Cultivating Community

McGill-Queen's Rural, Wildland, and Resource Studies Series
Series editors: Colin A.M. Duncan, James Murton, and R.W. Sandwell

The Rural, Wildland, and Resource Studies Series includes monographs, thematically unified edited collections, and rare out-of-print classics. It is inspired by Canadian Papers in Rural History, Donald H. Akenson's influential occasional papers series, and seeks to catalyze reconsideration of communities and places lying beyond city limits, outside centres of urban political and cultural power, and located at past and present sites of resource procurement and environmental change. Scholarly and popular interest in the environment, climate change, food, and a seemingly deepening divide between city and country is drawing non-urban places back into the mainstream. The series seeks to present the best environmentally contextualized research on topics such as agriculture, cottage living, fishing, the gathering of wild foods, mining, power generation, and rural commerce, within and beyond Canada's borders.

1 How Agriculture Made Canada
 Farming in the Nineteenth Century
 Peter A. Russell
2 The Once and Future Great Lakes Country
 An Ecological History
 John L. Riley
3 Consumers in the Bush
 Shopping in Rural Upper Canada
 Douglas McCalla
4 Subsistence under Capitalism
 Nature and Economy in Historical and Contemporary Perspectives
 Edited by James Murton, Dean Bavington, and Carly Dokis
5 Time and a Place
 An Environmental History of Prince Edward Island
 Edited by Edward MacDonald, Joshua MacFadyen, and Irené Novaczek
6 Powering Up Canada
 A History of Power, Fuel, and Energy from 1600
 Edited by R.W. Sandwell
7 Permanent Weekend
 Nature, Leisure, and Rural Gentrification
 John Michels
8 Nature, Place, and Story
 Rethinking Historic Sites in Canada
 Claire Elizabeth Campbell
9 The Subjugation of Canadian Wildlife
 Failures of Principle and Policy
 Max Foran
10 Flax Americana
 A History of the Fibre and Oil That Covered a Continent
 Joshua MacFadyen
11 At the Wilderness Edge
 The Rise of the Antidevelopment Movement on Canada's West Coast
 J.I. Little
12 The Greater Gulf
 Essays on the Environmental History of the Gulf of St Lawrence
 Edited by Claire E. Campbell, Edward MacDonald, and Brian Payne
13 The Miramichi Fire
 A History
 Alan MacEachern
14 Reading the Diaries of Henry Trent
 The Everyday Life of a Canadian Englishman, 1842–1898
 J.I. Little
15 Cultivating Community
 Women and Agricultural Fairs in Ontario
 Jodey Nurse

Cultivating Community

Women and Agricultural Fairs in Ontario

JODEY NURSE

McGill-Queen's University Press
Montreal & Kingston • London • Chicago

© McGill-Queen's University Press 2022

ISBN 978-0-2280-0914-6 (cloth)
ISBN 978-0-2280-0915-3 (paper)
ISBN 978-0-2280-0999-3 (ePDF)
ISBN 978-0-2280-1000-5 (ePUB)

Legal deposit first quarter 2022
Bibliothèque nationale du Québec

Printed in Canada on acid-free paper that is 100% ancient forest free (100% post-consumer recycled), processed chlorine free

This book has been published with the help of a grant from the Federation for the Humanities and Social Sciences, through the Awards to Scholarly Publications Program, using funds provided by the Social Sciences and Humanities Research Council of Canada.

Funded by the Government of Canada / Financé par le gouvernement du Canada | Canada | Canada Council for the Arts / Conseil des arts du Canada

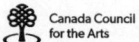

We acknowledge the support of the Canada Council for the Arts.

Nous remercions le Conseil des arts du Canada de son soutien.

Library and Archives Canada Cataloguing in Publication

Title: Cultivating community : women and agricultural fairs in Ontario / Jodey Nurse.
Names: Nurse, Jodey, author.
Series: McGill-Queen's rural, wildland, and resource studies series ; 15.
Description: Series statement: McGill-Queen's rural, wildland, and resource studies series ; 15 | Includes bibliographical references and index.
Identifiers: Canadiana (print) 20210334827 | Canadiana (ebook) 20210334886 | ISBN 9780228009153 (softcover) | ISBN 9780228009146 (hardcover) | ISBN 9780228009993 (PDF) | ISBN 9780228010005 (EPUB)
Subjects: LCSH: Agricultural exhibitions—Ontario. | LCSH: Agricultural exhibitions—Ontario—History. | LCSH: Women in agriculture—Ontario. | LCSH: Women in agriculture—Ontario—History. | LCSH: Agricultural exhibitions—Social aspects—Ontario. | LCSH: Agricultural exhibitions—Social aspects—Ontario—History.
Classification: LCC S557.C32 O66 2022 | DDC 630.74/713—dc23

Contents

Figures and Tables vii

Acknowledgments xi

Introduction 3

1 Getting on Board: Women and Agricultural Societies 18

2 Feeding the Family: Fair Food for All 57

3 Cultivating Beauty: The Flower Show 105

4 Providing Comfort, Refinement, and Respectability: Ladies' Domestic Manufactures, Fancywork, and Fine Art 137

5 Getting Outside the Inside Show: Female Horse and Livestock Exhibitors 183

6 From Church Booths to Baby Shows: Women on Display 211

Conclusion 245

Notes 251

Bibliography 317

Index 341

Figures and Tables

Figures

1.1 Example of Agricultural Society Fair Board Hierarchy, Nineteenth Century 25
1.2 Example of Agricultural Society Fair Board Hierarchy, 1930s 33
1.3 Example of Agricultural Society Fair Board Hierarchy, 1950s 48
2.1 The Crystal Palace at Picton Fairgrounds 68
2.2 Fairgoers Inspect the Food Exhibits 90
2.3 Domestic Science Judges at Rodney Fair 93
2.4 Prize Cooking at Schomberg Fair 95
2.5 Robin Hood Flour Advertisement, 1940 97
2.6 Robin Hood Flour Advertisement, 1945 98
2.7 A Family Practice at Shedden Fair 101
2.8 Canada's Flag Debate at Wallacetown Fair 102
3.1 Illustrated Florist 110–11
3.2 Prospect House in Peel County 115
3.3 Blossoms Attract at Aylmer Fair 130
3.4 Depicting Halloween at Erin Fall Fair 135
4.1 Prize Tatting at Schomberg Fair 163
4.2 Judges Inspecting Needlework at Waterdown Fair 165
4.3 Top Art Prize at Rodney Fair 166
4.4 Sarah Jane (Wheeler) Jackson's Prize-Winning Socks 170
4.5 Sarah Jane (Wheeler) Jackson's Prize-Winning Nightgown 171

4.6 Clara (Jackson) Robertson's Prize-Winning Work Shirt 172
4.7 Fanny Colwell Calvert's Prize-Winning Irish Crochet 173
4.8 Bernice, Irene, and Edna Rudd's Commemorative Quilt 174
4.9 Elizabeth Ann (Huxley) Donaldson's Crazy Quilt 176
4.10 Erin Agricultural Society Directorship Ribbon Sewn on to Crazy Quilt 177
4.11 Christena Nelena (McMillan) Neal's Quilt 179
4.12 Wellington County Fair Ribbons Sewn on to Quilt 180
5.1 Woman Feeding a Calf 185
5.2 Heavy Horse Stake Class at Erin Fair 199
5.3 Metcalf Calf Club 201
5.4 4-H Members Judging Livestock at Shedden Fairgrounds 202
5.5 Elgin 4-H Dairy Calf Club Showmanship Winner Sandra Dufty 203
5.6 4-H Aldborough Junior Calf Club Showmanship Winner Patricia McLean 204
5.7 Women Competing in Carriage Class at Erin Fair 206
6.1 Winners of the Oldest Persons at Smithville Fair 225
6.2 The Shedden Fair Queen 232
6.3 Baby Show Judges at Murillo Fall Fair 237
6.4 Baby Show at the Aylmer Fair 239
6.5 Big Crowd at Norfolk County Fair 243

Tables

Note: All individual classes are copied as they appear in the original prize lists. This includes original spelling, capitalization, and punctuation.

2.1 Erin Township Agricultural Society Fair, Erin, Wellington County: Classes for Vegetables, 1854 and 1901 78
2.2 Erin Township Agricultural Society Fair, Erin, Wellington Count: Classes for Fruit, 1854 and 1901 79
2.3 Peel County Agricultural Society Fair, Brampton, Peel Count: Classes for Fruit, 1871 and 1878 81
2.4 Brooke and Alvinston Agricultural Society Fair, Alvinston, Lambton County: Classes for Canned Goods, 1920 86

Figures and Tables

2.5 Collingwood Township Agricultural Society Fair, Clarksburg, Grey County: Classes for Canned Goods, 1930, 1935, 1940, 1945, and 1950 88–9
2.6 Central Agricultural Society Fair, Walters Falls, Grey Count: Classes for Baked Goods, 1890, 1900, 1910 94
2.7 Erin Township Agricultural Society Fair, Erin, Wellington County: Classes for Baked Goods, 1940 100
3.1 Peel County Agricultural Society Fair, Brampton, Peel County: Classes for Flowers and Plants, 1868 and 1880 118
3.2 Arran-Tara Agricultural Society Fair, Tara, Bruce County: Classes for Flowers and Plants, 1910, 1933, and 1950 123–5
3.3 Arran-Tara Agricultural Society, Tara, Bruce County: Classes for Flowers and Plants, 1965 132–3
3.4 Erin Township Agricultural Society Fair, Erin, Wellington County: Classes for Flowers and Plants, 1974 134
4.1 Erin Township Agricultural Society Fair, Erin, Wellington County: Classes for Domestic Manufactures, 1864, 1889, and 1910 142–3
4.2 Erin Township Agricultural Society Fair, Erin, Wellington County: Classes for Fancywork, 1889 149
4.3 Seaforth Agricultural Society Fair, Seaforth, Huron County: Classes for Ladies' Work, 1925 155
4.4 Arran-Tara Agricultural Society Fair, Tara, Bruce County: Classes for Ladies' Work, 1960 and 1970 160–2
4.5 Erin Township Agricultural Society Fair, Erin, Wellington County: Classes for Arts and Crafts, 1974 167

Acknowledgments

Agricultural fairs have always been an important part of my life. I spent most of my fall weekends as a child and into adulthood exhibiting livestock at these events. Moreover, I owe my genesis to the Canadian National Exhibition, where my parents met while exhibiting cattle, so I see this study as a chance to give back to the institution without which I might not have existed! I also wish to acknowledge the agricultural society members who volunteer so much of their time and other resources to ensuring that these valuable rural institutions not just continue, but flourish. A special thank you to the fair women who are the subject of this study, especially those who shared their stories with me.

This book would not have been possible without the encouragement of Catharine Anne Wilson, who supervised my doctoral research. Her award-winning studies of rural communities inspired me to pursue my doctorate, but it was her guidance, generosity, and kindness that made the completion of my dissertation, the basis for this book, possible. I will be forever grateful for her mentorship and for her friendship. I would also like to thank Adam Crerar and Belinda Leach for serving on the thesis committee and offering thoughtful assessments of my work, and Linda Mahood and Ruth Sandwell for serving as examiners and suggesting important revisions.

The research for this book was conducted with the financial support of an Ontario Graduate Scholarship as well as internal scholarships at the University of Guelph. My postdoctoral fellowship at the University of Waterloo and later my L.R. Wilson Assistant Professorship at the Wilson Institute for

Canadian History at McMaster University allowed me to complete the manuscript and publication process. I would like to thank my colleagues for their support during my time at these institutions. Bruce Muirhead deserves a special thank you not only for supporting this work but also for giving me the chance to pursue new research during my postdoctoral fellowship and for continuing to encourage my work as a research assistant professor at the University of Waterloo.

Many thanks to the staff at McGill-Queen's University Press for their support, especially Kyla Madden. Thank you to the anonymous reviewers of the manuscript who provided thoughtful criticism and encouragement. Too many scholars to name inspired this book but I do want to thank the Rural Women's Studies Association, whose members include pioneers of rural women's history, many of whom are now also my friends. Many other friends and relations contributed to this work through their encouragement. You know who you are, and I love you all!

Finally, this book is dedicated to my family. My parents, Jeff and Kenda, my siblings, Lee, Cristy, and Troy, my sister-in-law Mal, and my nieces, Leah, Callie, Lily, and Ellie, who have always been my greatest sources of support and unconditional love. I wish for many more years of fair-time memories together.

Cultivating Community

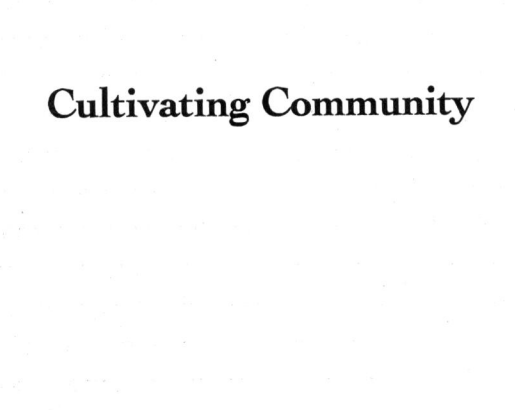

Chapter 6
Women and Agricultural Societies

In 1995, Mrs. Velda Dickenson, past president of the Women's Section of the Ontario Association of Agricultural Societies (OAAS), was awarded a special plaque for her many years of valuable service to the organization. The plaque was inscribed with a lengthy poem, read at convention time, which celebrated both her broader community service. The poem began by addressing the society members before thanking Dickenson for her dedication to the organization:

> To you, Velda, we do honor
> For all the work you have done,
> Your time, talent, and unselfish devotion,
> You have given, and many friends won.

The poem continued by describing elements of Dickenson's childhood, her participation in Junior Farmers and 4-H, and the "love of horses" which lured her to the fair each fall. It noted that she began with the Ancaster Agricultural Society before becoming a district representative and eventually the president of the Women's Section of the OAAS. Throughout her many years of service to the OAAS, she judged at fairs across the province and was especially fond of junior competitions because she hoped to inspire the next generation of fair exhibitors. The poem closed with the final stanza:

> And now this story comes to an end
> Memories have been brought back it is true;

Introduction

Agricultural fairs in Ontario have undergone significant transformations. Today, they are community events that draw local residents together to celebrate their talents, activities, and livelihoods. When agricultural fairs first began in the province, agricultural societies – the organizations that still host these events – were more intent on awarding premiums for livestock breeding and grain growing than they were about encouraging a wide variety of work-related and domestic pursuits. By the end of the nineteenth century, however, the agricultural fair was considered a family event that celebrated the pastimes of young and old, farm and town folk, and men and women alike. Indeed, according to an article published in the Toronto *Globe* in 1898, Ontario families' favourite fall pastime was attending "a Country Fair." The author told of a family whose eldest son drove one wagon loaded up with livestock while the father, mother, daughters, and younger sons took the "democrat waggon"[1] filled with "cheese-cloth-covered boxes of chickens, besides home-made bread and butter and fancy work." The family's drive along the country road was "full of silence and expectation of the day's pleasure," and the author commented, "Only those who are familiar with the country fair know its delights."[2] He explained that the fair was an institution onto itself, as "necessary to the farmer for a fall holiday as turkey is to the celebration of Christmas," that it was "a source of inspiration in the improvement of farm stock and produce," and "the gathering-place of many thousands of people in every township and county in the country."[3]

This description evokes an egalitarian household, travelling together to an event that represented opportunity for all its members. The husband, wife, sons, and daughters assembled to showcase their talents and contributions to the family and community. The livestock they transported epitomized the progressive farmer who had adopted improved animal husbandry, which resulted in superior stock that would serve him, his male heirs, and future generations of livestock breeders in the province. The women's domestic products and handiwork showcased their skill and commitment to creating goods that supplemented the household income, sustained the family, and gave evidence of their respectability and taste. Their silent contentment and hopeful expectation while travelling to the community's most significant social gathering emphasized the family's delight in the potential triumph their hard work might achieve.

Many families and individuals across the province of Ontario – and indeed Canada and beyond – travelled down country roads to enjoy agricultural fairs during this period. Still, the image of egalitarianism presented by the family in this story does not reflect the differences experienced by the women and men who attended. Common experiences exist at fairs, and the rural people who attend these events have some elements of shared experiences and consciousness. But fairs were and continue to be social institutions, and, as such, represent a host of cultural values and beliefs that often insist on the divergence of the sexes. Gender continues to shape men's and women's experiences in very different ways.[4] Historically, the categorization of women's versus men's work in rural households created a hierarchy of abilities – and thus relationships of power – not represented by the "separate but equal" mantra of the more progressive-minded people in the countryside.

This book explores how rural women (and men) presented their identities in the context of agricultural fairs in Ontario and how these gendered identities evolved from the mid-nineteenth century to the late twentieth century. It argues that fairs served as sites where sex-specific abilities and identities could be reinforced or challenged. While traditional domestic skills and handicrafts remained the domain of women throughout the period, women enlarged their sphere of influence on the fairgrounds by the mid-twentieth century. They took their places in the livestock show ring, on the athletic

Introduction

field, and in the boardroom – spaces that were previously closed to them. By focusing specifically on women's experiences at agricultural fairs, this book presents a fuller and more complex picture, one in which women's activities take center stage as essential components not only of the fair experience but also as important representations of rural life.

It is undeniable that women-centric exhibits were critical to agricultural fairs' growth and prosperity. The more traditional homecrafts showcase women's household provisioning and recreational pursuits and uncover how women understood the concepts of mutuality, hard work, thrift, and middle-class respectability and taste. They also help us better understand how women valued individual expression and sought empowerment through their work. The changes to and continuities of these exhibits also tell us a great deal about which items or skillsets were still valued (either for utilitarian or cultural purposes) and which were no longer deemed important. Some competitions also incorporated new technologies and trends that reflected contemporary society rather than anti-modern or outdated pastimes and products. Women's homecrafts were not static exhibits but rather dynamic displays of both past and present.

However, women did more than just participate in fairs by entering exhibits deemed appropriate for their sex. Agricultural fairs also allowed women to display new behaviours for public consideration, such as showing livestock, participating in athletic events, and occupying positions of authority and leadership. The range of experiences women had at Ontario fairs over the nineteenth and twentieth centuries are examined to illustrate how women used the agricultural fair's manifold nature to present different versions of womanhood, some of which supported socially constructed notions of feminine behaviour and others that worked to dismantle them. Women were not merely displaying items of work when they entered their exhibits; they presented objects that served as representations of their interests, values, and beliefs. These included a commitment to familial and communal cooperation, pride in the ability to provide for one's family, dedication to skill development, a belief that material things could embody respectability and refinement, and an enthusiasm for individual expression and artistry. Often their creations remained testaments to those associations long after their creators were gone. Moreover, by performing in a concert hall, in front of a grandstand, or in a show ring, women also conveyed self through action. Whether women managed exhibits, served pie to fairgoers, showed cattle, exhibited fancywork, or executed trick riding acts, what they did could enhance or reshape societal understandings of their abilities, who they were, and how they lived. Some women participated at fairs to showcase their loyalty to elements of traditional womanhood such as domesticity. In contrast, others increasingly moved outside of prescribed boundaries of femininity to display their strength, fearlessness, leadership, and knowledge. Fair women were not a unified group, and fairs were complex places where contradictory messages could be displayed to showcase an increasingly broad range of women's skills and talents. They enlarged both the literal and figurative space women occupied.

This is not to say that some assumptions about women were not stronger than others. Women, in general, have had rigid expectations placed on their behaviour. Societal ideas about appropriate conduct and opinions helped shape their identities so, unsurprisingly, many conformed to socially acceptable standards of behaviour on the fairgrounds. Women's identities were strongly tied to the household and included traditional domestic tasks such as fashioning clothing and goods, growing and producing foodstuffs, and cultivating respectable and moral environments in which their families could flourish. Rural men and women believed that women had an inherent ability to provide care, support, cooperation, and service to their families and communities. They also associated ideal femininity with beauty, taste, and refinement. And yet, because fairs were places "so amenable to disorder," they also allowed women to subvert gender roles and expectations.[5] The element of subversion at county and township fairs was not nearly as strong as it was at larger urban exhibitions, because local fairs did not provide the same opportunity to experiment with alternate identities and behaviours anonymously. Nevertheless, women still found opportunities to challenge the status quo. They used fairs to develop roles and responsibilities beyond the domestic realm by performing publicly in traditionally non-feminine ways, including carrying out feats of athleticism, entering the livestock show ring, and holding the highest office in an agricultural society – the position of president. Hegemonic forces helped shape women's experiences on and off the fairgrounds, but women could push the boundaries and move outside these larger societal powers.

The idea that fairs and exhibitions were used as instruments of cultural hegemony is popular among Canadian historians such as Keith Walden and Elsbeth Heaman.[6] As these scholars have demonstrated, exhibition organizers used these events to foster and reinforce cultural hierarchies of class, gender, and race. However, Walden and Heaman also argue that subaltern groups could co-opt these events for their own purposes. For instance, Walden highlights numerous cases where women used fairs to achieve increased mobility, destabilize notions of gender, and appropriate public spaces. Heaman argues as well that women used exhibitions to their advantage by demanding a broader public role for their sex. However, both of these studies focus more on large national and international fairs and exhibitions, and Heaman is critical of the women-centric exhibits that privileged the work of farm and leisured women and failed to consider the other types of work women were doing in the nineteenth century.[7] Walden and Heaman provide valuable analyses of women's participation at fairs and exhibitions, but more needs to be said about the township and county fairs in which women's exhibits tell us a great deal more than merely what "farm wives or leisured ladies" were making, or how women gained "a back door into the public sphere."[8] Unlike the provincial, national, and international exhibitions where women's exhibits were strategically selected and typically displayed by upper and middle-class urban men and women, county and township fairs directly involved rural women. Therefore these fairs provide a closer understanding of who these women were, what skills and characteristics they sought to showcase, and how they wanted to be remembered.[9] In addition, Walden's and Heaman's studies are confined to the nineteenth century. This book expands beyond the nineteenth century into the twentieth to highlight significant changes such as the expansion of women's domain beyond the confines of the agricultural hall and the midway booth and into the show ring and fair boardroom.[10]

Whether women used fairs as a medium to strengthen traditional concepts of womanhood or challenge them, ultimately the thousands of women who organized, administered, exhibited, performed, and attended fairs did so because such gatherings offered them opportunities to present themselves to their communities in their own way. In deciding to participate in these events, women did not necessarily accept all the principles espoused by agricultural societies. Still, they did have to believe that their involvement was worthwhile, either for themselves, their families, or their community. For many women, fairs were occasions where they created life-long friendships, supported community causes, developed and showcased valued skills, attained social authority, and simply enjoyed themselves. The joy, pride, and achievement they experienced motivated most of them to return annually.

Of course, agricultural societies also benefited from women's participation. Some men were initially wary of female members, but most came to accept and eventually celebrate the work women did to sustain their organizations and the fairs they hosted. Women increasingly moved into positions of authority in agricultural societies. By the 1970s, their ability to promote the fair to urban and rural audiences became especially important, because rural populations were noticeably declining.[11] Women's involvement in agricultural societies was necessary if fairs were to succeed, and by the end of the period covered by this study, most male members accepted, if not celebrated, this fact.

A Story about Rural Women

This is very much a women's history. We must tell women's stories, and many historical topics about women's experiences still require further consideration. We can and should say more about women's participation in agricultural societies and fairs, especially in the Canadian context. For instance, scholarly attention to agricultural societies has focused primarily on their activities during the eighteenth and nineteenth centuries, when women were rarely members and women's work was not officially recognized.[12] Focused as they were on agricultural progress through the improved breeding of livestock, the adoption of better-quality crops, and the "scientific" management of farmland, early agricultural societies likely had little desire to consider women's work in any serious fashion. By the second half of the nineteenth century, however, more exhibits highlighted domestic work, and by the twentieth century, these groups had a different attitude towards women's participation. Women had established themselves as important exhibitors and supporters of Ontario's agricultural fairs by this time, so in the 1910s and 1920s, agricultural societies began formally inviting women to join their organizations and assist in managing township and county fairs. This book emphasizes how women regularly seized opportunities for greater authority

Introduction

whenever they were available. Although many men continued to view women as subordinate, their fairground roles were never fixed and continued to transform over the nineteenth and twentieth centuries. Women did not congregate at the periphery of agricultural fairs during this period; they engaged with agricultural societies and fairs in great numbers, occupied myriad roles, and influenced the nature of the fairs.

Of course, one cannot study women's roles and behaviours in isolation. This study considers women's fair experiences within the larger context of gender relations and sexual difference between men and women to help us to uncover why and how women participated in these events. Still, the central focus is always on exploring the agricultural fair to construct a more nuanced portrait of the world as rural women saw it.[13]

Female actors are at the centre of this narrative, but so too are conceptions of rurality. Rural identity and community are also at the heart of this study, and therefore the term "rural" should be clarified. The word "rural" is a contested one. Historians studying rural society have had to decide how to define it: usually in terms of geography, in relation to population density and distance from urban centres, or as a cultural classification based on a particular way of life.[14] Ruth Sandwell provides a useful discussion of the considerable debate that surrounds the term. When presenting statistical data, she categorizes "rural" as defined by the Dominion Bureau of Statistics from 1871 to 1941, which distinguished rural people as those not living in incorporated cities, towns, and villages. Sandwell defines rural people as living in communities of fewer than five thousand people, but she also describes rural life as distinguished by "the dominance of the outdoors, the enormous amount of physical labour, and the centrality of the household."[15] Whatever categorization one uses, she finds that "Canada was a rural country, and Canadians were a rural people, until at least the Second World War."[16]

Many of the fairs investigated here took place in or just outside of villages and small towns whose residents were active participants along with individuals living in the countryside. Fair participants represented the occupational diversity of rural communities, and these individuals likely considered themselves different from one another in many respects. Even those in the same profession, such as farming, were different according to their class, ethnicity, race, and religion. I use the term "rural" to define all of these people, however, because they were all dependent on the hinterland in some form and understood their reliance on the land and the importance of family labour. I also accept the more expansive definition of "rural" that includes all people living outside major urban areas.[17] I recognize that rural people were not homogenous and that socio-economic differences are significant. Still, I believe that some meaningful generalizations can be made, particularly for residents from the agrarian-based communities most involved with fairs. Farming and farm life were central concerns of agricultural societies, and while they enlarged their interests over time, members generally remained dedicated to maintaining the fair's connection with agriculture. Furthermore, the women who participated in fairs often illustrated a broader rural tradition that scholars have argued valued conformity to community standards and cooperation, while also believing in independence and self-reliance.

Because fairs were premised on the belief that educational display and competition fostered improvement, this book addresses the literature on women's participation in rural reform. Rural reformers understood that women were the foundation of rural family life. They hoped to persuade women to adopt technology, science, business efficiency, and professional expertise to improve domestic tasks and help manage the home and the family.[18] Scholars have argued that such reformers met with varying degrees of success. At times women promoted change, and at other times they resisted it, according to their individual economic and social circumstances. What is clear, however, is that women were active participants in the changes taking place.

A focused analysis of women's involvement in agricultural societies, the organizations that were first tasked with improving rural life, illustrates how these groups sought to influence women to retain conventional forms of femininity and improve their domestic work. Unfortunately, it is difficult to know if what fair organizers promoted in terms of agricultural and domestic practices was implemented regularly in daily life. Furthermore, did their emphasis on the appearance of items, such as uniform, unblemished garden vegetables or attractively decorated cakes, actually make for better products? Like others who have studied the influence of exhibitions, this study is limited in the extent to which it can claim a causal relationship between organizers' stated ambitions and the evidence of rural households

Introduction

practicing those goals. For example, though fairs promoted classes for the manufacture of homemade canned goods, we cannot know if women canned more for their families than they otherwise would have because of fairs' efforts to encourage this practice.[19] What we do know, however, is that the practices promoted at fairs were ones that rural women generally supported and, as this book will show, provided women with opportunities to initiate, develop, or expand their skillset in a variety of pursuits, both domestic and nondomestic. Women's participation at fairs was voluntary, and whether they participated in or supported events that encouraged traditional boundaries of female behaviour or served to dismantle those boundaries depended on each woman.

The success women had in supporting or challenging prevailing notions of femininity was also related to their influence in a particular time and place. For instance, historian Darren Ferry has argued that women's growing authority in agricultural groups during the late nineteenth century reflected their increasing involvement in agrarian communities' public life as a whole. He also notes that male members still confined female members to the domestic sphere by incorporating women into their rituals and meetings in ways that reinforced their domestic role and reinforced conceptions of womanhood as defined by characteristics such as delicacy, loveliness, and sentimentality.[20] Indeed, even groups that expanded women's authority in rural communities often privileged more traditional forms of female participation.[21] This study agrees that popular perceptions of gender roles and responsibilities as well as the definition of agriculture as a profession undertaken by men influenced women's volunteerism in agricultural societies and fairs. Still, it demonstrates that over time, women and their male allies expanded the notion of agriculture to include women's work, thus allowing women claims to authority through agricultural pursuits.[22]

This study also contributes to our understanding of rural feminism(s) and agrarian patriarchy. Rural feminism (or –isms) is debated among scholars who study rural and farm women in the late nineteenth and twentieth centuries. Some argue rural society was a hotbed for conservative and patriarchal discourse, while others contend it was a place where necessity promoted egalitarian forces that gave women social and economic power unequalled in urban settings.[23] For instance, Nancy Grey Osterud recognizes that "conjunction and disjunction" existed in women's and men's relationships, but that women never retreated into a separate women's sphere. Instead, women sought "mutuality in their marriages, reciprocity in their performance of labour, and integration in their modes of sociability," thereby enlarging the "dimensions of sharing in their relationships with men."[24] Although the cooperative nature of cross-gendered relationships changed over time, the reality of farming allowed women to shape their work and encourage all household members to see their lives and interests as relational.[25]

However, the American historian Deborah Fink challenges the assumption that it was easier for women to achieve equality with men on the farm because of the proximity of their work, and instead posits that agrarian idealism was male-oriented and women occupied the subordinate roles of mothers and wives. Fink argues that cultural assumptions about women's domestic roles remained unchanged in the countryside, so women were never perceived as equal with their husbands.[26] Scholars writing about Canadian farm women support this conclusion. They argue that agrarian communities were fundamentally conservative ones where women had limited property ownership, decision-making, and leadership opportunities.[27] While one should be careful not to overemphasize a universal "common culture of reciprocity and respect among women and men"[28] in rural communities and recognize that significant power asymmetries limited the roles rural women held and the opportunities they could pursue, this study finds that cooperation among the sexes was central to the functioning of agricultural fairs and that women actively worked to dismantle the boundaries limiting them.

The nature of women's activism in agricultural societies and fairs is complicated, of course. Indeed, the degree to which feminism existed in the countryside has been especially contentious among scholars studying Ontario Women's Institutes (WI). Historian Monda Halpern argues that many "farm women were indeed feminist, and that this feminism was more progressive than most of us would presume" by demonstrating how WI women regularly worked towards their interests and challenged the privileges and priorities of men.[29] In contrast, Margaret Kechnie contends that the WIs did not satisfy the needs of farm women but served to "satisfy the goals of state-sponsored male farm leadership" and "elite and middle-class town women," arguing further that these groups reinforced the idea that women

Introduction

were domestic helpmates rather than agricultural partners.[30] Other historians, such as Linda Ambrose, have recognized that the leadership of Ontario WIs may have wanted to impose an ideal of domesticity but that WI members worked towards different aims. Ambrose contends that the WIS allowed rural women to enlarge their public sphere through valuable opportunities to develop leadership skills and political acumen, influence community projects, seek education on various topics of interest, and socialize.[31] Scholar Louise Carbert also finds that the WI did not ignore the full range of responsibilities and labour farm women undertook in the home and encouraged domestic and market-oriented projects.[32] Beyond Ontario, Women's Institutes also played a critical role in defining and meeting rural women's needs, creating a female cultural space, and challenging the boundaries of acceptable femininity.[33]

Women's participation in agricultural societies and fairs stimulate similar discussions. Was women's involvement representative of other patriarchal institutions where they were regulated or ignored, or did these groups and events allow women space for greater mutuality and equality between the sexes and the ability to achieve greater authority than was otherwise possible in the wider society? Did women consciously use agricultural societies and fairs to seek out sisterhood and empowerment? Or did they simply fill a void in these organizations and events that, purposefully or not, enabled them to have more power over time because of their talents and organizational skills? Can we use agricultural societies and fairs to judge women's broader experiences in the countryside? I argue that, yes, women's involvement in agricultural societies was constrained by a patriarchal system, but that nonetheless a degree of cooperation and mutuality was achieved among the sexes that characterized similar relations in rural households and allowed for an appreciation of women's contributions. Agricultural societies benefited from women's organizational skills and hard work, but women also benefited from the opportunities available at fairs to increase their economic, social, and cultural authority. Whether or not they sought it, women often found sisterhood and empowerment through these experiences. They did not attain equality with men (equality had not been realized outside the fairgrounds either, of course), but by the 1970s, at least some women had become agricultural society presidents, a significant achievement in the history of these organizations.

The Sources and Parameters

This book focuses on the period between 1846, when the provincial association for agricultural societies was first created and women's exhibits were expanded, and 1980, the end of a decade that experienced significant growth in the number of women serving on fair board executives and managing fair exhibits and displays. In terms of geography, I have selected examples from various locations around the province. Most sources focus on the central and western regions of Ontario, especially the regions of Peel and York and the counties of Bruce, Elgin, Grey, Norfolk, Wellington, and Hamilton and Wentworth. Archival information found in Middlesex, Peterborough, and Simcoe Counties was also useful. I focused on areas that allowed for an abundance of primary source possibilities and were easily accessible and geographically suitable. This analysis includes other counties and regions of Ontario, but the reports on their fairs and the depth of archival research are generally not as deep. At one point, 354 agricultural societies operated in the province; today 217 remain, each with its own rich and vibrant history.

I do not profess to have completed an exhaustive study. The most prominent regions in this study were also regions in Ontario that exhibited a high degree of ethnic homogeneity, especially in the nineteenth and early twentieth centuries. Over the twentieth century, rural communities became more diverse after increased immigration, mainly European, brought newcomers who were not of British descent. The historical records surveyed for this study did not provide for a sufficient analysis of the different ethnic and racial groups, such as First Nations, Franco-Ontarians, and Eastern-European immigrants, that also experienced fairs and participated in agricultural societies. I believe that additional research on this topic will reveal a great deal more than what has been discussed within this book's parameters, and I look forward to seeing other work on this topic.

This study uses a variety of sources, such as newspapers and journals; government publications and annual reports, including the annual reports of the Ontario Association of Agricultural Societies (OAAS); township and county agricultural society minutes, membership and account books; general correspondence; and published fair prize lists found in public archival and private collections. Other useful sources included published and unpublished diaries and photographs. Diaries provide a rich source of evidence

Introduction 15

for historians studying rural life. Although they are not a large part of my primary source base, they offer some critical insight, including the relationship between what women did at fairs and their daily chores, rituals, and pastimes. Photographs are beneficial for providing evidence of and illustrating some of women's late-nineteenth and twentieth centuries' fair exhibits and activities.

Another key feature of this work is the incorporation of a material cultural analysis in Chapter 4. Artifacts are a defining feature of my chapter on women's domestic manufactures, fancywork, and fine art because these types of items have been preserved and can provide exceptional insight into their creators' lives. Artifacts can be intimidating sources for historians, who may question their ability to detect and appreciate objects' minute details.[34] The inherent complexity of objects, their ability to serve as tools and "signals, signs, symbols," and the fact that "much of their meaning is subliminal and unconscious" is also challenging.[35] Many approaches to material culture exist,[36] but historians bring unique skills to the task. By situating objects in their historical context, we can find deeper stories that allow us to understand better the social, cultural, and economic environments in which they existed.[37] Objects are especially useful when studying rural women. I had difficulty finding the voices of female fair exhibitors in the nineteenth and early twentieth centuries, so when I uncovered items women had shown at fairs or made from materials awarded at fairs, these objects allowed me to consider new questions for my work. We often consider objects for their utility or aesthetic qualities, but we sometimes forget that they can also act as "companions to our emotional lives or as provocations to thought."[38] The power of objects to connect people both to ideas and to other people was evident in the artifacts used for this study.[39] Interaction with objects as historical sources forces one to value what things reveal about "meaning making, identity formation, and commemoration."[40]

Oral history was also important for this research. I interviewed eleven women who exhibited at or were involved with agricultural societies during the period of this study.[41] These women were involved with fairs in the Regional Municipalities of Halton and Hamilton-Wentworth and Dundas and Wellington Counties. When I began this project, I did not intend to conduct interviews or incorporate oral history into my methodology. When I extended my study into the postwar period, the opportunity to use oral history was exciting. Although this is not a large sample of women, these interviews provided rich information about their fair experiences that I could not have garnered from other sources. For example, much of the material I use for earlier periods of the study necessarily relies on others' assessments of rural women and their work rather than women's own valuations. Oral history allows for a shift in focus where women tell their own stories. In the same way that material culture provides a crucial tool for studying rural women, oral history can also be transformative because it gives historians another way to access rural women's everyday experiences, including relationships with neighbours and kin and the special moments of their lives, such as winning a baking competition or a dairy calf showmanship class.[42] Oral history brings recognition to groups of people who have been largely unnoticed in the historical record.

Oral history also has its challenges, including a required closeness between researcher and subject that can be intimidating. However, the relationship between interviewer and interviewee can also be richly rewarding for scholars who are open to learning and collaborating in the reconstruction of history and appreciating how the use of a human voice can "breathe life into history."[43] Of course, oral history faces the fact that "we do not have memory as much as remembrances, or even performances of remembering, where what is remembered is shaped fundamentally both by the meaning of the initial experience to the individual in question and by the psychological – and inextricably social – circumstances of recall."[44] The present inevitably shapes what is recalled. Historians who engage in oral history recognize that "there remain many contested and unresolved questions about how memory can meaningfully be interrogated by historians."[45] However, for historians, it is essential to understand how people defined themselves, including what words they used and with whom they did or did not identify.[46] Despite its challenges, oral history provides the opportunity for rural women to comment on their experiences and the meanings they held and continue to hold. Admittedly, the stories in this study, including those of the women interviewed, were mostly told from the perspective of the dedicated female agricultural society member, exhibitor, entertainer, or long-time attendee. The negative experience of a woman who joined an agricultural society for one year and did not return or a woman who vowed never to exhibit again after a demoralizing loss is absent from the historical record. Future researchers

Introduction

studying agricultural societies and similar groups should be encouraged to reach beyond a "self-selected group" to fully represent what these organizations offered and what they did not.[47]

The book's chapters are thematically organized, but internally they demonstrate chronological development. Chapter 1 examines women's involvement in agricultural societies' organizational aspects, their roles and responsibilities, and how their participation was shaped by societal forces, male leadership, and their own initiatives. I provide an assessment of the reasons why women volunteered in these organizations. Chapters 2, 3, and 4 focus on traditional women's exhibits. Chapter 2 focuses on food exhibits, chapter 3 on flowers, and chapter 4 on domestic manufactures, fancywork, and fine art. All three chapters demonstrate how the items women exhibited at fairs conveyed societal values and assumptions about women's work and character and broader social, cultural, economic, and technological change during the period. Chapter 2 provides an extended discussion of women's contribution to the overall fair display, but chapters 3 and 4 also address how women helped enhance presentations made on the fairgrounds. Chapter 5 departs from the previous three chapters and considers women in what was traditionally defined as men's areas of expertise, specifically horse and cattle competitions. I examine the reasons for women's involvement in these competitions and the reactions to their involvement. Chapter 6 provides an analysis of some of the other ways women involved themselves on the fairgrounds. This includes how women used fairs for fundraising, earning a living, entertaining and being entertained, and winning accolades or meeting new friends. The conclusion provides some closing thoughts on the arguments presented.

1

Getting on Board
Women and Agricultural Societies

In 1965, Mrs Velda Dickenson, past president of the Women's Section of the Ontario Association of Agricultural Societies (OAAS), was awarded a special plaque for her years of valuable service to the organization. The plaque was inscribed with a lengthy poem about Dickenson's time with the association and her broader community service. The poem began by addressing the society members before thanking Dickenson for her dedication to the organization:

> To you, Velda, we do honor,
> For all the work you have done,
> Your time, talent, and unselfish devotion,
> You have given, and many friends won.

The poem continued by describing elements of Dickenson's childhood, her participation in Junior Farmers and 4-H, and the "love of horses" which lured her to the fair each fall. It noted that she began with the Ancaster Agricultural Society before becoming a district representative and eventually the president of the Women's Section of the OAAS. Throughout her many years of service to the OAAS, she judged at fairs across the province and was especially fond of junior competitions because she hoped to inspire the next generation of fair exhibitors. The poem closed with the final stanza:

> And now this story comes to an end
>   Memories have been brought back it is true;

We ask you to accept this parting gift,
A token of our love and devotion to you.

Dickenson and thousands of women like her gave their time, talent, and devotion to help organize and manage local fairs across Ontario. Some of these women received acknowledgement for their work. For Dickenson, this recognition came in the form of a plaque, which the OAAS presented to her in front of hundreds of her peers at their annual convention in Toronto.[1]

However, Ontario agricultural societies in 1965 were not the same organizations they had been a century earlier. In the nineteenth century, women were rarely members of these societies. Although some women paid for membership, no records survive of women serving on executive committees or directorships during this period. The women who assisted men on fair boards acted as "helpmates" by arranging fair banquets or aiding with women's exhibits; they were not recognized for their service in any official capacity. In the twentieth century, this changed. In the first three decades of the century, some women officially joined fair board committees. By the 1930s, "lady director" positions became common and the OAAS established a separate "Women's Section." Agricultural societies created more space for women on fair boards, partially in response to women's demands and growing concerns about decreased membership. Perceptions of women and their ability to serve in public positions were changing in the early twentieth century. Women were eager to expand their authority and enter new roles, including leadership positions within agricultural societies.

The power asymmetry between men and women persisted because of long-held assumptions about women's roles and capabilities. Still, female agricultural society members remained dedicated to serving their communities, sustaining public spirit and sociability, and advancing women's work. Historians and sociologists studying rural women have often found that women worked together for their mutual benefit. Their volunteering activities fit into the broader context of their working lives and their understanding of rural identity and community as dependent on cooperation.[2] Scholars have also noted that these women often acted in ways that can be characterized as "social feminism," meaning that they prized female values and competencies that they believed necessitated their influence in social matters. They highlighted their experiential differences, unlike equity feminists who claimed equality by virtue of their similarity to men.[3] Some scholars have portrayed these women as feminists because many of them were committed to improving women's lives and breaking down patriarchal structures and values by asserting their right to participate in the public sphere.[4] Dickenson represents one of the thousands of women who used agricultural societies to create a larger public role and a better life for themselves, their families, and their communities.

Setting the Stage: 1792–1846

When the first agricultural society in Ontario was established in 1792, women were not included. The Niagara Agricultural Society began under John Graves Simcoe's patronage. Simcoe was Upper Canada's first Lieutenant-Governor, and he hoped that these groups would help improve farming practices in the province.[5] In the late eighteenth century, agriculture was broadly defined as "Tillage; husbandry,"[6] however, the word held a much deeper meaning for agricultural improvers. Agriculture was not merely a practice – the tilling of soil or animal husbandry with a view to harvesting crops or livestock – it was a science and a profession that was explicitly understood to be undertaken by men. Late eighteenth- and early nineteenth-century improvers in Canada argued that agriculture was central to the advancement of human society and that scientific practices were central to the creation of better "husbandmen."[7] Late nineteenth-century improvers also insisted that agriculture was "the back-bone and sinew of every country; it had made nations, caused them to become mighty, and when neglected they have collapsed; it must take the lead in all employments; it is the impetus that sets everything in motion."[8] Agricultural societies promoted agriculture as the profession to which all others were "subservient," in part because they perceived it as men's work.[9] Of course, many women performed farm labour, and some were singled out for their prowess,[10] but the idea that "Agriculture, from Adam down" had been the noble pursuit of men was common.[11]

Simcoe wanted to transplant a model of agricultural societies that existed in Britain to Upper Canada to facilitate the improvement he believed was necessary for success.[12] This encompassed a range of technological, cultural, and institutional changes meant to encourage farmers to accept modern capitalist modes of agriculture.[13] Agricultural societies were voluntary associations which disseminated relevant information and advice on progressive practices for land use and animal husbandry, thereby stimulating further settlement and production. Beyond better farming techniques, some improvers also espoused self-improvement in terms of "conduct, propriety, and respectability."[14] Agricultural improvers who believed that the material and spiritual were inseparable also opened the door for women to have a more significant role because they recognized women as the family's moral guardians. For some improvers, the home underpinned the political economy of improvement. A wife's moral fortitude and thrifty management within the home were as necessary as her husband's labours outside of it.[15]

Still, the early membership of agricultural societies was almost exclusively male, so it is not surprising that historians studying these groups have not considered women in any great detail. Male members focused on strengthening middle-class masculine authority, and women were often beyond their purview.[16] The public sphere was considered a masculine one. While women were occasionally able to step into the civic spotlight, generally men wished to exclude them from public roles.[17] In the 1800s, some rural women established a public presence by serving in voluntary associations such as temperance societies and church groups. Still, they were unable to attain leadership positions in agricultural institutions.[18] Farm journals published articles supporting ideas about the inherent differences between the sexes and women's inability to perform public and professional roles. Any attempt women made to deviate from their domestic responsibilities was considered a revolt against nature and an abuse of energy that would result in social upheaval.[19] Despite a small number of women registering as members of agricultural societies (mainly widows and those unmarried), no women served on these organizations' executive or directorship throughout the nineteenth century.[20]

Initially, agricultural societies focused on hosting competitions associated with the most obvious pursuits of farm men. Fair exhibitors won premiums for showing livestock and field crops and for essays on progressive methods of clearing land, crop rotation, fertilization, and general cultivation. Agricultural societies later added implement classes to emphasize the tools men needed to do their best work. At the same time, male livestock was privileged over female livestock because sires were believed to have a greater influence on progeny than dams.[21] Dairy product and textile exhibits were also added, but they did not garner nearly as much attention as classes meant for masculine pursuits.[22] Agricultural societies perceived women's work as subordinate to men's. They believed agriculture was a male profession, but they recognized women could be essential helpmates who, under proper supervision, were capable of contributing to some forms of farm production. They believed women excelled at dairying, for example, because the work was associated with their feminine skillset. As this book will later illustrate, rural women eventually took advantage of these competitions to showcase their abilities and contributions to their households and the broader community. However, early on, agricultural and fairs typically highlighted men's success in the countryside, not women's.

Behind the Scenes: 1846–1910s

The limited appeal of the earliest agricultural societies and the fairs they held meant that during the first half of the nineteenth century, their existence was precarious at best. Often groups were defunct mere years after they began. By the 1850s, however, more than forty county societies and at least 150 township societies existed throughout the province.[23] By the twentieth century, more than 350 county and township agricultural societies operated in Ontario.[24] Government funding aided the growth of these groups since 1830,[25] but agricultural societies soon realized that their long-term success depended on hosting robust annual fairs that appealed to the broader rural community.

The first provincial association of agricultural societies, the Agricultural Association of Upper Canada, was established in 1846 to circulate information about advanced farming practices and technology through publications, experimental work, and public meetings. The association's most important function, however, was its management of the Provincial Exhibition. The first Provincial Exhibition held in 1846 stimulated interest across Ontario and set a standard for what township and county fairs should display, including home manufactures, dairy products, and vegetables and fruits, all goods typically produced through women's work.[26] The importance of women in the household economy, and therefore in the material success of a farm family, made agricultural improvers realize the need to incorporate "women's work" into fairs if they were to achieve their goal of improving "every branch of Rural and Domestic Economy."[27] Agricultural societies had pragmatic reasons for including women, such as to increase gate receipts, but they were also committed to having all rural people adopt progressive ways. Beyond the benefits their work had for the family economy, nineteenth-century women bore responsibility for their families' moral welfare. Reformers argued that if women developed a "spirit of intelligence, order, neatness, and taste" through their honest industry and hard work, the agricultural class, in general, would be advanced.[28]

The ethos of competition was fundamental to agricultural fairs from the beginning. Fair supporters believed that the pride aroused by competition was central to encouraging exceptional performance.[29] Although a public show of pride was considered unwomanly because self-display was associated with sexual immorality and disorder,[30] women who organized and spoke at temperance, anti-slavery, and women's suffrage assemblies had established precedents for being in the public eye, and fair organizers believed all individuals benefited from the education that competition implicitly bestowed.[31] Agricultural societies sought women's patronage once they agreed that women's work was essential to the domestic economy and that women's presence had a refining influence on the fairgrounds.[32] Men who had been indifferent to women's participation at fairs came around to the idea that women had a "refining and subduing influence" and that the "intercourse of rational beings in a refined and cultivated society" should be encouraged.[33] Women were already attending fairs across the province, and reformers believed the benefits of including women outweighed the cost of having them compete in public.

Agricultural improvers also believed that women needed more education in the domestic realm. As early as 1849, an article in the *Canadian Agriculturalist* advocated for teaching women how to conduct their domestic tasks in accordance with the "scientific principles" of the day. The author, identified as "Amanda," argued that women were obligated to educate themselves about advances in domestic affairs. She explained that a "knowledge of chemistry and dietetics, in a cook, is invaluable to a family" and that an understanding of "the laws of health, and life, and mental philosophy, is absolutely necessary to the proper rearing of children." As she explained, "The suffering I have seen and experienced for want of knowledge, and the almost incredible advantage gained by the application of a few practical ideas, makes me very desirous for others, as well as myself, that we should have "more light."[34] Like other moral reformers, rural reformers wanted to cleanse and enlighten their communities.[35] They wished to shed "more light" on women's work because they believed improving it was necessary for improving rural communities. These ideas moved beyond popular farming journals and became entrenched in rural organizations such as the Dominion Grange and, later still, in rural women's organizations such as the Women's Institutes of Ontario and the United Farm Women's associations.[36]

Despite some groups' exclusionary policies, by the late nineteenth century, most organizations recognized that women were central to rural life. Women's participation in the Grange and the Patrons of Industry was a significant reason for the groups' rise in the late nineteenth century.[37] Although agricultural societies did not include women in their meetings, increased gender inclusion in the countryside likely contributed to the expansion of women's exhibits in the 1880s. Fair advocates argued that, although chiefly for the farmer, fairs also provided useful instruction for women because they would "see patterns of fancy work that they can look at and copy when they get home, or they can see some nice way of putting up fruit; or perhaps in a chat with some exhibitor of butter they may learn of a better way of treating their cream to make good butter."[38] Women's exhibits highlighted their enterprise, skill, and refinement and showcased their home, garden, and farm work. The steady increase in classes encouraged more women to compete, further stimulating interest. Local newspapers noted how women did their part to make fairs a success. The *Brampton Times* reported that at the 1871 Peel County Fair "in the ladies' department the fair sex contributed rather more than their share to the general effect of the exhibition, proving that their commitment for its success increases with each recurring year."[39]

By the 1880s, however, women remained excluded from fair boards, as shown in Figure 1.1. Fairs were managed by men who served as elected officers or directors. Each board had an executive committee with an elected president, vice-president (some boards had first and second vice-presidents), a secretary and treasurer (or a secretary-treasurer – both were paid positions), and roughly a dozen directors (this number varied over time and place). Typically, directors acted as superintendents of various management committees (cattle, horse, field crops, etc.), which had two or three members

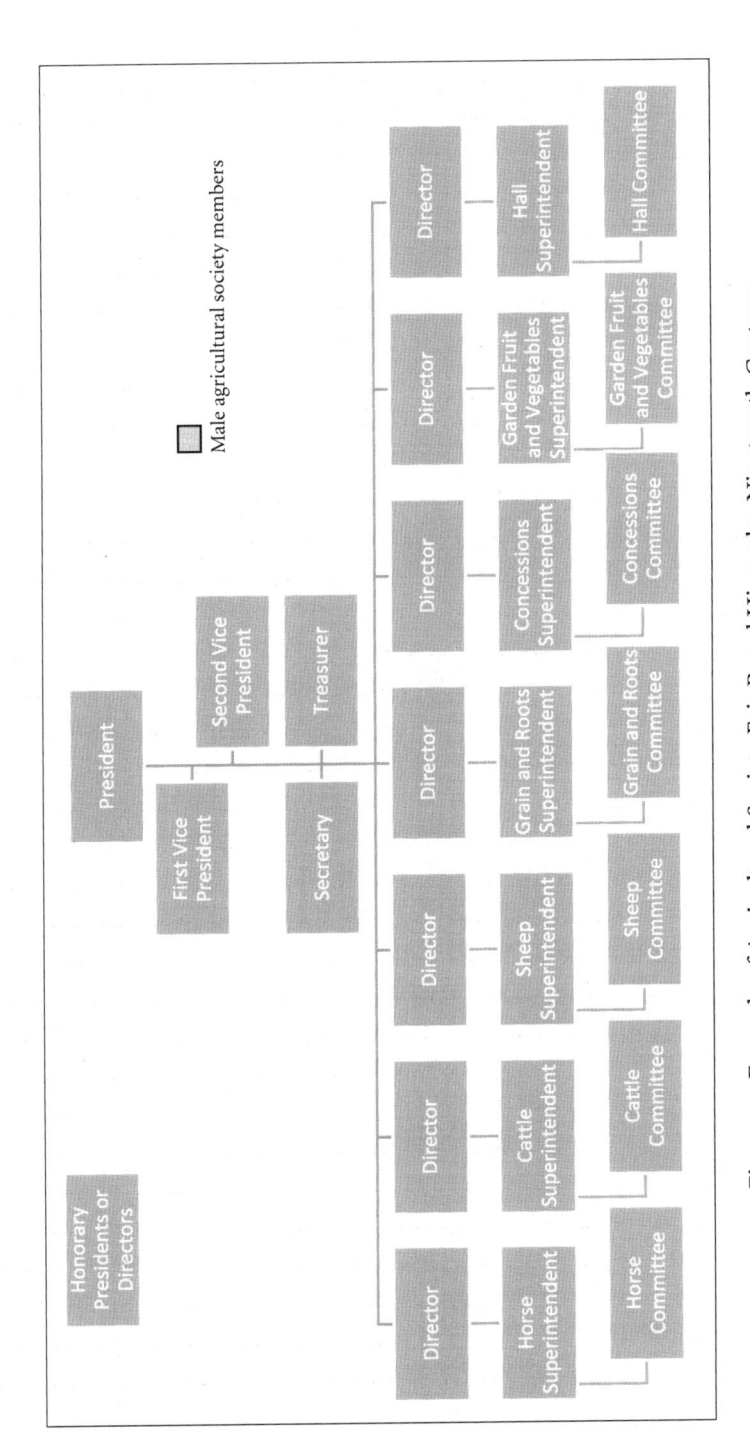

Figure 1.1 Example of Agricultural Society Fair Board Hierarchy, Nineteenth Century

serving. Some fairs also recognized honorary presidents or directors, who were usually past presidents or wealthy patrons of the society. The main difference between township and county societies in the nineteenth century was that county societies were usually required to elect at least one director from each district's township, and township societies simply selected from their local membership. Altogether, these male officers and directors directed fair programming.

The membership of agricultural societies fluctuated over time. Some organizations were more successful in gaining and sustaining membership than others, but rarely did they have more than a few female members.[40] For example, between 1868 and 1896, only nine women registered as members of the Brant Agricultural Society, with no more than one woman listed in any given year in a membership roll that often had upwards of one hundred members recorded annually. Of the nine reported women, five had never been married.[41] Fair prize lists during this period also illustrate the absence of women in the official record. Nineteenth-century Erin Agricultural Society prize lists recorded all-male fair executives and directorships.[42] Even after the First World War, only men served on the fair board, and the directors in charge of all departments, including "Ladies' Woollen" and "Fancywork," were men.[43] When women were members, they were not allowed to hold positions on the board. Instead, they helped male directors organize women's exhibits without any official recognition for their service.

The topics discussed at agricultural society meetings reflected men's interests. Meeting minutes and published reports show that these groups usually conversed about livestock and field crops, the state of current farming practices, and developments in agricultural markets and trade. Men sometimes discussed the significance of industrial manufactures but rarely mentioned domestic work. Beyond discussions that advocated for increased and improved dairy production in the province – specifically improved butter and increased cheese production – or the desire for more homemade and commercially-made cloth, any reference to items women made was absent.[44]

Nevertheless, agricultural societies did seek more participation from women.[45] In 1887, Toronto hosted an international convention for fairs and exhibitions. One of the topics discussed was how agricultural societies could get women to take a greater interest in fair work.[46] Mrs Noe, Miss Mary R. Heron, and Miss Kate Connelly formed a committee to report on this question. They recommended that agricultural societies secure women's cooperation by allowing women full authority to revise prize lists, purchase supplies, arrange exhibits, make entries, issue premiums, select judges, and allot exhibit space. They also advised agricultural societies to award healthy premiums to women's exhibits equal to other departments, which would compensate women for their labour and expertise.[47] The recommendations made it clear that women wanted more autonomy to manage their affairs, more space for their exhibits, more money for their work, and more respect and authority. It appears, however, that few agricultural societies acted on these recommendations. Although women wanted to have a more significant role and had a clear vision of what was needed, they remained stifled.

In 1902, Mr G.C. Creelman became the superintendent of Agricultural Societies. Creelman already served as the superintendent of Farmers' Institutes and Secretary of the Fruit Growers' Association, and in his new role, he was tasked with uniting "in some measure the work of the institutes and the fairs" and developing "the latter along educational lines rather than in circus features."[48] Creelman was critical of groups that focused mainly on managing fairs rather than thinking about additional ways to improve rural communities. He especially disapproved of fairs that employed "special attractions" that were unrelated to agricultural exhibits. Creelman proposed several reforms, including co-operating with Farmers' and Women's Institutes to "furnish lectures on agricultural topics, and give demonstrations in domestic science and butter making."[49] In 1904, he emphasized that fairs should secure "a women's building or tent on each fair ground for demonstrations and lectures on domestic science,"[50] and in 1908, he encouraged programming that educated both sexes.[51]

Creelman believed strongly in reforming the countryside, and he saw women as useful allies. Margaret Kechnie notes in her history of the Women's Institutes of Ontario that Creelman aggressively promoted the expansion of the WI to help facilitate rural change.[52] Reformers were concerned about rural depopulation, and they argued that discontented women led to dissatisfied families. They sought to ease women's toil by convincing them to make domestic improvements, such as adopting "scientific" methods in the home that would elevate their standard of life.[53] These methods included new time- and labour-saving technology, prudent and efficient business practices in the home, and professional expertise.[54]

Agricultural societies understood that they had a role in facilitating these changes, but male members were hesitant to give women too much authority. Rural reformers urged women, like men, to accept the family farm and home as an interconnected business enterprise and apply the rules of capitalist and scientific efficiency to its management.[55] Still, they did not feel that women were capable or deserving of leading agricultural institutions. While women might be "the glue that held the rural world together," they were not the ones tasked with determining the shape of that world.[56]

Women's own opinions of agricultural societies before the First World War are hard to find. During fair time, newspaper reporters might interview the occasional female judge, exhibitor, or visitor, but such reports were rare and provided little detail. For instance, in 1909, Mrs Campbell, a Wallacetown Fair judge, was quoted in a local newspaper merely stating that the display that year was "the best yet, and amongst the finest she has seen at the smaller fairs."[57] Usually, male newspaper reporters summarized their own opinion of women's exhibits or quoted from male agricultural society officers and directors. It was also common for reporters to limit their discussion of women's exhibits because they proudly professed ignorance about such "feminine arts." If women aided men in organizing domestic displays, they were never mentioned. The idea that fairs were the "annual reunion of the old boys"[58] was common, despite the reality that women attended and participated in great numbers. Women's silence reflected their limited power. Even though subsequent chapters will show that women's exhibits attracted exhibitors and fairgoers and improved the overall fair display,[59] the prevailing attitude was that men were public figures and that women should remain behind the scenes.

Moving beyond Membership: 1910–1940s

The first half of the twentieth century signalled a change for women in agricultural societies. Although this period did not see a full extension of women's rights, assumptions about women's roles were shifting. In Ontario, most women won the right to vote in 1917, and they subsequently expanded their social, economic, and political opportunities in the interwar years.[60] By the Second World War, the OAAS had created a Women's Section of the organization with its own president and executive. Here female delegates and district representatives met annually to discuss how to advance their interests on the fairgrounds. By the 1940s, many agricultural societies in Ontario allowed women to serve on fair boards, usually as "lady directors" who managed women's exhibits and growing children's departments. Overall, women's agricultural society activities during this period remained subordinate to those of their male counterparts. Still, women achieved more authority in areas considered to be of interest to women and made use of new opportunities to promote their work and serve their communities.

The first decades of the twentieth century saw the expansion of rural reformers' efforts to modernize farmsteads and agricultural and domestic practices further. The Progressive Era saw the proliferation of ideas about rural women's need to embrace modern conveniences. Reformers advocated science, technology, and professional expertise as the solutions needed for reducing women's drudgery, elevating their standard of living, and maintaining their devotion to rural life.[61] Some of the reasons for these measures were declining rural populations, labour shortages, and growing urban dominance, which stimulated farmers and rural people to take political and social action.[62] Rural women were also in a position to facilitate change. When most women won the vote in Ontario in 1917, they were part of a new electorate that had the power to influence politics. *The Globe* placed advertisements in journals such as the *Farmer's Advocate* to "appeal for progressive, right thinking" farm women who would take advantage of the vote by buying a newspaper subscription that kept them "informed on the questions of the hour."[63] *The Globe* had both an urban and a rural readership, and it published articles that encouraged rural women. In 1919, one journalist argued that previously women had not received their fair share of improvements on the farm, but "a new era" had emerged wherein women began to assert themselves and "man, with the consciousness in him of the past errors of his sex," was willing to accept that future expenditure must be made on "a more equitable basis.[64]

During the first decade of the twentieth century, agricultural societies began to broaden women's roles. The first action many groups took was inviting Women's Institute (WI) members to participate. The growth of WIs in Ontario in the early twentieth century was impressive. The first branch of the WI was formed in Stoney Creek in 1897.[65] By 1904, 150 branches with approximately 25,000 members existed,[66] and in 1914 the WI grew to 888 branches with almost 30,000 members.[67] These groups made significant contributions to fairs around the province. For example, in Elgin County in 1909, the Rodney Women's Institute organized a committee of women who planned and managed the ladies' exhibits.[68] These women had the authority to determine the prize money awarded for the individual classes, although male board members likely determined the department budgets.[69] In 1913, the Southwold and Dunwich Agricultural Society asked WI members to convene a committee to arrange for the Shedden Fair dinner.[70] By 1915, the Shedden WI members were also in charge of the flower committee, and the group sold candy and ice cream at the fair.[71] Other agricultural societies soon sought the aid of local WIs to organize women's exhibits, arrange dinners, and supply food on the fairgrounds.[72] Although women likely had helped most agricultural societies with these tasks previously, the official recognition of their participation and the authority awarded them – however limited – suggests men welcomed women's assistance in carefully defined areas. Church groups and other voluntary organizations in rural communities had relied heavily on women's volunteer work for their practical day-to-day operations and fundraising events, and agricultural societies recognized this. Often women's volunteer work coincided with their domestic roles, perceived nurturing natures, and responsibility for socialization.[73] Agricultural societies were no different in harnessing women's talents and contributions.

Women's role in fostering rural sociability was important. During the First World War, many women volunteered for community groups that fundraised for the war effort and boosted patriotic sentiment and sociability.[74] Even during the war, fairs were considered a special "holiday and a social time" for rural communities.[75] The primary function of fairs may still have been agricultural improvement,[76] but most people accepted that "there is no reason why the county fall fair should not be educational as well as a promoter of sociability and a good place to visit."[77] Women's position in agricultural societies did not change substantially during these years, but they were credited with nurturing a "spirit of neighbouring" and acting as "a great civilian army" that routinely took on activities that benefited their communities.[78] While men advanced the agricultural interests of their community, nation, and empire,[79] women were expected to promote the social interests of the community and nurture future citizens.[80]

Women also exerted agency by dictating the terms of their participation. For example, the Shedden WI had provided meals at the local fair for many years, but on 17 September 1921, their secretary reported that "after finding that the booth privileges were not satisfactory, they decided they would neither serve meals nor run a booth that day."[81] The details of this incident were not given, and the next year the WI resumed its meal service after securing the ground privileges it requested.[82] The WI's return to the fairgrounds demonstrates its ability to negotiate with the agricultural society and have its needs met. WI members could leverage their close relationship and long-standing service to agricultural societies, along with their experience and organizational ability in providing food services, as a way to ensure they received fair, if not preferential, treatment. WI members also often helped male fair board members (some of whom were their husbands) organize women's exhibits, so maintaining a good relationship between them was important to agricultural societies.

The Shedden WI and the fair board met throughout the years to make sure that both sides were clear about their roles and responsibilities.[83] In 1925, the WI again made it known that its participation depended on its ability to secure space "free of charge" for its booth.[84] The agricultural society agreed to this, and in return, the WI served food and sponsored an Institute Special Prize, worth $5, to be "divided equally between the Elgin Girls and Junior Farmers entries."[85] Participating at the fair allowed the Shedden WI an opportunity to raise funds and awareness for its causes. In 1925, it raised a profit of $48.10 by serving meals at the fair, which members used to buy batten and lining for a quilt made for a needy family.[86] The relationship between these two groups illustrated their symbiotic nature, but also that this symbiosis was under constant negotiation.

In the 1920s and 1930s, some agricultural societies demonstrated a greater commitment to include women by creating "lady director" positions on fair boards specifically to advise on women's exhibits, usually under the supervision of a male superintendent (see Figure 1.2). Women's new voting rights and the public service they provided during wartime likely influenced this shift, along with agricultural societies' recognition of the improvements WI members and other women had already made to fair management. At the Ontario Association of Fairs and Exhibitions (later renamed the Ontario Association of Agricultural Societies) Convention in 1920, Mrs Laura Rose Stephen of Huntingdon, Quebec, spoke about female directors on fair boards. She was of the "strong opinion" that "no Fall Fair Board is complete without women representatives" and that the events of the war and its aftermath had enabled women to showcase their ability to participate in every sphere of society.

> It seems unnecessary for one to expatiate even briefly on the advisability of women acting with men for the furtherance of fair work. Women have made such advances into preserves held sacred for men, it would seem that no position is really now secure to men alone ... We have women lawyers, doctors, judges, ministers, and members of parliament. No door seems now closed fast to women. The unusual civic conditions of the past six years have placed women side by side with men in workshop, office and laboratory. Men have had, through circumstances, to see women from a different viewpoint and, I will be pardoned when I say the vision of man has broadened and his conception of woman's worth and possibilities has vastly increased. Women have demonstrated that they have education, brains, nerve power and physical endurance sufficient to make them useful and needed in almost every sphere of world activity.[87]

Stephen contended that a close, mutual relationship existed between men and women on the farm and that women were intimately linked with the business and knowledgeable about every branch of agriculture. Therefore, she argued, "to place woman on an agricultural fair board is not lifting her to a new and bewildering position, but to one into which she fits and can immediately do good work."[88] Stephen foresaw objections from some men about the inclusion of women, but she dismissed these objections as shortsighted and "dragging on the neck of progress."[89] Stephen described how women would improve fairs by creating more attractively displayed exhibits, enlisting qualified judges, arranging better food and meal services, and bringing about new and improved ideas more generally. Her speech's overall

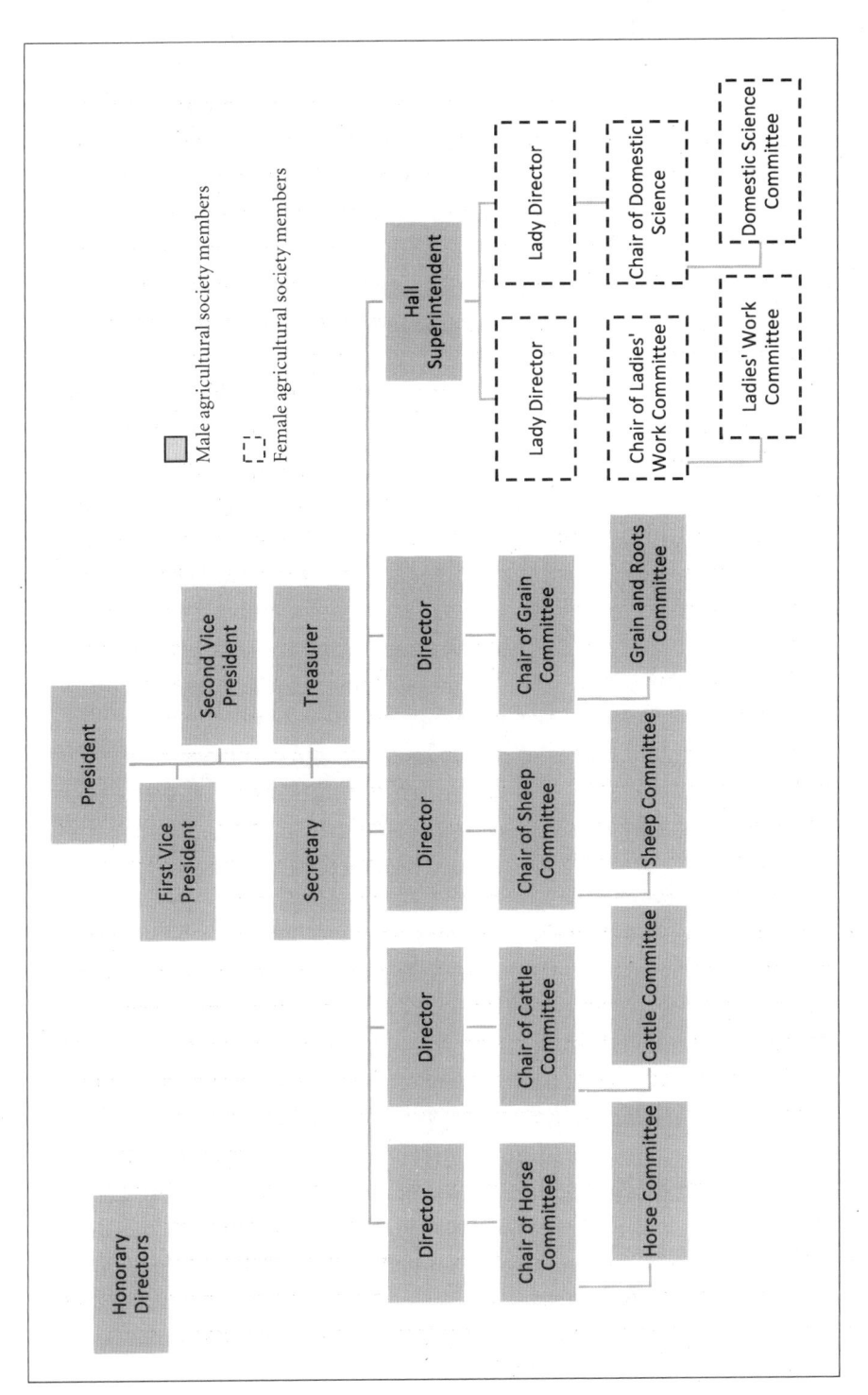

Figure 1.2 Example of Agricultural Society Fair Board Hierarchy, 1930s

message was that women had proven their usefulness to agricultural societies and that fair boards needed to accept female organizers if these events were to flourish in the future.

The 1920s signalled a shifting and sometimes volatile landscape in Ontario's rural communities. Farm prices had risen steadily after 1896 until they peaked in 1920, after which post-wwi European recovery ushered in a rapid decline in prices from 1920 to 1923. In 1920, the number of agricultural societies also peaked at 354 and then began to decline. This period witnessed the amalgamation of some county societies with township societies, and several new township societies merged as well during these years.[90] Agricultural society records of the period are inconsistent and do not reveal a complete picture of membership numbers. The societies expressed worries over stagnation and decline as their numbers decreased.[91] Agricultural commodity prices peaked again in 1925 and held until 1929,[92] but throughout the 1920s, fair organizers maintained a keen desire to improve fairs and create greater interest among farmers, and to attend to "the many things that make home attractive and prepare our men and women of the future for the more serious problems of life."[93] At the same time, women expressed the desire for leadership roles on fair boards. These factors likely influenced the noticeable increase in female directors at that time. For example, the Beeton Agricultural Society in Simcoe County accepted its first "Lady Directors" at the annual general meeting in 1925. The women took responsibility for revising the prize list for women's exhibits. The fair board secretary and two male directors supervised this work.[94] In 1928, the Harriston Agricultural Society in Wellington County first allowed female directors by naming each of the local WI Presidents to the fair board to represent women's interests.[95] 1928 was also the first year that the Normanby Agricultural Society in Grey County reported a "Ladies' Committee," consisting of three women.[96]

Fairs typically reflected ongoing changes in the rural landscape, and they adjusted the types of competitions they offered accordingly. Still, there were times when the male organizers of these events believed they knew what was best, even if their ideas conflicted with what women told them. For instance, in 1928, Miss E.A. Slicter of the Women's Institute and Mrs Mark Senn of Caledonia told the mostly male delegates at the annual OAAS Convention that women's exhibits needed more attention. Senn noted how fairs failed to update prize lists for women's competitions. For example, she insisted that bread and butter classes, "two of the former standbys," were no longer needed.[97] She also said that women welcomed the change whereby "creameries and bakeries were doing the work as well or better and relieving the farm household of much drudgery." George Robertson of Ottawa disagreed. He argued that "if the women gave up baking and dairy work entirely, the men were in for a bad time."[98] The exchange between Senn and Robertson illustrates how women and men did not always agree. Of course, Mrs Senn did not represent all female competitors (homemade bread remained a popular class item for years to come). Still, her remarks and the response they received are worth consideration. By engaging women in these discussions, agricultural society officers and directors recognized a need for women's expertise in creating a more modern, inclusive fair. At the same time, however, they limited women's ability to make change. Although they asked for women's advice, they did not necessarily take it.

The continued decline in the number of agricultural societies in the 1930s and the economic difficulties that resulted from the Great Depression suggest that including both women and men was more critical than ever.[99] Most fairs reduced admission prices to stave off decreased attendance during the 1930s. Still, the lost revenues, as well as the debt some agricultural societies had incurred to improve their fairgrounds, meant that many groups were struggling.[100] At the same time that agricultural societies struggled, women's political and social authority increased, and their contributions to household economy were more important than ever. One reporter credited women's support for fairs as the reason they survived during these difficult years and suggested that "more active participation by the womenfolk was what the fall fair needed to put it on its feet again as an important and interesting annual event." He noted that "though for a few years inclined to languish, [the fair] has 'come back' in vigorous fashion, and to the wives and daughters of the farm must be given a great deal of the credit."[101] Although agricultural societies across the province were inconsistent at providing women with greater responsibilities, it became difficult for them to ignore the benefits women brought to their organizations. Slowly more and more groups increased the number of women involved. For example, by 1930, the Collingwood Township Agricultural Society's fair board had an equal number of men and women serving as directors, and the next year women outnumbered men eleven to nine.[102] By 1931, the Elgin County Fair had twelve women serving on the board as "Lady Directors."[103]

Agricultural societies needed women, and women knew it. Female members found themselves in a position to ask for more. At the 1930 Ontario Association of Fairs and Exhibitions Convention, Miss K. Goodfellow of Long Branch told the delegates that women's participation at fairs was necessary because "women are the back-bone of our country! What would our Fairs be like without their handiwork? Where would our men be if our cooks ceased their work, thinking it not worth the effort?" She advised agricultural societies that female members should receive "every encouragement in their work" and that agricultural societies should employ qualified female judges and provide networking opportunities for women.[104] Goodfellow pointed out that there was a double standard at play, and only when women were given equal opportunities to men would fairs advance all rural residents' interests.[105]

Women reacted against what they understood was a second-class membership. Even societies that allowed female members to serve as associate directors or "lady directors" excluded them from being individually recognized in prize lists. For example, the 1934 West Elgin Agricultural Society had female associate directors, but none were listed in official prize lists as management committee members, even for women's exhibits. In the 1935 Erin Agricultural Society's prize list, the category of "Lady Directors" was recognized, but instead of listing these women by name, as the male directors were, they simply identified them as the "wives of the Officers and Directors."[106] When women were included, they almost always served with men who were singled out as the superintendent or director in charge.[107] Other agricultural societies continued to exclude women altogether. In 1938, J.A. Carroll, the Superintendent of the OAAS, reported that "There is at least one Society which boasts of the fact that they have no women around them at all."[108] Some groups had significantly increased their female participation, which was heartening, but their continued failure to recognize women by name or allow women more control over their exhibits demonstrates the limited power women held.

Women's roles in agricultural societies varied in the 1930s, but a notable transformation resulted when the OAAS created the Women's Section of the association. In 1937, female members organized the first "Women's Meeting" at the OAAS Annual Convention in Toronto, and in 1938 they created an official "Women's Section." The creation of the Women's Section was credited to Ethel Brant Monture, who was the first president of the Women's Section and a woman considered "synonymous with movements for community betterment."[109] She complained of being the only woman who attended the convention and "of going home with her clothes smelling of tobacco smoke."[110] Monture had felt earlier conventions had "nothing of interest for [women]" because male organizers had given "no thought … that their section was important." But Monture also knew "when men took visitors to the fair, they always went first to the women's section," so she believed that interest in women's exhibits existed.[111] When Monture approached J.A. Carroll about hosting a women's meeting, he challenged her to find enough women to attend. Monture travelled the province to garner support. In 1937, seventeen women met to set up a tentative committee. She explained that they were "good women, and capable. But the men did not take kindly to the idea of a separate women's section or even to the idea of lady delegates, and I was not popular at the time." Still, she persevered, and in 1938, two hundred female delegates attended the first official Women's Meeting. Monture commented later that women's enthusiasm had "never waned. Five years after the organization of the section, the *Farmer's Advocate* had an article on the renaissance of fairs. They said they did not know why, but we women could guess."[112]

As the president and chair of the first official meeting in 1938, Monture recognized the historic significance of the event, noting that it was "quite the historical occasion" for women to meet for the first time in the OAAS' history. She contended that the meeting would lay the foundation for "a mighty structure [to] grow which will do much to foster not only the welfare of women in Agricultural Societies but all work of Societies, and indeed of agriculture in country life."[113] J.A. Carroll spoke to the women assembled and expressed his gratitude for their attendance, but he also cautioned them about the challenges ahead. He admitted that he was "not very clear" on the best way to promote women's interests.

> Many Societies, as you know, will soon be celebrating their hundredth birthdays. During a century some of them have become quite con-

firmed in their habits, and therefore, we cannot expect too radical changes. Anyone may become a member of an Agricultural Society by paying $1.00 membership fee. In the case of married couples it will be a matter of battling it out as to whether the one dollar entitles the husband or wife to membership. If two dollars are paid, you will have two votes at Agricultural Societies meetings. I may be criticized for saying so, but I believe in the interests of the work that women should be more aggressive and should take more part in members' meetings. If a group of women show that they are sufficiently interested to come to the annual meeting, with power to vote, I feel more attention will be paid to them than otherwise.[114]

Carroll recognized that many women were interested in society work but allowed their husbands to be the public face of the family on the board, while others were members but did not attend meetings regularly, likely because of men's disinterest in discussing women's activities. Furthermore, Carroll explained that agricultural societies were inconsistent in providing official titles for women on their boards, and male members often refused to allow women to become full-fledged directors because only a limited number of directorships were given, and they wanted to reserve them for men. He noted that women were "usually" permitted to attend board meetings and "say all they wish to about the management of *their* particular department" [italics added]. Carroll expressed his desire to see women interested in all fair departments, but he pointed out that women had a special responsibility regarding domestic departments.[115] He made clear where women belonged and what their proper responsibilities were.

Still, he regretted that some agricultural societies had "no recognized channel of getting information or suggestions from women" and that they preferred to rely on unelected officers' and directors' wives for their input or simply forgo consulting women at all, which he believed resulted in inferior domestic exhibits. Carroll advised all societies to create a women's committee to manage women's work. Despite his limited understanding of how women could contribute, Carroll was an important ally for female members, and he assured them that "With respect to the provincial organization, be assured you are welcome here at all times to participate in [the] convention."[116]

Of course, women advocated strongly for their participation in agricultural societies as well. Mrs J.K. Kelly, a future president of the Women's Section, spoke in 1938 and assured women that they had "a very definite place in fairs." Kelly was critical of what she perceived as a shortage of entertainment geared to women's interests, the small amount of prize money they received, and a lack of co-operation between the sexes, which she attributed to men's disinterest.[117] Kelly believed that "the men feel they are much more important than women" and thus did not offer women's work the same appreciation as their own.[118]

These concerns prompted women to establish a separate Women's Section in 1938. A separate section of the OAAS allowed women to continue to be a part of the organization while developing a subculture that combatted male dominance.[119] The Women's Section gave its members a comfortable space where they could build confidence in their skills and enjoy other women's camaraderie without fear of male judgement.[120] Whether identified as "Patronesses, Honorary Directors, Women's Committee, Associate Directors and Lady Directors," Kelly and women like her felt that women should hold their own meetings to discuss matters related to women's exhibits.[121] Kelly advocated for fairs to have a female convener of women's work because women understood this work best and could make efficient decisions. She also outlined female directors' duties, including their responsibility to understand what was made and grown in their communities, showcase those items, organize and display entries submitted for evaluation, and encourage more women to participate.[122]

Kelly was particularly pleased by the creation of female district representatives in 1938, who attended the OAAS Convention and reported on fairs in their region. Agricultural societies selected district representatives at their annual district meetings. District meetings brought together neighbouring agricultural society members to learn from one another about shared challenges and opportunities. Once women's divisions were created, women designated their own representatives to discuss concerns specific to women's work and "to advance the best ideas in each district."[123]

Like agricultural improvers from the nineteenth century who believed the home was fundamental to the political economy, women involved in agricultural societies thought they had an essential role in improving country life. In 1938, Miss Bess McDermand, the first female superintendent of the Ontario Women's Institutes after George Putnam's retirement in 1934, explained that WIS remained useful allies for agricultural societies because their members were committed to the betterment of the countryside. She noted that each WI branch concerned itself with "making women better home-makers." Therefore, these groups had "a relationship to Agricultural Societies" because the "home affects the business of farming more than the home affects business of any other kind and so we have this close association of bettering agriculture and of bettering homes. They march together."[124] McDermand had grown up on an Ontario farm before leaving to receive post-secondary education, and she believed that the home and farm were united.[125] For her, advancements in farming were a "means to an end." Farmers made more money so that their families "may have a better living, but before a better living is possible the money must be turned into healthful food, useful clothing and comfortable shelter."[126]

The significance of the 1938 OAAS Women's Meeting did not go unnoticed. The *Globe* reported in an article entitled "Rural Ontario Women Make Entry into Field Hitherto Held by Men" that "History was made for rural women of Ontario, Thursday afternoon, when representatives of many parts of the Province – members of the Women's Committee of the Ontario Agricultural Association of Agricultural Societies... [gathered] for their first sessions, and discussed, with enthusiasm, rural matters pertaining particularly to women's interests. Women's status in the societies; duties and obligations of women directors, fall fairs, Women's Institute exhibits, judging and judges, all of vital concern to rural women, were subjects presented by able speakers, and discussed with animation by the delegates."[127] OAAS President W.J. Hill admitted that previously they had always elected male delegates, noting that fewer than a dozen women attended the Annual Convention of 300 or 400 people. Yet, he acknowledged that fairs "would not be much of a success were it not for the assistance of the women."[128] Mrs Monture argued women wanted "to work along with the men for the betterment of the community." She recognized that "in some quarters we are not very welcome," but asserted, "we are going to be there just the same!"[129]

The *Farmer's Advocate* also published an article about the first Women's Meeting, titled "Rural Women Invade Man's Realm in Fair Work." It described how women met to assess their status in agricultural societies and how their fair contributions needed to be recognized.[130] "News of the uprising must have got abroad," the article went on, surprised that more than sixty-five women had assembled. The reporter covering the event told how, before the meeting, Mrs Monture had been advised by an "elderly delegate ... perhaps from Bruce or Glengarry" and of Scottish background that "they would be juist as weel off at hame washing the dushes."[131] However, the *Farmer's Advocate* also published another article that insisted the "Women stole the show" that year and that "women members of the various boards can be of great assistance in year-round activities of any live Society."[132] Female members did not let the criticism they received deter them. Unlike research that suggests female volunteers tended to belittle or minimize their efforts, the women assembled at the OAAS Convention were confident that their contributions were vital to fairs' success.[133] Like other rural women volunteers, those in agricultural societies were proud of their work. Despite some negativity about the meeting, the women at the convention had established a "new order" where at least one woman would attend agricultural society district and annual meetings every year.[134]

Despite women's passion, however, they continued to struggle for respect. In 1938, Mrs O'Leary of Lindsay recalled how she had asked a man "knowledgeable about judging" what he would expect in a female judge, and he had replied that "she would be good-looking and have red hair!"[135] O'Leary told the story to highlight some men's very gendered and condescending perspective. This disregard for women also continued in official programs and prize lists. Even by 1941, the West Elgin Agricultural Society simply identified their "Lady Associate Directors" serving in the "Home Department" as "M.W. Page and Associate Lady Directors" in their official prize list.[136] Page, the department supervisor, was singled out for recognition, but the female committee members were not.

The 1938 OAAS Women's Meeting did not cause all agricultural societies to change established practices immediately, but the creation of the Women's Section was a significant result of changes that gave female members the confidence to organize in the first place. Many took advantage of changes in the OAAS constitution and sent at least one female delegate from each local society to the convention in Toronto. In 1939, *The Globe* reported that a "group of smartly attired, alert rural women from many parts of the province" met at the King Edward Hotel to discuss how they might aid in making agricultural fairs more representative and valuable for their communities.[137] The description of rural women as "smartly attired" and "alert" illustrates the way that women's appearance was central to how others perceived them. Still, the reporter also identified one of the speakers, Miss Ina Hodgins of Carp, as a "farmer, stockbreeder and judge at Ottawa Valley Fairs." These descriptors were typically reserved for men, so by acknowledging Hodgins's skill and authority in these positions, the reporter admitted women's abilities went beyond their appearance.

Hodgins herself made a point of describing the interconnectedness of farm men and women. She asserted that "Behind the man who is exhibiting something of good quality you will generally find a woman of ambition and determination encouraging him."[138] She also explained that women wanted to exhibit because they liked "to do something on our own account. Most of us can do some one thing just a bit better than another can. Naturally we, too, like to do little tricks in public and walk off with a pretty colored ribbon." Hodgins described fairs as "places which give us one grand chance of bringing our light from under a bushel and setting it up on the hill ... It may be a cake, jelly, butter, the best embroidery or the most artistic rug."[139] Hodgins's examples confined women's work to traditionally female activities. Although she was involved in animal breeding and farming and was clearly committed to extending women's influence, her list of women's contributions remained squarely within the domestic realm.[140]

During the Second World War, women had another opportunity to demand more authority in agricultural societies. They acted as a reserve force in many facets of rural life during the war, especially in harvesting crops,[141] and agricultural societies also required their help. Although male organizers continued to dominate fair boards, women were enlarging their sphere of authority. In 1940, the OAAS granted women equal voting status. Dr J.J. Wilson of Burks Falls noted that "previously women were without status on the board or in the association," but they now had equality with men.[142] In 1941, the president of the OAAS, William Walker, asked convention delegates, "What would our Fairs of to-day be without [women's] valuable help?"[143] revealing the role women played in keeping agricultural societies and fairs afloat.

As in the First World War period, members of agricultural societies acknowledged women for their wartime efforts during the Second World War. Walker congratulated them for creating special classes for war work, and although he noted that some other fair exhibits received fewer entries because women "were hard at work in the war effort," he acknowledged that "That is as it should be. We are proud of them."[144] Mrs Kelly noted in her 1941 OAAS Women's President's Address that women had a responsibility to uplift community morale, and she believed "a progressive, aggressive Agricultural Society" was a vital part of the task and a "necessity in any community."[145] The women who spoke at OAAS Conventions during the war emphasized women's ability to care for their families, assist in agricultural production, and cooperate to ensure that the social and economic issues in their rural communities were met.[146] Women also encouraged fair organizers to create prize lists that included war work and discouraged using rationed commodities or employing women in superfluous tasks.[147]

In 1943, J.A. Carroll praised women at the OAAS Convention for their productive agricultural work during the nation's time of need. Furthermore, he encouraged agricultural societies to work harder to have their fairs represent women's changing roles on the farm by offering classes for work he argued had previously been undertaken by men, from dairying (he neglected women's longstanding history in this domain) to driving a team of horses: "Are women proud of the fact that more cows are being milked by them now? We think they ought to be. Could this not be reflected at the fair by a competition of milk maids? Women are using horses much more now. How about hitching and driving competitions? What we have in mind is that while it may be necessary to cut down on some of the classes for fancywork and other exhibits that women worked at formerly, should we not add classes emphasizing the productive war work on which women and girls are now engaged? Would it be possible to have classes for them, to demonstrate what they have been doing on farms?"[148] Carroll wanted fair programs to represent what women were doing in the countryside so that those activities would be recognized, encouraged, and improved.

Carroll continued to advocate for more leadership opportunities for women, but he remained careful to describe how they might "appropriately" achieve this goal. He mentioned that some societies still limited women's influence, and he tried to advise women on how to best facilitate change. Carroll paternalistically warned women that "it would be useless or unwise to scold them [men], and a much better plan would be to become more aggressive without being arrogant." He noted that he believed women could offer their services during agricultural societies' annual meetings, and asked "Why should the annual meeting not be conducted on a plan similar to this convention? That is to have a separate session for women, and for men and women to meet together for the election of officers and programme features in which both are mutually interested."[149] While Carroll valued women's service, he believed women had to conform to the rules set before them. In a telling piece of advice to female OAAS delegates, he suggested that "If conditions should favour such a move [to attend the annual meeting], we think the men would not object, in fact would be very pleased, if the ladies were to say: "Now we would like to *help* you to improve attendance and interest at the annual meeting, and we are *prepared to serve* refreshments if there are no objections. Women in the home and in the community have a direct responsibility and usually assume it for social activities. Why not apply this principle to Agricultural Societies? [Italics added]"[150] Carroll knew changes were needed, but he was unwilling to consider that the system itself required changing. He suggested women had to take the initiative and prove their usefulness to male board members in a way that did not challenge their traditional roles. Although supportive of women, his recommendations illustrated his ideological struggles surrounding women's involvement.

Ultimately, women's view of their roles exceeded that of Carroll's. They believed that they could more than manage women's exhibits, and they understood their broader worth to these organizations. The importance of women's involvement during wartime also gave them the leverage they needed to make change. At the 1945 OAAS Convention, Mrs Kelly argued that there were "a number of Fairs which would have been cancelled had not the women encouraged and given extra assistance." She felt that women deserved the opportunity of reaching their "proper place in the set-up of Fair management."[151] During the war, many agricultural societies created women's divisions modelled after the Women's Section of the OAAS. Kelly advised women to attend their agricultural society's annual meeting to report on their departments and elect officers before adjourning to their own meeting to discuss their division and select a convention delegate and district representative.[152] Executive committees continued to be dominated by men, but with the creation of separate women's committees, women could elect their own honorary presidents, vice-presidents, secretaries, and directors. The number of women on fair boards generally increased during the war as well. The South Brant Agricultural Society had twelve male directors on the board for the 1945 Burford Fair but nineteen female directors to take care of the women's division.[153] Men continued to hold the balance of power in agricultural societies by holding executive committee offices and supervisory roles, but the women's sphere within these organizations was expanding.

Ladies on Board: The Postwar Period

The postwar period was when women finally broke through the "glass ceiling" of fair board management. Some women were even elected to the highest agricultural society office – president. Tremendous change occurred in rural communities: the number of farm families decreased significantly, and more women worked in the paid workforce. Indeed, numerous sociological studies during the 1970s and 1980s showed that women's off-farm labour was increasingly necessary to support their households.[154] Of course, many of the women who worked off the farm also provided vital on-farm work, including assisting in managerial tasks, bookkeeping, and providing supplementary labour.[155] This is to say nothing about the extensive domestic and caretaking duties these women continued to perform.

Those women who remained were committed to maintaining the family farming business, and many supported agricultural societies that were also responding to the demographic shifts.[156] Agricultural societies looked for ways to elevate the status of farming in urban people's minds and stimulate more interest and a greater appreciation for agriculture among the expanding urban population. They wanted to educate visitors about the value of rural life. Women's domestic work and handicrafts had often served as a connection between rural and urban women, so it is unsurprising that agricultural societies increasingly relied on women to manage exhibits that interested all sorts of fairgoers. Women took advantage of this fact to expand their influence, and while traditional women's work continued to be their focus, more women were transcending boundaries that had previously seemed impassable.

The "Home Department" had always been women's domain, but many agricultural societies had had male superintendents and supervisors as liaisons between female committee members and male boards. In the postwar period, agricultural societies finally recognized women as the ones in charge of these departments. In 1950, the OAAS Women's president, Miss Ina Hodgins of Carp, reminded female OAAS delegates that they needed to "keep in mind the objectives of the women's section of the fair."[157] Hodgins believed women should focus on raising and teaching new and improved practices and standards for handicrafts and other women's work while also encouraging a "love for the beautiful," introducing "new classes related to the special needs of the times."[158] Although Hodgins urged women to "not be afraid to make a change. What was impossible last year may be possible now,"[159] she was committed to maintaining women's focus on their traditional roles as organizers of the domestic, garden, art, craft, and children's work.[160] Mrs A.L. Dickson, the secretary-treasurer of the Perth and District Agricultural Society, argued that it was "high time that women on fair boards have a voice equal with men." However, she also contended that "We women do not want to run the fairs; we just want to feel that we are a necessary part of the fair programs and that we should be allowed to run our part of it without reservations."[161] During this period, it was common for women to limit their participation to more traditional pursuits. Women wanted greater control over the exhibits under their purview, but they never attempted to supplant men's authority over traditionally masculine departments such as draft horse or livestock shows.

Women were reluctant to manage departments unconnected with the domestic realm, but they did seek more representation on executive committees and greater input in fair administration. It was not until the 1960s that female directors in some agricultural societies could vote on fair business at the annual general meeting. Before then, women had to wait in the hallway while men elected the executive officers and attended to other business matters. Women could only join the men for the social program that followed, usually to serve refreshments.[162]

Women continued to feel that the best way to represent their interests was through separate women's divisions in which they elected their own executive officers and generated a level of authority for themselves. The creation of separate women's divisions (later renamed "homecrafts" divisions) allowed women to elect executive officers, directors, and committee members and run their departments as they saw fit. Figure 1.3 demonstrates the divided roles in agricultural societies based on gender at this time. For example, in 1951, female members of the Paisley Agricultural Society organized a separate women's division. Although the agricultural society had allowed female directors in 1944 to help manage the school and women's exhibits, the creation of a formal women's section gave women the ability to hold separate meetings focused on their concerns. Men and women met together to discuss general preparations of the fair, but "the ladies retired to the kitchen or another room" partway through meetings to discuss their matters while the men continued with their business in the boardroom.[163] In the same way that the Women's Section of the OAAS met separately, women in local agricultural societies believed that their work was better served (and their time better spent) when they met independently and focused on improving the departments under their control.

Nevertheless, many women remained dissatisfied with the limits placed on their ability to manage their affairs. Some women believed that holding office on fair executive committees would enable greater control over their departments, create more transparency, and offer women a chance to voice their opinions. Women's initial foray on executive committees usually involved secretary or treasurer positions. Mrs Koelher of Dundalk remarked that, as the first female secretary of the Dundalk Agricultural Society in 1948, she had been "a rare bird indeed." By 1956, however, fifty-two agricultural societies had female secretaries.[164] By 1959, sixty-five of the 260 existing agricultural societies had either a female secretary or treasurer. Some organizations, such as the Teeswater Agricultural Society, had women serving as first or second vice-presidents.[165]

The election of female agricultural society presidents was especially significant. In 1959, the Smithville Agricultural Society elected its first female president.[166] In 1960, seventy-two agricultural societies had female secretaries and another female president, Mrs O.R. Sproule, had been elected to the Wyoming Agricultural Society's fair board.[167] F.A. Lashley reported that Sproule won the position because of "her capabilities and keen interest in the general programme of the fair."[168] Although female secretaries and

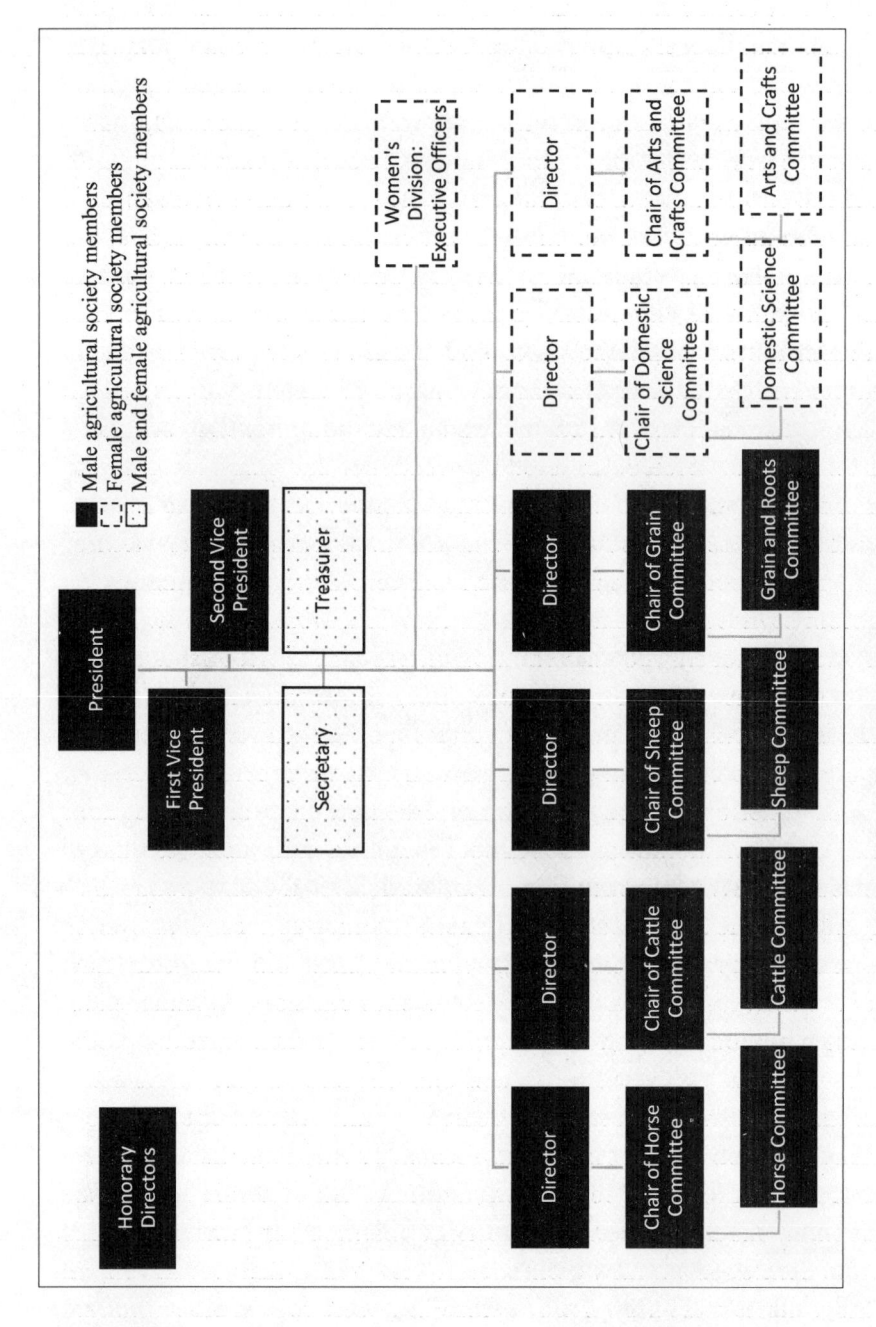

Figure 1.3 Example of Agricultural Society Fair Board Hierarchy, 1950s

treasurers became common, they were not equal in status to presidents or vice presidents because they were paid positions. They were employees, and thus their work was directed by the organization's leaders. Presidents, however, were selected based on their ability to represent the entire agricultural society. The women who held these positions had convinced all members – men included – that they had the leadership skills, public speaking ability, knowledge, and confidence to excel in the role, and, perhaps most importantly, the respect not just of their fellow agricultural society members but also of the community.

Jeanette Jameson became the first female president of the Rockton Agricultural Society in 1975, and she remembered her election as unusual for the time. Jameson had grown up on a farm, one of three daughters, and she was used to doing whatever work needed to be done, regardless of the task. "I was [treated like] a boy, I had no choice [laughs]," she explained. "I don't think I got out of anything on the farm." Jameson completed eighteen 4-H clubs in her youth, which included homemaking clubs but also calf and sheep clubs. Jameson's broader interests in agriculture, coupled with the fact that she "married into a fair family," helped facilitate her involvement in the Rockton Agricultural Society as a director in 1959.[169] Jameson was a teacher at the time, so she was first appointed to the school fair committee and later became the department's chairperson. When she was elected president of the society in 1975, she explained that "There were no women on the board when I became president. I was sitting there because I was the head of the school fair department and somebody said … Jeanette should be president." She felt the men on the board sometimes looked "suspiciously at the things [she] was doing" but noted her response was, "you just sort of turn your head and go on." Despite this, Jameson believed that she never felt "condemned in any way," and she thought her fellow officers never "looked down on [her] or anything," but she did recognize that "it was a big change for Rockton, for sure."[170]

Fellow Rockton Agricultural Society member and past president Eleanor Wood believed that acceptance of Jameson's role was partly because her husband had previously served in the position. She reasoned that members viewed the Jamesons as a couple, and because Neal was "bedrock" in the association, it was likely safe to "venture out into Jeanette."[171] Wood clarified that, though friends, it was not uncommon for male board members to question "a woman's point of view."[172] Jameson, however, had "fire in her belly," as well as a "practical" view to how things should be done, one reason she believed male members were quick to support her direction.[173] Despite Jameson's recognition that she experienced some of the gendered assumptions of the time, she asserted that she never thought of herself in those terms. "I never considered myself female or male, like, you know, because I didn't grow up that way. There was no division in my house. When I was teaching, I became a principal of a school in the early 60s, so you know, that was not normal for that time either, so it was not ground-breaking at all. If there's a job to be done, you do it."[174]

Not all women wanted to move beyond their traditional roles. Wood explained that when she got involved in the Rockton Agricultural Society in the 1960s and before Jameson was elected president, "Ladies kept to themselves, they didn't aspire to be on the board." Wood explained that they had no problem letting the men run "those other things," and they focused on looking after their own affairs. In time, however, women's committees felt that they needed greater control, especially in terms of finances. As Wood explained, it was frustrating for women to constantly have to seek approval from the main fair board to move forward with their projects: "you needed a new stapler, and they [the main fair board] wouldn't give you any money for that, and you'd have to go, sort of on bended knee for that, and you began to think, maybe we should have our own representative on there, and it took a while before anybody wanted to step into that den of iniquity [laughs], where the men were because men will handle a meeting, they will go at each other's throats and be done with it when it's over, but let them take a woman apart at a meeting, and she's never going to get over it, and that, I think, was a fear ... Women didn't speak up like they might have ... I think eventually we found women strong enough to do that, and that's when it happened."[175] Wood suggested that women felt ill-prepared to deal with fierce debates and disagreements that could occur at fair board meetings. She understood men to have a "thicker skin" than women, and she believed the heated discussions that often took place when fair board members disagreed intimidated and alienated some women. Wood suggested it took time before women were willing to speak up and stand out.

One might conclude that because women had difficulty attaining positions of power in agricultural societies that considerable tension existed between the sexes, but the women I interviewed never expressed any significant concern over this imbalance. Most women emphasized that men and women worked together to organize successful fairs. Often women served in agricultural societies alongside their husbands, male friends, and neighbours, and they believed these men had their best interests at heart. When Glenda Benton, past president of the Homecrafts Division and long-time member of the Georgetown Agricultural Society, described the relationship of the men and women on the fair board during this period, she explained that it was one of friendship. She noted that the sexes worked well together and that the "men were always there when we needed them."[176] On the other hand, women may not have wanted to be seen as troublemakers. As sociologists Peggy Petrzalka and Susan E. Mannon have argued, social cohesion was often more important to rural women than personal profit or advancement. Many women were willing to limit their ambitions if they believed it was for the common good.[177]

Another reason women shied away from taking on additional tasks was the amount of responsibility they already had at home. Many rural women did not have the time to attend meetings, especially if their husbands were also members and they had young children to attend to. Margaret Lovering of the Ancaster Agricultural Society had gone to the fair since she was a child, but it was not until later in life that she was able to get involved in the agricultural society. For her husband to be involved, Lovering had to stay at home and take care of their six children. Before she joined the fair board as a member of the Crafts Committee in 1972 (she later served as the Chairman of the Crafts Section and the Secretary-Treasurer for the Women's Section), Lovering helped run the school children's fair, but as she explained, "Well, I had small children, I had six children you know [laughs], and it was very hard, and when I farmed them out to a lady in Alberton, she didn't think that I got back quick enough, but as soon as the judging was over I was back home, and I said, that's it! Until my husband became president of the fair, and they asked me to join the fair then."[178] Even though her husband and the Lovering family had been involved for many years, she explained, "I just couldn't with small children, because, how can you drag a bunch of children to the fair and try to help?" Although Lovering explained that a lot of women "took their kids with them ... they didn't have six children [laughs]."[179] Both married and unmarried women were involved in agricultural societies, and many married women also had husbands who were members. However, when a couple were members, the husband's involvement often took precedence over the wife's because women were expected to take care of responsibilities at home.

Another issue that frustrated women was how some women's divisions were still beholden to the main fair board for funds, funds the women themselves usually raised. A conflict could arise when women felt the men did not provide them with their fair share. Long-time member of the Erin Agricultural Society, Myrtle Reid, had been a "Lady Director" when the women organized a separate women's division in 1953 to help pay for the newly built coliseum on the fairgrounds that housed the local ice rink. The women began a series of fundraising activities, including managing a refreshment booth during weekend hockey games in the winter and catering bonspiels and other events at the rink.[180] Because of the money they raised, the society could afford the payments and upkeep of the new building.[181] She recalled that the "Ladies' Division" had to turn over its fundraising efforts to the main fair board even though the women thought they should have some say in how that money was spent. Reid explained, "I think there was always this nagging that ... the women were giving over their proceeds from our fundraising to the guys and maybe not getting enough recognition for it, and then it came about that the women wanted to be represented on the main board to find out what they were doing ... We all wanted to know what the guys were planning."[182]

June Switzer, also a long-time member of the Erin Agricultural Society and a past president, explained that even in the 1980s, if women wanted to do something different, they had to bring their request to the board of directors. "At that time, all the board were all men," Switzer described, and "depending on who was president at the time, you got totally dismissed, and it was never brought up," or "other times it was great." Many of these women also had husbands serving who might forward their ideas. Still, if a president or other board members were not inclined to listen, the women's committee had difficulty facilitating their agenda. Switzer explained how "one president

I worked with – nice guy – but he was very old school. You women just do your thing, don't suggest that we change things to make this [or] that better for you."[183] Other women's sections did create their own treasury to finance their competitions and had more autonomy. Glenda Benton recalled that the Georgetown Agricultural Society's fair board had to borrow money from the women's division one year to pay prize money for exhibits outside of the women's department.[184] Still, many women continued to have their authority restricted, and increasingly they found this intolerable.

By the 1970s, a new generation of women was challenging the status quo. Second-wave feminism was influencing women in rural areas,[185] and women in agricultural communities were establishing new organizations in which they could have a more meaningful role.[186] In 1975, for example, over 300 farm women came together to form Women for the Survival of Agriculture (WSA). Members were concerned with the lack of communication between producers and consumers, unfair criticisms of the farm community, and "the inequalities that exist for farm women in regard to property and marital rights."[187] One of the founding members, Dianne Harkin of Winchester, explained that their meetings were "alive and exciting," and instead of talking about children and recipes, they "just talk farming." WSA also sponsored other women to attend farm business courses.[188] Harkin, and women like her, believed that women were responsible for telling farmers' stories because they themselves were farmers, and not just farm wives.

The same change in mindset was seen at fairs. At the Canadian National Exhibition, women had already begun to challenge their unequal status on the board,[189] and women involved in county and township societies also demanded equality at all levels of fair administration. When June Switzer joined the Erin Agricultural Society in 1972, she became unsettled by the tradition of referring to female members and directors by their husbands' names in prize lists and reports. She and her friend Mary Wall realized this needed to change: "Mary was one of the first I knew who kept her maiden name even though she was married, and it suddenly dawned on us that we were doing all this work – and at that point in time, the women did all of the hall stuff and a whole lot more ... we had more responsibilities really than a lot of the other directors had – and we're thinking, 'and I'm Mrs. Craig Switzer? I don't think so! I have a name!' and that was in the early 1980s."[190]

Switzer no longer wanted to be referred to by her husband's name; she wanted her own recognition and the ability for someone to say "Hey, June." Switzer explained that the idea of recognizing women by their first as well as last name was accepted widely; many believed it was long overdue.[191]

Women's increased responsibilities on the fairgrounds also reflected the changing influence of agricultural societies as tools of farm reform and innovation. Agricultural societies had once been an important part of the Ontario Department of Agriculture's focus on agricultural improvement, but by the 1970s the annual report by the Minister of Agriculture and Food only briefly mentioned their activities. By 1980, the published report on agricultural societies was less than 130 words in length.[192] This neglect reflected the societies' declining importance in directing policy in the province. Although the societies continued to profess their commitment to agricultural improvement, their influence had weakened over the course of the twentieth century compared to corporate and government-appointed experts, as well as the large number of specialized agricultural and livestock associations that had begun in the late nineteenth century. These other groups had mostly taken over the education of farm people. Simultaneously, rural populations were decreasing, and more entertainment options and greater mobility meant that fairs had significant competition for fairgoers' attention. Agricultural societies were concerned about how they would remain relevant in the future if their allure had diminished. They decided that their role would become one of community building and educating an urban population that was increasingly removed from agriculture,[193] and women's exhibits at fairs were praised for their ability to "create enthusiasm and a good community spirit" among local and visiting populations alike.[194]

Agricultural societies' most important tasks in the postwar period became their ability to foster community, encourage rural sociability, and educate urban populations about agriculture life – all things their female members had long been doing. Women's work continued to connect farm, town, and city women's interests, and their exhibits attracted a diverse group of fairgoers. Baked and cooked goods, arts and crafts, flowers; these were all items that had the power to bridge geography and encourage people from both urban and rural backgrounds to participate at the fair. Fair organizers recognized that urban visitors were needed for a successful fair, and they wondered if fair time should be known as "Rural Fellowship Day," a day "when rural people can exhibit their produce and improve public relations with our urban neighbours."[195]

Conclusion

The women who devoted so much of their volunteer time to their agricultural societies did so because they realized they were needed and they enjoyed the work involved. When asked why she continued to serve on the Erin Agricultural Society's fair board, June Switzer explained: "Partly because I know that I'm useful and that I'm helpful ... you get a sense that [you] can do this job, and I feel that it is really necessary that somebody does this type of thing, and so I'm still there, and it's still fun."[196] Margaret Lovering expressed similar feelings when she explained that she enjoyed being a part of an agricultural society. She felt her service was appreciated and needed: "They're always calling, Marg would you do this, and Marg would you do that [laughs]."[197] Rather than complaining about the amount of work they were required to do, most women were pleased that their agricultural societies needed them. Of course, both Switzer and Lovering were long-time members; other women may have simply stopped volunteering if they found they did not enjoy the work or receive the same benefits.

Fair women also faced challenges, including limited authority to influence fair administration and indifference to their interests and activities. Still, they responded to opportunities to increase their control, such as creating their own divisions within agricultural societies. Women's divisions helped to develop a subculture that allowed women a space to fight the subordination they may have felt by their exclusion from other roles and opportunities. Ultimately, however, those who continued to participate did so because they believed the benefits they received outweighed the costs. One will never know how Velda Dickenson felt when she accepted her plaque in 1965 and heard the poem written for her. Did she focus on the "many friends won" and their "love and devotion"? Was she satisfied that women had separate but important roles to play? Or did she long for more women to hold office in the upper echelons of agricultural society management? Whatever her feelings, or those of the women around her, most female agricultural society members were committed to serving despite the inequalities that existed because they valued the contributions they made and believed they were doing vital work to advance themselves, their families, and their communities.

2

Feeding the Family
Fair Food for All

Food is intrinsically connected with the agricultural fair. Midway concession stands evoke popcorn, cotton candy, french fries, foot-long hotdogs, or even the more recent invention, deep-fried Mars bars. In homecrafts buildings, fairgoers expect to see homemade bread, pies, jams, jellies, and other foodstuffs. Anyone who recognizes the fair's educational purpose associates food with the livestock in the show ring and the uniform groups of apples, tomatoes, beets, and potatoes exhibited in the agricultural hall. Food presented at agricultural fairs comes in many forms, but they all remind fairgoers that rural people are deeply invested in feeding their families, communities, the nation, and the world.

In the nineteenth and twentieth centuries, both men and women contributed to the abundance of foodstuff found on Ontario fairgrounds, but their contributions were not valued in the same way. Men's food production was recognized as a part of the larger agricultural economy, in which commodity crops and livestock were grown to supply national and international markets. Women's food production, however, was contextualized within their domestic roles and their obligations to their families and local communities. Just as the previous chapter showed that women were expected to serve men on fair boards, this chapter demonstrates that many people saw women's food provisioning as supplementary to men's wage-earning abilities on the family farm. The idea that women's contributions were secondary to men's shaped both the types of food competitions they entered and the prestige afforded them. Fairs were inherently political and performative. By focusing on women's achievements in competitions for garden vegetables, fruit, dairy and domestic produce, fair organizers and newspaper reporters portrayed women as caretakers of the family and the community rather than producers for the nation and the world. Still, they could not ignore women's significant contributions to household economy. And despite attempts to limit women's food provisioning to households, food exhibits such as dairy products highlighted their production for commercial markets as well. Additionally, changes in women's food exhibits over time illustrate how new products and technologies – whether new varieties of fruit or new cultural concepts such as the TV dinner – were altering the nature of women's work in the home, the garden, and the barnyard. Whether women chose to accept or resist elements of the transforming foodscape are reflected in their exhibits at Ontario agricultural fairs.

Another important element of food exhibits was the way they shaped and enhanced the fairs' overall appearance. Food represented the bounty of the countryside. Female exhibitors and organizers worked tirelessly to achieve orderly, tasteful, and skillfully presented food displays as part of their mandate to refine the fairgrounds. Fair organizers supported women's desire to showcase their "love for the beautiful" in addition to their hard work, thrift, and material contributions to the household and community.

Exhibiting Food: Showcasing the Bounty of the Countryside

Exhibits that showcased the abundance of foodstuffs grown in the province were an essential feature of fairs. They also emphasized women's ability to provide for the provisioning and comfort of their families. Visual knowledge was central to the fair experience, and fairgoers – including women who visited or participated in food exhibits – learned about and assessed the items on display as well as the work necessary to produce them. Moreover, food exhibits often reflected contemporary ideas about how best to display things. As Keith Walden explains in his study of the Toronto Industrial Exhibition, "a culture of the eye" dominated late-nineteenth-century exhibitions and "the desire to see and the need to see became more acute."[1] This included aspirational visions such as items that might be found in a middle-class parlour or a bountiful collection of fall fruit and vegetables that acted as a feast for the eye.

Both women and men contributed to the food shown at fairs, but it took time before female exhibitors gained public recognition for their work. Traditionally, the male head of the household was recorded in fair winners' lists, even for products made by female household members. Initially, this may have been because many agricultural societies only allowed members to compete, and most of them were men. Still, this tradition persisted long after competition became open to non-members. Furthermore, although agricultural societies usually allowed family members to exhibit under a single membership fee, women's entries were still identified as having been made by male household heads. This makes it difficult to say with certainty who was most responsible for the food items exhibited at fairs. Still, with the long history of women making butter, cheese, bread, and other food products as well as their work in the garden, orchard, and hen house, it is safe to conclude that women often made the items exhibited in these categories.[2] For example, even though the official winners' list for the Erin Fair reported that John Thompson and Henry Smith won prizes for "two beautiful loaves" in 1865, it was also noted that the "home made bread, of which there were no less than seventeen entries, showed what the thrifty wives of Erin could do."[3] Prize lists reported men's names, but anyone reading those reports would have understood that particular winning entries showcased the talents of a Miss or a Mrs.[4]

The practice of privileging male names also reflected the idea that rural families were working units where everyone produced for the entire household's benefit. However, family solidarity was more of an illusion than a reality. Even though all family members might have contributed to the household economy, male heads of household were legal owners of the land and the yields that came from it.[5] Writers in nineteenth-century farm journals advocated for farm women's rights and lamented the fact that many men considered themselves the breadwinners and ignored how women's work sustained the family.[6] However, most rural people accepted that men occupied the breadwinner role and were the public face of the family.[7] In her study of the impact of technology on farm women in the twentieth century, Historian Katherine Jellison notes that although women gained some degree of economic and political power over the century, rural patriarchy remained. Even in the postwar period, "men were still the 'farmers' and women their 'helpers,' a perception reinforced by women's assertion that they were not 'farmers' but 'farm wives.'"[8] This idea is supported by other studies that conclude that Canadian farm families were typically patriarchal institutions where women's work was undervalued and women themselves rarely recognized as equal contributors in production.[9] The public appreciated the work women exhibited at agricultural fairs, but women's participation remained constrained and their contributions were not always fully recognized.

Newspaper reporters did not wholly ignore women's fair contributions, and it was not uncommon for them to acknowledge that "No part of the exhibition is more instructive than that which comes under woman's supervision."[10] The ability to feed the family was fundamental to women's caretaking identities, and therefore prizes for food competitions were particularly prestigious. The *Farmer's Advocate* advised men that when selecting a wife, they should insist that she had culinary knowledge and skills because "however little we may in the days of our health and vigour care about choice food and cookery, we very soon get tired of heavy or burnt bread and of spoiled joints of meat."[11] Cooking was said to be "an art upon which so much of our daily life depends," so women were expected to perform it well.[12] Women's talent in the kitchen was central to their domestic skills and thus their success as wives.

Thousands of women exhibited at fairs over the years, and their participation illustrates that they prided themselves on their reputations as good bakers, cooks, and provisioners. These women believed that competition in a public arena was a legitimate means of recognizing their merit. In 1868, the *Brampton Times* asserted that, at fairs, the "prizes are worth contending for, aside from those offered by the committees. 'The best bread maker in the country' is an honor that would sit gracefully on any woman,' and that the 'finest butter neatly stamped' ... is certain to be looked at, and ... to be inquired for."[13] Newspaper reporters also approved of the industrious character of female exhibitors.[14] Women's industry in producing foodstuffs, along with other domestic work, provided "a view to comfort, utility, and the sustentation of the family."[15] Food, which was "designed for the delectation of the palate or satisfaction of the appetite,"[16] allowed women's primary role as caretakers to be put on display and celebrated as something all women should aspire to and take pride in.

Vision's privileged position in determining merit was rarely questioned. Fair organizers and judges promoted the idea that one could judge food's quality without tasting it; this was especially true of fruits and vegetables, often judged on their appearance alone. By the twentieth century, the use of judging scorecards became standard, and they emphasized fruit and vegetables' form, shape, colour, uniformity, and condition as the basis for evaluation. For example, one was to judge a vegetable's quality by its "Smoothness, texture, [and] freedom from blemishes" and determine its uniformity by ensuring that the specimens were "one size and colour."[17] Sometimes judges sampled fruit for its flavour, but again, most assessments privileged the item's size, colour, texture, form, and freedom from blemishes.[18] General fair goers relied on sight as their primary means of acquiring knowledge, especially since they were discouraged from getting close enough to taste, touch, or smell the entries. In her research on agricultural exhibitions in Nova Scotia, Sara Spike employs what anthropologist Cristina Grasseni calls "skilled visions" to explain how exhibition organizers used sight to educate the public on the ideal model for any given thing.[19] Imparting visual knowledge was a fundamental purpose of fairs, whether through the confirmation of a purebred Shorthorn bull or a three-tiered wedding cake.[20]

Newspapers reporting on fairs confirmed that visually pleasing displays were an effective way to communicate knowledge to fairgoers. The reporter for the *Hillsburg Beaver* contended that at the 1907 Erin Fair, the assembled crowd did not care "a continental for 'educational features,' or would stand for them for any length of time." Instead, the reporter noted the people wanted "good, well-arranged exhibits, sharp judging, no dragging or delays, and what we might term "movement."[21] The *Ontario Farmer* agreed that fairgoers obtained useful information when they kept "their eyes and ears open." The article suggested that there were two classes of people at fairs, the observant and unobservant, and that only an enquiring mind found "enough to engage its best attention and waken its fullest energies on such occasions, while a dull sleepy mind will go and come like a door on its hinges."[22] Such rhetoric illustrates how rural reformers expected fairgoers to keep the spirit of improvement alive, and agricultural societies realized it was their duty to educate fairgoers in an easy way.

When food exhibits displayed knowledge attractively, reporters praised women for their contributions. The judges at the 1863 Minto Township Fair in Harriston found the local butter entries superior to and more tastefully arranged than those at the Provincial Exhibition. They also insisted that it was "the ladies [who] were instrumental in getting up the best part of the show."[23] The *Brampton Times* gave credit to female exhibitors at the 1870 Peel County Fair for the "butter and cheese, and staff of life ... flanked on either side by deliciously clear jars of honey, bottles of home made wine, and [the] most luscious and tempting looking samples of table grapes, peaches, pears, &c., &c., which taken as a whole, formed a very pretty picture."[24] The "taste" women's food entries exhibited was more a comment on women's ability to discern good quality and a high aesthetic standard than it was a comment on the foods' flavour.

Food items also conveyed flavour and texture visually. Fairgoers appreciated displays "of toothsome edibles" that whetted the appetite of every passerby, and large exhibits of "bread, butter, honey, pies, cakes, pickles and preserves [that stood] out in tantalizing array to the hungry visitor."[25] While other competitions in the "Ladies' Department," such as home manufactures, fancywork, and fine art, also exhibited good taste, food exhibits visually communicated "tasty" as well.

The popularity of food exhibits continued to grow over the years, and, even during the uncertainty of the Second World War, they flourished. At the 1944 Woodbridge Fair in York County, crowds jammed into the exhibition hall to see the bountiful foodstuffs grown and produced in the county. A reporter from *The Globe and Mail* provided a detailed account of the display by explaining that "it was almost impossible to move in the mob that slowly surged past the fencing which separated the envious eyes from the tables on which were lined the cakes with deep frosting, pies of delicate autumn hues, cookies of tempting crispness and mountainous loaves of bread with bulbous eaves along their sides."[26] The luxury of plenty combined with a richness of colour, while the perceived quality of the taste and texture conveyed the idea that Woodbridge was a fertile agricultural region, blessed with skilled women who used the land's productiveness and their superior talents to feed their families and communities.

Feeding the Family 63

The seasonality of fairs limited the variety of fresh food displayed. Fairs were usually fall harvest celebrations, so the produce exhibited reflected what was at its peak of maturity.[27] Fall fairs also gave women time to preserve summer and fall fruits and vegetables, craft dairy products, and prepare other foodstuffs. While fall fairs could not display all of Ontario's variety of produce, the agricultural economy's essential products were usually represented. The Toronto *Daily Globe* noted that the first Weston Agricultural and Horticultural Society's Fair in 1876 was too late in the season for an extensive range of fruit, so exhibitors showed only late-harvest fruits such as apples in great numbers. Despite this limitation, more than one hundred exhibitors participated in the fresh fruit competition that year.[28] Apples' seasonality, along with their suitability for the Ontario climate and popularity among residents in the province, resulted in a wealth of apple exhibits over the years.[29]

The day of the week that organizers held the fair could be critical for women's exhibits. Throughout the nineteenth century and well into the twentieth century, fairs were not held on weekends. Agricultural societies adopted this practice to avoid the Sabbath, but some realized that Mondays posed problems too. In 1951, Mrs Alex Robinson of Cochrane, the director of the women's section of the OAAS, reported that one fair in her region held on a Monday suffered "a bad slump" in baking entries.[30] Robinson warned agricultural societies about the consequences of hosting fairs immediately after the Lord's Day, because the more devout women in the community would not have the time to prepare their exhibits.[31]

Another challenge faced by food exhibit organizers was the weather during the growing season. Unlike domestic manufactures or fancywork, garden fruits and vegetables were dependent on good growing conditions. Unusually dry or rainy weather could result in a poor harvest and thus an inferior exhibit. In 1883, the *Brampton Conservator* reported that cold weather and severe frost had damaged fruits and flowers that year, so "no one expected to see a full and rich exhibit in these departments" at the Brampton Fair that year.[32]

Poor weather during the fair itself was another challenge. Many reports cited rain as the most frequent cause for poor attendance. In 1877, it was reported that the significant drawback of the Peel County Fair "was the rain and the mud ... had the weather been fine the attendance of visitors would have been fully three times as many as it was, and the entries even greater in number than they were."[33] The reporter acknowledged that, unfortunately, agricultural societies could not remedy poor weather, and therefore "must be content and hope for better days in future."[34] In 1884, the Peel County Fair again suffered downpours at fair time, but the local newspaper reporter saw the humour in the way they affected the women who attended: "As soon as the rain ceased, the crowds from down town began pouring in through the gates. The ground was wet, soft and soapy, and in less than half an hour after the crowd arrived it was in the most lovely, charming, slippery, muddy state that eye ever beheld." The "lovely, charming" state of affairs left the reporter in particular awe of the women who "all honor to them for their pluck and patience, came sliding through and into the main hall."[35] "Such crushing, crowding and squeezing you never saw before," the reporter declared. Despite this, however, he contended that "everybody enjoyed it immensely. There was more fun than at a country paring-bee."[36] Whether or not the fairgoers truly enjoyed the chaos created by the wet and soggy conditions we cannot know. The women's exhibits may have received more attention as fairgoers sought shelter from the rain, but the lost revenue had society members praying for good weather the following year.

Fair organizers also had to protect food exhibits and maintain their quality over the course of the fair. As local fairs grew in size and duration in the twentieth century, the women organizing the domestic science exhibits found it especially difficult to keep food in good condition. At the 1934 OAAS Convention, Miss May Needham, future president of the Women's Section of the OAAS,[37] advised domestic science committees that in order to contend with "flies, dust, and germs," suitable display cases were needed to keep exhibits sanitary.[38] Pests and rodents were another issue. In 1929, Miss Goodfellow, a Department of Agriculture home economist who was regularly called on to judge at local fairs and later became secretary-treasurer of the Women's Section of the OAAS,[39] recalled when she had been judging the fancywork at a fair and noticed a mouse running through the domestic science entries. She expressed her relief that she was not judging that exhibit, but later that evening the domestic science committee asked if she would take up the task, revealing they were disappointed with the appointed judge's placings. "I did it," Goodfellow explained, "but it wasn't as pleasant a task as if I hadn't seen that mouse."[40] Other "pests" also caused issues. The brazen fairgoer, for example, was a problem that only a proper set of barriers could deter. It was reported that the butter at the 1854 Provincial Exhibition was considered excellent, thanks to the "house-wife" who, after "applying a thumbful to her mouth declared [it] 'splendid.'"[41] At the 1932 Springfield Fair, no children entered homemade candy in the children's department because the previous year their entries had been stolen. The organizers set up barriers, including a locked entrance and a high woven wire to guard the food and ensure children they should "have no fear of any further loss," but the damage was done.[42]

Women believed that new and improved products helped secure displays and encouraged progressivism and modernity. Needham advised all female delegates at the 1934 OAAS Convention to adopt the use of cellophane as a first step in protecting food exhibits because she saw cellophane as the best defense against pests, even better than glass-covered cases, which she believed were a good idea but allowed flies to enter during the judging process.[43] In the years following the Second World War, a lot of attention was paid to which fairs were making improvements. At the 1949 OAAS Convention, district representative Miss Agnes Yuill from Middleville mentioned how, in continuing to grow and prosper, Renfrew Fair made sure that all baking was under glass,[44] while Mrs J.H. Booth of Port Arthur expressed satisfaction that the Lakehead Exhibition in Thunder Bay was arranging for a special building with large glass cases so that all food exhibits would be protected.[45] Mrs Guthrie Reid of Teeswater expressed her pleasure with some of the fine new buildings, arenas, and community halls built on fairgrounds in her district, but she stated explicitly that glass cases for food exhibits were still needed.[46] The next year the same concern was conveyed by Mrs Allan Koehler who reported that her only criticism for the district was the lack of proper display cases.[47]

If agricultural societies did not properly care for food, they risked discouraging exhibitors who expected to consume or sell their entries afterwards. Agricultural societies also encouraged donors to sponsor special prizes. Donors gave first-place exhibitors a generous monetary sum or valuable item in return for the winning entry. If the entries were damaged or inedible, sponsors would be less inclined to donate the following year. Various community groups also raised money by auctioning off prize-winning exhibits. The exclusivity associated with purchasing (and eating) a winning food item, and an atmosphere of good-natured competition between peers often meant that high prices were paid. During the First World War, an auction at the Wallacetown Fair for the "prodigious loaves of bread, and likewise buns, for which W. McLaudress offered special prizes, caused considerable rivalry, especially among the citizens of St. Thomas, who know a good thing when they see it."[48] The society was able to raise $18 for the Red Cross from exhibitor-donated items.[49] When food was kept safe and edible it was a valuable asset.

Wastefulness was a point of contention for those who sought to improve food classes. This included waste from improperly protected exhibits, as well as excessive amounts of food required for individual entries. Thrift was a positive characteristic among rural women because it not only represented a moral virtue associated with Protestant values such as "self-discipline, hard work, sobriety, honesty, diligence, and industry,"[50] but was directly related to popular ideals of agrarian production, whereby the man and woman both contributed to the farm enterprise through their adoption of scientific methods and careful economy.[51] Careful economy was especially important during the Great Depression.[52] At the 1934 OAAS Convention, Needham recommended that the quantity of food required per entry be reduced. She suggested that classes for large quantities of butter, such as the class for 20- to 30-pound crocks of butter, be eliminated.[53] She reasoned that farmers were making less butter, and that it was "most unfair to pry to the bottom of a 30 lb. crock, to test for award, thus rendering the exhibit unfit for winter preservation." She also suggested there was no need for large collections of canned fruit and vegetables, as one sealer was enough.[54] Needham's concern with waste was illustrative of the value of thrift during the difficult years of the Depression. Unlike the late-Victorian penchant for overabundance at fairs,[55] the 1930s were a time when farmers and townspeople alike had to be efficient and frugal. Although most agricultural societies offered classes that were similar to those offered before the Depression, sometimes the amount of an item or the number of items required per entry was reduced, as evidenced in domestic science and dairy produce exhibits.[56]

The way in which food exhibits were displayed was also important. At fairs, a contemporary fashion for taxonomy influenced people's belief that the "completeness" of displays and their rational organization was necessary for educational purposes. The reporter for Toronto's *Daily Globe* was upset with the lack of finished displays on the first day of the 1876 Guelph Central Fair. He beseeched agricultural societies to "put a stop to the constantly growing evil of allowing exhibitors to deter putting their articles in place until near the close of the first day."[57] He argued that all fairgoers, whether they arrived for the first day of the fair or the last, had the right to see a complete show, and he added that visitors were "not put in the best of humour by finding everything on deck in a state of provoking confusion."[58]

Beyond incomplete exhibits and tardy entries, the *Daily Globe* reporter also criticized disorganized displays. He was distressed by the jumbled exhibits that defied classification, noting how one was "constantly stumbling across articles up stairs which under any regard for the eternal fitness of things should have been found alongside a similar class of goods down stairs, and *vice versa*." He explained that "Just when one is congratulating himself and thanking his stars that he has got one department finished at last, in spite of digs in ribs from moving boxes and collisions with excited assistants, he finds that that department is 'in parts' like Daniel Deronda, and that the 'parts' are pretty much all over the building."[59] Disorganization hindered organizers' ability to present entries to their best advantage. The Victorians believed that proper systems of classification were essential for the clear communication of knowledge.[60] The reality, however, is that most organizers were restricted by the facilities at their disposal. Women would have wanted their exhibits arranged intelligently and neatly, but most fairs housed various women's exhibits, including foodstuffs, in one large building or shelter. Often these structures contained everything from home appliances and horse harnesses to cakes and crazy quilts. Of course, not everyone was critical of fair organizers' ability to construct displays. A reporter for the *York Herald* praised the women's exhibits at the Union Fall Fair in Markham for the taste and skill they displayed, as well as the management for ensuring the entries were "in excellent order."[61]

Fair organizers advocated for spacious halls to be built to showcase exhibits. The Toronto Township Agricultural Society secured the use of a large new drill shed in 1868 so that "the prolific display of grain, fruit, dairy produce, manufactured articles, ladies' work, &c.," could be "neatly arranged,"[62] while in 1892 the Bayham Agricultural Society secured a spacious building that allowed indoor exhibits to be displayed to their best advantage.[63] As agricultural societies matured, they often tried to secure funds to renovate

Figure 2.1 The Crystal Palace at Picton Fairgrounds
Built at the Picton Fairgrounds in 1887, this was one of many exhibition halls in North America that were scaled-down versions of Joseph Paxton's 1851 Crystal Palace in London. Loyal Orange Association of British America, "Picton, Prince Edward County, Ontario. A souvenir presented to the delegates of the Grand Orange Lodge, June 1904." 917.13587 P37 BR, Baldwin Collection, Toronto Reference Library. Image courtesy of the Toronto Public Library.

or erect permanent exhibit halls.[64] The Picton Agricultural Society in Prince Edward County raised funds to build a "Crystal Palace" in 1887, which was modeled on the 1851 London Crystal Palace. The size and type of permanent structures on fairgrounds could, however, vary enormously.

Large exhibition halls were a matter of pride for fair organizers. Throughout the late nineteenth and into the twentieth century, agricultural societies sought ways to improve their fairgrounds, and often an impressive hall or arena was their most obvious sign of advancement. In 1950, Mrs J.H. Booth, the district representative for the Dryden, Lenora, Rainy River, and Lakehead Fairs, proudly reported at the OAAS Women's Meeting that a new building for women's work that included "large cases with glass windows for needlework" was built at the Lakehead Exhibition Grounds to celebrate their sixtieth anniversary.[65] Agricultural societies in Northern Ontario were often younger than many of their southern counterparts, and they were eager to highlight the improvement and progress they had made to illustrate that their fairs could be as progressive as southern ones.

Exhibitors' adherence to instructions also led to orderly and uniform exhibits. Exhibitors who did not follow the rules were usually disqualified from competition. In 1850, Mr Thurtell, a judge at the county fair held in Guelph, remarked that a considerable number of excellent butter entries could not compete because they were not "rightly put up"; in this case samples were put in crocks rather than packed for export.[66] Even by 1934, organizers were advised to always "state exactly what is required" so that competitors understood that "if it calls for six buns, that means "six," not "thirteen," and if for light fruit cake, it does not mean "dark cake"; if for sandwiches for afternoon tea, that does not mean for a "lunch pail."[67]

When an entry was faulted for something not necessarily visible to fairgoers, mistrust and suspicion ensued. Needham told OAAS delegates in 1934 that because exhibitors sometimes used stained pans to bake their bread, buns, biscuits, and other goods, their exhibits were discoloured. While fairgoers could not see the discoloured underside of the items, judging did, and therefore penalized such entries. She warned that "If cooks were more careful about the pans, used for exhibits for the Fair, they might get more prizes."[68] An exhibitor's unseen error and made the results difficult to understand when sight was the main method of fairgoers' evaluation.

Occasionally, food exhibitors tried to deceive judges. At the 1936 OAAS Convention, judges described some of the "trickiest methods" used by exhibitors. One judge recalled how a woman entered a jar of raw peas that were simply covered with cold water rather than cooked, in order to preserve their green colour. Another woman entered "home-preserved" fruits, which were discovered to have been purchased and "dumped into a sealer and taken to the fairgrounds."[69] These women believed that seeing was believing and that their entries would be judged on appearance alone, but their tricks were revealed when the jars were opened for sampling.[70] The idea that only appearance mattered also gave rise to a number of dishonest honey exhibitors. *The Globe and Mail* reported a particularly contemptable stunt in which honey had been replaced with alternate liquids having the "clear and not too thick" appearance judges desired. The result was that "Curious judges who sampled honey shown in glass jars got a mouthful of castor oil at one fall fair – kerosene at another."[71] Actions such as these illustrate not only that women were not afraid of competition, but that some were willing to cheat in order to win, risking their reputations in the process.

To legitimize results and thus maintain the integrity of what was on display, agricultural societies outlined the rules for exhibitors in their prize books. In the "Rules and Regulations" of the first Central Agricultural Society Fall Exhibition held in Walters Falls, Grey County, in 1889, organizers made it clear that "Upon the discovery of any fraud, or deception, either in the production of ownership, concerning any article exhibited, the Directors will have power to withhold any prize awarded and may prohibit any such party from exhibiting in any class for two or three years, and may publish the names of such persons or not, as may be deemed expedient."[72] The rules also guaranteed exhibitors that "No person shall be a Judge in any class, in which he is an exhibitor or interested in or has any relations exhibiting," and they emphasized that the "Judges are expected, in the execution of their duties, to be careful to act with the most rigid impartiality, and make their entries in a clear and conspicuous manner."[73] Fair organizers reassured exhibitors that the measures in place ensured impartial competition, and they warned would-be cheaters of the consequences for dishonesty.

Despite the drama that could occur, most fairs were conducted yearly without any great controversy. Even criticism of judges was limited, because generally fairgoers and exhibitors understood that most judges tried their best to award prizes fairly.[74] Of course, some people questioned judges' knowledge or judiciousness, but for the most part, exhibitors recognized that the merits of their work would ultimately be rewarded. June Switzer of the Erin Agricultural Society explained how subjective judging could be. She noted that, despite the introduction of fairly strict point systems and standard classifications, judges had their own assumptions about how exhibits should be made and presented, and exhibitors understood this.[75]

June exhibited baked goods and other items at fairs because she wanted to showcase her talents, engage with her community, and enjoy the company of those she met there. When asked about the skills she developed because of her participation, she explained: "when it looks really nice it also counts

… even if it's just a cinnamon bun, the presentation, the type of plate you put it on and the way you ice it, or whatever, or don't ice it, has a bearing on everything and you can look in a showcase and see, oh, that one really looks nice in the way they presented it; so little details like that … paying attention to details, paying attention to rules, that if there is a rule that it's supposed to be this size, well, you can be disqualified, and they won't look at it even if it's wonderful if it's not the right size. Life lessons. Life lessons."[76] Visual knowledge was central for how women learned about and assessed their work at fairs. Women were expected to follow instructions, pay attention to details, and present skillfully made entries that reflected their success as caretakers of their families.

Food exhibit organizers were similarly committed to orderly and tasteful presentations of exhibits. An abundance of foodstuffs from the countryside was a marker of agricultural societies' success in engaging with producers. The colourful display of produce at the 1960 Erin Fair was enthusiastically described by the local paper: "On entering the coliseum, the eye was immediately drawn to a giant purple and gold cornucopia; and a horn of plenty it was, for flowing from its bell was every conceivable vegetable the soil could produce."[77] Women's ability to use new techniques and materials for showcasing that bounty illustrated their dedication to improving and modernizing the fair and the countryside. By displaying their work, women were not only teaching others what proper homemade goods looked like, but they were also presenting themselves as contributing to the material and moral advancement of rural society.

Dairy Produce

Dairy products were among the first items included at fairs that showcased farm women's economic contributions and fetched some of the highest premiums on the fairground. Most dairying in the nineteenth and early twentieth centuries was done by women, who thereby made a significant economic contribution to the rural household, so agricultural improvers were interested in its development.[78] Dairy products highlighted values such as industry and thrift; traits that were a respected part of rural women's identity. The evolution of dairy produce classes over the course of this investigation, however, illustrates that farm women produced goods that allowed them to contribute economically to the family, but when those products were no longer beneficial, either because they had a reduced market-share or the labour and time involved became too costly, women were unafraid to develop other forms of production.

Dairying was important because it helped safeguard families who also relied on income from commodity crops whose prices depended on unpredictable export markets. Some dairy produce such as butter and cheese could be exported, but dairy products generally had a strong local demand and were important to family provisioning.[79] In 1851, homemade butter production was almost 16.1 million pounds; by 1891 it was 55.6 million pounds.[80] Butter was always produced in greater amounts than cheese because it was more easily made and marketed, and before the rise of cheese factories, it was also more profitable.[81]

Agricultural societies valued dairy exhibits because they encouraged farmers to improve dairying for the future. Reports of the inferiority of Canadian butter on the British market resulted in lower prices than were paid for British butter, and complaints that it was "badly made, badly packed, and, last but not least ... put into bad tubs" prompted improvers to chastise producers.[82] Most agricultural societies adopted dairy produce classes early on to try to improve the production of butter and cheese, and by mid-nineteenth-century most fairs had at least one class for each of these products, if not more.[83]

Butter production was said to require great care and judgement.[84] In 1859, the *Canadian Agriculturist* published an article about an award-winning female butter maker who advised the use of the purest of ingredients, clean instruments, milk kept at the proper temperature, proper churning, washing, and salting procedures, and finally, ensuring the product was correctly packaged and stored.[85] Agricultural societies were hesitant to give women credit for some types of agricultural production, but they celebrated women's role in dairying. Idealized images of milkmaids were popular in the nineteenth century,[86] but rural men and women understood that dairying was onerous work that deserved praise for its importance to the farm economy.[87] The dairy produce judge for the 1850 Eramosa Fair, Mr E. Passmore, declared that he "was sure that if he were a young man, and looking for a wife, he would come to Eramosa for one – particularly if he were wanting one who could make good butter."[88] In Guelph in 1854, the size and quality of the dairy display was a significant part of the County Show's success. The *Guelph Advertiser* reported that "it was amongst the butter that the greatest interest was evinced, and in which there was the greatest competition." The large numbers of entries were reportedly "so good that the judges wished they had prizes for one half," and that the entries were "a sight to make a dairy maid feel proud, and bore most honorable testimony to the industry of the farmers' wives and daughters."[89] The judges stated that the best entries were by women clearly committed to advancing their skills beyond the ordinary. The praise women received for their work was not simply local boosterism; their ability to produce superior dairy products was respected and encouraged.

Exhibitors usually received admiration for their dairy exhibits, but agricultural societies were often critical of the dairy industry. A reporter noted that even though the cheese shown at the 1848 Nichol Fair was very good, Canadian farmers generally did not devote much attention to dairying.[90] Some agricultural society and newspaper reports criticized the quality of Canadian dairy produce, particularly butter, and argued that these poor products decreased the overall value of dairy commodities and caused the "the blush of shame on the cheek of every producer."[91] When the entries for dairy produce were few, farmers were also criticized for their indifference.[92] Dairy exhibits – good or bad, small or large – allowed agricultural societies and fairgoers the opportunity to discuss the industry and its future, as well as women's efforts in improving this line of work.

Butter was more popularly shown at fairs because homemade cheese production was waning in the late nineteenth century.[93] Domestic cheese production in Ontario was halved during the 1870s and continued to decline over the years, while the domestic production of butter plateaued for three decades after 1881.[94] Even before many women gave up cheese-making, homemade cheese classes were generally seen as deficient. It was reported the butter exhibit at the 1860 Perth County Fair in Stratford was large and of good quality, but that there were only about half a dozen cheese entries, generally of inferior quality.[95] The 1890 West Elgin Fair had twenty-seven butter entries and fifty-one entries in bread, but there was only one exhibitor for cheese, Mr W. Ostrander, and this was for "factory cheese."[96] Although the 1894 Markham Fair reported that the "cheese, both homemade and factory, is fully up to the high standard which Canadian cheese has attained,"[97] most fairs by then had limited entries, largely confined to factory-made rather than homemade cheese. The decline of cheese entries at fairs supports historians' assertions that the rise of cheese factories in the province in the last quarter of the nineteenth century took a toll on homemade production.[98] Production in factories was often less wasteful and provided a more reliable product, and greater amounts of cheese could be produced by a smaller workforce.[99] In 1910, the total value of homemade cheese in Canada was $154,000, while the value of homemade butter was over $30 million for that same year.[100] By the twentieth century, many fairs, such as Erin, had discontinued the cheese competition altogether.[101]

Butter exhibitors made among the highest premiums in the nineteenth century, which indicated the value agricultural societies placed on the product. At the 1878 Brampton Fair, the most valuable award given was a horse rake worth $32, which was awarded to the winning "Span with wagon" entry in the horse show. In the cattle show, the top prize was a feed mill worth $35, awarded to the best herd (one bull and four females).[102] Other prizes generally ranged between $1 and $5, but women competing in the "firkin of butter" class that year had the chance to win a sewing machine presented by T. Smyth, valued at $50. The competition attracted forty-eight entries and was reported by the local newspaper as among "the keenest in the whole fair."[103] Special top prizes donated by individuals and businesses were common for butter competitions. At the 1869 Peel County Fair, the best twenty-five pounds of butter received an additional $20,[104] and at the 1884 Toronto Fair, Mrs William Dolson of Alloa, Peel County, won $100 for the best tub of butter.[105]

Valuable prizes placed pressure on judges to select the right winner. At the 1853 Eramosa Fair, the judges "felt in a fix amongst the ladies," because their butter entries were so good.[106] The West Riding and Vaughan Agricultural Society hosted a successful butter exhibit that had eighty entries. The judges reportedly had difficulty deciding on the winners, and it was wondered whether "ever such a show of butter was seen anywhere."[107] Butter competitions could be highly spirited. The reporter covering the 1912 Dufferin County Fair in Orangeville wrote that the "county has some famous butter makers and there was great rivalry among the exhibitors to capture the cash prize of twenty-five dollars for the best ten-pound butter prints, given by Mr James Curry of Toronto, a former resident of Dufferin county."

Eighteen women competed and the winner, Mrs Charles Crombie of Mono Township, exchanged her entry for Curry's generous prize.[108]

By the late nineteenth and early twentieth centuries, however, it was clear that homemade butter was waning. The steady decline in competition at the 1892 Puslinch Fall Show was reasoned to be the result of the many creameries that existed in the region.[109] After the First World War, new technology and a desire for greater product uniformity strengthened creameries' market share.[110] By the post-war period, butter classes had declined significantly. At the OAAS Convention in 1950, district representative Mrs A. Drysdale reported that despite growth in other competitions, butter entries were down in her district.[111] Mrs Kestle of Exeter also indicated that dairy produce was no longer a viable competition in her district and that most fairs decided to host non-competitive dairy displays sponsored by milk producers and creameries, rather than a traditional competition.[112] When homemade dairy entries were offered, it often surprised organizers. Mrs W.C. Huckle of Bracebridge noted that she was surprised that one fair in her district still attracted entries for homemade bread and crocks of butter.[113]

By 1950, declining homemade dairy production meant that women no longer made butter and cheese in large amounts. Butter could be bought more easily than produced, and the dairy farmers were expanding their herds and mechanizing in order to ship milk to be processed elsewhere rather than processing it themselves.[114] Instead of making butter, they focused on increasing milk production and improving the quality and components of their milk through breeding better cows. Dairy farming remained an important agricultural sector in Ontario, and dairy cattle competitions grew over the years. Domestic science classes for other foodstuffs also continued to grow, and the number of female exhibitors increased, but dairy products were no longer the stars of the show.

Garden Vegetables and Fruits

The garden was central to many rural households. Agricultural journals, such as the *Farmer's Advocate*, argued that farmers should "have the best garden of any class of citizens,"[115] and garden vegetable and fruit classes proved popular at fairs. Men and women contributed to the production of garden produce, but because it is difficult to identify the work of women in early fair winners' lists, it can be difficult to assess how many entries were the result of their efforts. Even when women were more commonly recognized for the entries in "Ladies' Work," entries for garden vegetables and fruit classes were considered the result of a household effort rather than individual effort, and fewer women were named as competitors. Still, by the late nineteenth century, more women were being identified in these classes and competing alongside men for top prizes.

Women's participation in garden exhibits reflected both their provisioning role and their ability to earn income for the family. Unlike wage work in garment-making or teaching, where women earned less than men simply because of their sex, garden products had a market value that was "not dependent on the hands which raised them."[116] An article published in the 1872 *Farmer's Advocate* contended that many women had become excellent small-scale market gardeners, noting that one female market gardener who grew a variety of fruit "lost nothing in refinement of feeling by her out-door work," and was happier and healthier because of it.[117] This idea was also encouraged by the *Ontario Farmer*, which argued that even if gardening only returned a moderate profit, "in the re-establishment of impaired health, it was great."[118] While one must question the idyllic images presented by male reformers writing about women's market gardening, their discourse at least shows that men accepted and encouraged female gardeners.

Women's association with household gardening, however, may be why agricultural societies initially focused on field crops, typically cultivated by men, rather than small-scale gardening. Agricultural societies privileged men's agricultural pursuits, and so it is unsurprising the most early and mid-nineteenth century fairs turned their attention to the cultivation of crops, such as wheat, barley, oats, and corn, as well as field roots, such as turnips and potatoes.[119] Many types and varieties of garden vegetables, fruit trees and bushes, flowers, and other plants grew in Upper Canada,[120] but agricultural societies wanted to concentrate their efforts on improving the grains, seeds, forage, tubers, and roots that were standard cash crops for farmers or used as livestock feed.[121]

The production of garden vegetables and fruits had never been entirely gendered. Men, women, and children often worked together to produce these items for the family. In Upper Canada in the early nineteenth century, it was typical for men to plough the garden and for women to plant and care for it.[122] The division of labour in the garden meant that female heads of household were largely responsible for its management, and children could contribute significantly to its maintenance.[123] And, of course, some individuals took on all or most of the gardening duties in the family, either out of interest or necessity, regardless of their gender.

Prize lists are useful records for analyzing the growth of garden exhibits and changes in popular varieties of produce. An 1854 report of the Erin Fair identified two departments in which vegetables were listed: "Grains and Seeds" and "Roots." Of the twenty-two classes listed in those departments, fourteen could be broadly classified as vegetables, but some root vegetables such as turnips and beets would have also been used for livestock feed, not simply for human consumption. Likely only eleven vegetable classes in the "Roots" category were meant, to some degree, for household consumption. Only one fruit class was offered, categorized under the "Roots" department, which was for the "Best half-bushel Apples," but a discretionary prize was also awarded for an entry of "Isabella" Grape[s]."[124]

By 1901, thirty-two vegetable classes existed in departments for "Grain and Seeds," "Field Roots," and the added departments of "Garden Roots and Vegetables" and "Fruits and Flowers"; more than double the number from the 1854 fair. Twenty-five of these vegetable classes were largely for household use. The expansion of fruits was even more impressive, with seventeen classes added (sixteen listed in the "Fruits and Flowers" department; "Citrons" were recorded in the "Garden Roots and Vegetables" department). This amounted to more than six times the number of fruit classes as found in 1854.[125] Although more kinds of vegetables and fruits were shown, the main reason for the expansion of classes was the increased number of plant varieties included in competitions. In 1854, the Erin Fair offered one class for onions; in 1901, five classes existed for different varieties. In 1854, one class for the best half-bushel of apples existed; by 1901, the competition had grown to ten classes for specific varieties, such as Golden Russet, Northern Spy, Greening, Baldwin, Snow, Colvert, and St Lawrence.

Items such as onions, potatoes, turnips, and apples were popularly shown across the province because they were well suited to regional growing conditions and they were often produced as surplus commodities sold to the

Table 2.1
Erin Township Agricultural Society Fair, Erin, Wellington County:
Classes for Vegetables, 1854 and 1901

1854	1901
Best bushel of Pink-eyed Potatoes	Early potatoes
Best bushel of any other kind (potato)	Elephant potatoes
Best bushel Swedish Turnips	Pearless Savoy (potato)
Best bushel any other kind (turnip)	Any other late variety (potato)
Best dozen Onions	Swede Turnips
Best dozen Carrots	Heaviest Swede turnips, properly trimmed
Best six Cabbages	Turnips, any other kind
Best dozen Tomatoes	Mangold Wurtzel
Best dozen Parsnips	Sugar Beets
Best dozen Beets	Field carrots
Best dozen Mangel Wurzel	Pumpkin
	Squash
	Beets
	Parsnips
	Table carrots
	Red onions
	Onions, any other kind
	Potato onions
	English potato onions
	Top onions
	Cauliflower
	Cabbage
	Heaviest cabbage
	Tomatoes
	Celery
	Radish*

*Radishes were listed under the "Fruit and Flower" category in the fair prize list, but have been included in the vegetable list here to illustrate item specific classes. Many times fairs or newspaper reports miscategorized classes in prize lists.

Table 2.2
Erin Township Agricultural Society Fair, Erin, Wellington County:
Classes for Fruits, 1854 and 1901

1854	1901
Best half-bushel Apples "Isabella" Grape*	American Golden Russett (apples) North Spy (apples) Greenings (apples) Baldwin (apples) Winter apples, any other kind, correctly named Snow (apples) Colverts (apples) St. Lawrence (apples) Fall apples, any other kind, correctly named Winter apples, collection of 6 varieties, named Fall pears Winter pears Crab apples Plums Black grapes Grapes, any other color Citrons*

*Citrons were listed under the "Garden Roots and Vegetables" category in the fair prize list, but have been listed with fruit here to illustrate item specific classes. "Isabella" Grapes were awarded a discretionary prize by the judges for their merit, despite having no official class in which to be entered.

local community or exported to other regions of the country and overseas.[126] Agricultural societies focused on these products because of their larger significance to the province. Apples were especially important to the horticultural industry in Ontario. They were promoted as *the* fruit to grow in Ontario, and their popularity was evidenced by the numerous apple classes offered at most fairs by the end of the nineteenth century. The number of apples exhibited was impressive. At the 1879 Grand Union Exhibition of the West Riding of York and Township of Vaughan Agricultural Societies, 113 of the 132 entries in the fruit classes were for apples.[127] The number of classes for different kinds of apples also grew significantly. The 1871 Peel County Fair offered two classes for apples; by 1878 there were seventeen. During this time, fruit classes at the Peel Fair expanded from eight official classes to more than forty-two classes for different varieties of seasonal fruits, such as apples, pears, plums, peaches, and grapes.[128] Some classes required a large quantity of fruit to be shown, such as ten varieties of apples for a single entry, and were meant for farmers who were capitalizing on the growing export market for Canadian apples.[129] In 1919, Canada's horticultural reputation was cited as having "been made mainly from the fine apples that have been produced in this country."[130]

Horticultural classes were open to all sorts of exhibitors, but the increased number of classes that required large amounts of fruit per entry, or many different varieties of a single fruit per entry, suggests that agricultural societies wanted to promote classes in which large-scale operators – men – would compete. The horticultural industry expanded significantly in the south-central regions of Ontario in the late nineteenth century.[131] Research on the best varieties of apples for cultivation in Ontario was conducted by the provincial Department of Agriculture and the Ontario Agricultural College,[132] and fair judges for vegetable and fruit classes were often men who worked at these institutions. Rural families benefited from horticultural research,[133] but its aim, and the addition of fair classes for newly developed varieties of apples and other fruits and vegetables, was to support and encourage leading male horticulturalists – not female market gardeners.

But despite the increasing promotion of fruit and vegetables grown by large-scale growers, women continued to exhibit. At the 1892 Erin Fair, Miss M. Overland won the potato, onion, and celery classes, placed second in the "short horn table carrots" class, and placed second in the Colvert and St Lawrence apple classes. Mrs Smart was the second-prize winner for the squash class, while Mrs Bennie won the Snow apple class, and placed second in the classes for American Golden Russett apples. Of the thirty-two different names listed as winners for the "Garden Roots and Vegetables" and "Fruit" classes, however, only three were women's.[134] For an agricultural society with a history of not recording the names of female exhibitors, one also has to wonder how many husbands took credit for their wives' work. On the 1925 Erin Fair winners list, almost thirty years later, Miss M. Overland

Table 2.3
Peel County Agricultural Society Fair, Brampton, Peel County: Classes for Fruit, 1871 and 1878

1871	1878
Winter Apples	10 varieties apples
Fall Apples	4 varieties cooking apples
Variety of Pears	4 varieties table apples
Sample of Peaches	Snow apples
Quinces	Duchess of Oldenburg apples
Sample of Egg Plums	Fall Pippin apples
Sample of Plums	Gravenstein apples
3 Clusters of Colored Grapes	Spitzenburg apples
Best collection of table grapes	Ribston Pippin apples
	St. Lawrence apples
	Rambeay apples
	Baldwin apples
	Rhode Island Greening apples
	Golden Russett apples
	English Russett apples
	Northern Spy apples
	Cultivated Crabapples
	10 varieties pears
	5 varieties pears
	3 varieties pears
	Flemish Beauty pears
	Louise Bonnie de Jersey pears
	Duchess de Angouleme pears
	Vicar of Winkfield pears
	Three varieties plums
	Greengages
	Egg plums
	Lombard plums
	Dessert plums
	Cooking plums
	Peaches
	Colored grapes
	Concord grapes
	Isabella grapes
	Delaware grapes
	Adirondac grapes
	Clinton grapes
	Rogers No. 4 grapes
	Rogers No. 15 grapes
	Rogers No. 19 grapes
	Hartford Prolific grapes
	Arnold's Othelle grapes

was still winning many of the garden classes, but was joined by ten new female prize-winners.

Most women won for vegetables rather than fruit; fruit classes were generally for apple production.[135] Jessie Milton, a long-time Georgetown Agricultural Society member, had a large garden from which she and her husband entered produce in fair competitions.[136] Many women may have also shared garden responsibilities with their husbands, but entered their exhibits in their husbands' names. Women were unafraid to compete alongside men in competitions where men were the main exhibitors. Women's garden produce was seen as an "economical investment" that could be made into a "range of culinaries"[137] for their families, but it also provided an additional source of income. When Jessie Milton's father died in 1936, her mother carried on farming with the help of her three children. During the Second World War, Milton recalled that they could not find hired help, so along with some "shares" from neighbours, she and her sister and younger brother helped their mother with field work and milking. Milton remembers that things were difficult for her mother, but the best source of income for her family was the Northern Spy and McIntosh apples they sold from the orchard her father had planted in 1919.[138]

During the Second World War, women's food provisioning – in the field, in the garden, and in the kitchen – were all discussed at OAAS Conventions, and women were encouraged to pay attention to the value of the family garden.[139] In wartime, women had an even greater role in food production, and were credited with saving agricultural harvests during periods of male labour shortage.[140] As in the United States, Victory Gardens and community canning projects in Canada acted as community builders,[141] while gardening clubs became a popular way to advocate for improved gardening practices and plant selection. Girls' garden clubs with club project titles such as "Vegetables to Keep Us Fit," "Every Farm Home – a Vegetable Garden," "Gardens for Health," and "Strength for Victory" were led by home economists and other women to teach and encourage young women to do their duty to provision for the family and the nation.[142] Agricultural societies sponsored display competitions at fairs to encourage young women to showcase their work and illustrate how women's gardening was seen as a service to their family and their community.[143] Women did their patriotic duty by growing

Feeding the Family 83

food that would help win the war by keeping soldiers and workers fit, and future generations healthy.[144]

In the postwar period, more attention was paid to new varieties of garden vegetables and fruits that were available to the household. As with ornamental gardeners, food gardeners benefited from the increasing array of plant varieties available in seed catalogues.[145] The gardening column in the *Farmer's Advocate* told gardeners to purchase the newest and best varieties of plants, since new strains were introduced each year that made their hard work more rewarding.[146]

Garden competitions remained popular because families appreciated the benefits of a home garden. Rural women exhibited produce because it showcased their skills. Women's gardening efforts continued to be associated with local, small-scale production, but during wartime they received more recognition for their ability to feed the nation. Women also appreciated the ability to grow fruits and vegetables that could be used to make other fare, such as canned goods and preserves, which they believed was important for the health and economy of their households.

Canned Goods

While everyone exhibited raw vegetables and fruits, the domestic science exhibitors were mostly women. Canned vegetables and fruit showcased women's ability to add value to produce and their skill and thriftiness in the kitchen. In 1855, the *Canadian Agriculturist* published a poem that conveyed the delight a well-stocked household pantry provoked:

FAMILY JARS.

Jars of jelly, jars of jam,
Jars of potted beef and ham,
Jars of early gooseberries nice,
Jars of mince-meat, jars of spice,
Jars of orange marmalade,
Jars of pickles, all home-made,
Jars of cordial cider wine,

Jars of honey, superfine,
Would the only jars were these,
Which occur in families![147]

Canning was a task that mainly women performed.[148] Farming journals advised women that with a small outlay of time and labour, they could grow fruit and make a good profit selling canned fruit, jams, and jellies.[149] The *Farmer's Advocate's* Minnie May advised women that fruit such as pears, quinces, peaches, plums, apples and cherries, were profitably grown, and that when preparing jars of jellies, jams, catsup, and other sauces, they needed to make them look tempting so that they sold easily.[150]

Agricultural societies' encouraged women's canning by introducing a variety of "Domestic Produce," later renamed "Domestic Science," classes at fairs for canned fruits and vegetables, jams, jellies, marmalades, vinegar, catsup, salad dressings, and canned meats such as chicken, pork, and beef. Preserving food for the family was an important part of women's household labour in the nineteenth century. Many rural women had the space to grow vegetables and fruit or collect wild berries, which gave some families enough food for most of their needs throughout the winter and early spring.[151] With the invention of the Mason jar in mid-century and improvements in home canning processes, women canned more than ever before and with better results.[152] Women's canning expanded even further in the twentieth century when cheap glass became more readily available.[153]

The Women's Institutes of Ontario and the Department of Agriculture encouraged women to can as an economical way to provision their household.[154] Agricultural societies offered classes for canned goods, held demonstrations, and constructed displays for educating rural women on the most effective preserving methods. Because exhibits showcased the finished product rather than the method of preserving, demonstrations were also used to advance women's learning.[155] Agricultural societies and other rural reformers recognized that women made the majority of household purchases, and that therefore they needed to be educated about consumer choices.[156] In the minds of reformers, rural women needed to become more progressive homemakers, efficient consumers, and market-oriented producers.[157] Rural women often dismissed reformers who criticized their practises and labelled

Feeding the Family 85

them ignorant and backward,[158] but when new technology or methods were presented at fairs, they were more open to receiving the message.

During times of scarcity, canning was particularly important for keeping families supplied with a variety of food while saving money. During the Depression, home canning was so widespread that in the United States, sales of glass jars soared above those in the preceding decade.[159] The sales of factory canned goods, which had doubled between 1919 and 1929, correspondingly dropped with an increase in home-canned products.[160] Farm families relied on self-provisioning rather than just cash income throughout this period.[161] This illustrates how growing food and making value-added products remained important for the economy of the household.

Many women prided themselves on their preserves. Jessie Milton recalled that her family had a large garden and orchard, and her mother always preserved food for the family. She described how her mother never purchased canned goods until the 1940s, explaining "that would be a very embarrassing thing, to be caught with a can."[162] Glenda Benton recalled how her family canned vegetables and fruit, but also meat: "My dad would kill our own beef and mother would can the beef right in the jars, the sealers, because there were no freezers ... so that would be our beef in the winter."[163] Benton and Milton, like other women, were skilled at canning throughout their lives. Benton explained that she started exhibiting at the Georgetown Fall Fair in baking and preserves because she was making around twelve different kinds of preserves every year anyway, so "all of a sudden I thought, well, why don't I show them at the fair, and that's what I did."[164]

The popularity of canned goods was clearly evident in fair prize lists. In 1900, most township fairs held about a half-dozen classes for different jellies and canned fruits and vegetables. By 1920, however, fairs such as the Brooke and Alvinston Fall Fair in Lambton County were offering more than thirty-one classes for pickles, preserves, and canned poultry and meat (see Table 2.4).[165] Exhibits of canned goods were popular because unlike baking, canned items could be made months in advance, easily transported, and displayed with little concern about sanitation or spoilage.

The popularity of canned goods increased in the 1930s, and while some fairs reduced classes during the Second World War, classes expanded again in the postwar period. The Collingwood Township Fair in Clarksburg is a

Table 2.4
Brooke and Alvinston Agricultural Society Fair, Alvinston, Lambton County: Classes for Canned Goods, 1920

1920

1 sealer canned raspberries, red
1 sealer canned cherries, red or black
1 sealer canned rhubarb
1 sealer canned pears
1 sealer canned plums, green or yellow
1 sealer canned peaches
1 sealer preserved strawberries
1 sealer preserved apples
1 sealer preserved citron
1 sealer preserved peaches
1 sealer grape jam
1 sealer black raspberry jam
1 sealer black currant jam
1 glass apple jelly
1 glass red currant jelly
1 glass raspberry jelly
1 glass other native fruit jelly
1 glass orange or grape fruit marmalade
1 glass native fruit marmalade
1 sealer canned tomatoes
1 sealer canned corn
1 sealer canned peas
1 sealer canned beans
1 sealer other vegetable canned
1 sealer canned chicken
1 sealer canned mince meat
1 bottle mixed pickles, sour
1 bottle mustard pickles
1 bottle sweet pickles
1 bottle other variety pickles
1 bottle tomato catsup

Feeding the Family 87

great example of this. In 1930, the fair had fifteen classes for a variety of canned vegetables, fruits, meats, jams, sauces, and dressings. In 1935, classes grew to twenty-three with the addition of items such as "Homemade Catsup" and the separation of previously grouped items. In 1940, classes for homemade soup and cold meat sauces were dropped, but more classes for jam and new classes for canned beans and tomato juice were added, totaling twenty-four classes. The number of classes was reduced to eighteen in 1945 because organizers amalgamated classes, so that exhibitors had to enter three types of canned fruits together in one class rather than multiple classes. By 1950, twenty-seven classes existed, including one for canned pineapple. Women's knowledge and creativity was tested with the addition of a class for "A Home Canned Meal" (see Table 2.5).[166]

In the postwar period, canning continued to be part of rural women's household provisioning. Even in a period when canning was not an economic necessity and the number of canned goods for purchase increased, rural women continued this practice because it saved money.[167] Preserving was associated with industry and thrift – personal traits that continued to be a source of pride for rural women. Photographs of canning exhibits show that they were interested in what other women in the community were making (see Figure 2.2). Female judges, exhibitors, and fairgoers alike understood the hard work that went into each jar on display, but they remained committed to preserving foodstuffs because of their value for the rural household and community more broadly.

Baked Goods and Other Culinary Fare

Rural households considered that baking and cooking were necessities, and well into the mid-twentieth century women continued to make a variety of baked goods and home-cooked meals for their families' consumption, as well as for sale or barter in the community.[168] Many women enjoyed cooking and baking, while for others it was simply a chore.[169] At fairs, baking and cooked goods usually fell under the label of domestic produce, domestic science, or culinary arts. This section focuses on baked items because even though some cooked food such as casseroles or meat loaves appeared at fairs, baked goods were much more prevalent because they could be displayed longer and more easily. They were also popular because they illustrated

Table 2.5
Collingwood Township Agricultural Fair, Clarksburg, Grey County:
Classes for Canned Goods, 1930, 1935, 1940, 1945, and 1950

1930	1935	1940	1945	1950
One pint sealer, canned corn, any method	Pint Sealer Canned Peas	Pint Canned Strawberries	Pint of Sandwiche Filling (home prepared)	A Home Canned Meal
One pint sealer, canned peas, any method	Pint Sealer, Canned Corn	Pint Canned Raspberries	One Pint each, canned Raspberries, Peaches and Cherries	Pint Sandwich Spread
One pint sealer, canned tomatoes, any method	Pint Sealer Canned Tomatoes	Pint Canned Peaches	Pint Canned Strawberries, Pears and Plums	Pint Canned Raspberries
One pint sealer, canned chicken	Pint Sealer Canned Chicken	Pint Canned Pears	Pint of Canned Tomatoes (whole)	Pint Canned Pineapple
One pint, each canned Raspberries, Strawberries and Cherries	Raspberry, 1 pint	Pint Canned Cherries	Pint of Canned Tomato Juice	Pint Canned Peaches
One pint each canned Peaches, Pears, Plums	Strawberry, 1 pint	Pint Canned Plums	Glass of Strawberry Jam	Pint Canned Pears
One pint each canned Gooseberry, Black Currant and Citron	Cherries, 1 pint	Pint Canned Rhubarb	Glass of Raspberry Jam	Pint Canned Strawberries
Jam, 2 var., 1 pt. each, "nature fruit," named	Peaches, 1 pint	Pint Tomatoes (whole)	Glass of Conserve	Pint Canned Gooseberries
Marmalade, 2 var., each named	Pears, 1 pint	Pint Tomato Juice	Glass of Jelly	Pint Canned Cherries

Soup Mixture, homemade, 1 qt., attach recipe	Plums, 1 pint	Glass of Orange Marmalade		Pint Canned Corn
Salad Dressing, homemade, 1 pt., attach recipe	B. Currants, 1 pint	Pint of Pickled Beets	Pint Strawberry Jam	Pint Canned Peas
Pickles mixed, 2 var, sweet and sour, 1 pt.	Jam, 2 varieties, 1 pint each	Pint Pickled Onions	Pint Raspberry Jam	Pint Canned Beans
One quart mustard pickles	Marmalade, 2 varieties, 1 pint each	Pint of Nine Day Pickled Cucumbers	Pint Conserve	Pint Canned Tomatoes
One quart ceilar Cold Meat Sauce	Jelly, 2 varieties, 1 pint each	Pint Mustard Pickles	Pint Jelly	Pint Tomato Juice
One quart Ceilar Preserved Rhubarb	Preserved Rhubard, plain, 1 pint	Pint Ripe Cucumbers	Pint Orange Marmalade	Glass wild Strawberry Jam
	Soup Mixture, homemade, pint	Pint Tomato Catsup	Pint Pickled Onions	Glass Raspberry Jam
	Salad Dressing, 1 pint	Pint Chili Sauce	Pint Pickled Cucumbers (9-day pickles)	Glass Grape Jam
	Mixed Pickles, sweet, 1 pint	Pint Salad Dressing	Pint Mustard Pickles	Glass Black Current Jam
	Mixed Pickles, sour, 1 pint		Pint Ripe Cucumbers	Orange Marmalade
	Mustard Pickles, sweet, 1 pint		Pint Tomato Catsup	Glass Apple Jelly
	1 pint Cold Meat Sauce		Pint Chili Sauce	
	1 pound Homemade Soup			Pint Pickled Beets
	1 pint Homemade Catsup		Pint Salad Dressing	Pint Pickled Onions
			Pint Canned Peas	Pint Ripe Cucumber Relish
			Pint Canned Corn	Pint Tomato Catsup
				Pint Chili Sauce
			Pint Canned Beans	Pint Dill Pickles
				Pint Salad Dressing

Figure 2.2 Fairgoers Inspect the Food Exhibits
This image shows that many women were interested in observing their peers' canned goods. Although men and children also viewed these exhibits, women were the intended audience. "Exhibitors at the Fall Fair, 1950." Port Hope Archives, 2003-47-3-2157. Image courtesy of Port Hope Archives.

women's ability to produce sweet treats. Long-time exhibitors in baking classes testified to the enjoyment women received from showing their baking. *The Globe and Mail* reported in 1938 that "Mrs. Freeman Green, 84-year-old resident of Ridgetown, made her seventy-second consecutive entry at the Ridgetown Fair." Green started exhibiting when she was twelve years old, when she "made her first entry with a loaf of bread, and since then has never missed a year."[170] Considerable skill was needed to bake well, and women respected those who won top prizes. Baking classes evolved over the years, showing that women were willing to adopt new products, although they still prepared were perennial favourites.

In the nineteenth century, unlike working-class women in urban centres who commonly bought bread, rural families tended to make their own.[171] The task was laborious, especially for yeast breads.[172] Attention to detail was also important in order to maintain yeast cultures, while proper kneading required a significant degree of physical exertion.[173] Articles in farm journals suggested that bread making was a subject that interested many, and recipes and methods were described for making the best varieties.[174] Diaries from women in nineteenth-century Ontario suggest that baking good bread was a significant achievement. Matilda Bowers Eby expressed satisfaction when, at the age of nineteen, she recorded that she had been baking all day and produced the "most beautiful" looking bread, also admitting, "I almost flatter myself that I will become a good baker yet; something not every woman can brag of."[175] Minnie May, the popular columnist for the *Farmer's Advocate* in the nineteenth century, also agreed that bread making was important and stood "at the head of domestic accomplishments, since the health and happiness of the family depend incalculably on bread, 'the staff of life.'"[176]

Bread was the first type of baked good shown at fairs across Ontario in the 1860s. Like butter and cheese, it was considered a staple that showcased rural women's skills. At the 1864 Guelph Township Fair, the first bread class, for the "best 6 lb loaf of bread" drew seven entries, which was considered a good showing for a new class.[177] At the Eramosa Fair that same year, it was agreed that the bread class was "an excellent idea" when twenty-two entries were made, four times as many entries as the year before when the class had first been offered.[178]

As with butter, bread classes received high premiums and special prizes. At the 1869 Peel County Fall Fair, the winner of the class for "the best two loaves of homemade Bread" won a Balmoral Stove valued at $28.[179] At the 1878 Peel County Fair, women made twenty-three entries of "hop yeast bread" and thirteen entries for buns. The newspaper congratulated the winners and warned that anyone "who would not be content to make a square meal on such bread, buns and butter as were shown at Brampton Fair this week ought to be sent to break stones for the corporations and fed a week of saw dust."[180]

Initially, bread classes consisted mostly of loaves and buns.[181] Other baked items, such as biscuits, pies, cakes, and tarts, were added to some agricultural societies' prize lists in the 1880s and 1890s, but generally they were limited until the twentieth century. Fewer classes likely reflected agricultural societies' preference for less perishable exhibits, as well as their focus on other pursuits. Also, baking was laborious, and although bakers received some help in the form of leavening agents in the mid to late nineteenth century, it was still a difficult task.[182]

Although judges could taste the baking exhibits, fairgoers based their judgement on appearance alone, so it was important that exhibitors knew "how to make the staff of life palatable as well as attractive"[183] (see Figure 2.3). Although some reports suggested that judges initially selected their winners only for their appearance, generally sight, smell, and taste all contributed to an entry's placing. Some women complained that because judges typically chose the winners before the fair opened to the public, and without explanation, appearance was what everyone relied on to make their own judgements.[184] The baking at the 1889 Nassagaweya Fair showcased "toothsome" exhibits, leaving fairgoers hungry for a bite.[185] The *Globe* reported that at the 1894 Markham Fair, the "bread and baking shown are of a very high quality, and afford a clue to the healthy appearance of the people of this neighbourhood." The newspaper noted that the "Snow-white bread, light buns and pies, cakes and tartlettes that would not incommode a newborn babe were constantly surrounded by good housewives, who always appear to enjoy looking at other housewives' work."[186] The "snow-white" bread and "light" coloured buns spoke to society's understanding of goodness and purity. Purity Flour was a brand that sponsored fair competitions and whose advertisements in farm and home journals depicted women in white outfits and gloves, seated embroidering the words "purity" at a table full of delicate-looking cakes and buns.[187] Ideals of purity and wholesomeness were associated with women's domesticity. At the 1899 West Durham Fair in Bowmanville, women entered 132 items in the domestic produce classes, and the display of goods was described as "exceedingly fine."[188] Baked goods were especially appealing food exhibits; from the more lavish cakes, tarts, and pies, to the more quotidian breads and buns, these items encouraged fairgoers to imagine the wonderful textures and tastes each entry had. The creation of delicious-looking food enticed passersby, who praised women for their "mouth-watering" entries.[189]

In the twentieth century, fairs continued to offer bread competitions. Women were told that if "good bread is of such great importance, there is

Feeding the Family 93

Figure 2.3 Domestic Science Judges at Rodney Fair
The local newspaper noted that "One of the largest displays in the exhibits building at the Rodney Fair this year is in the domestic science section, where hundreds of mouth-watering cakes and pastries are being shown. Some types of exhibits are more fun to judge than others, and the baking exhibit is about the best. Judging test cake samples are, from left to right, Mrs. Wilfred McMillan, domestic science bookkeeper; Mrs. Andrew Cipu and Mrs. Phillip Schliehauf; Mrs. Ruth Little, women's division president; Mrs. Dave McPherson, Dutton, judge; and Mrs. A. Plyley." "Judge's Decision Is Final," 1959. *St. Thomas Times-Journal* fonds, C8 Sh3 B1 F11 1, Elgin County Archives. Image courtesy of Elgin County Archives.

surely no other accomplishment in cookery over which a woman or girl should be so proud as over her ability to make it – make it fit to take a prize at a fair."[190] The number of other baked goods classes was also growing. At the 1890 Central Agricultural Society's Fair in Walters Falls, Grey County, these classes items featured "Home-made Bread, 2 loaves," "Pie, Frosted," and "Fruit Cake, plain, not iced." The homemade-bread class excluded professional bakers, and the first prize winner was awarded "1 pair Gentlemen's Plush Slippers" valued at $1.75; second place received "1 pair Ladies' Slippers"

Table 2.6
Central Agricultural Society Fair, Walters Falls, Grey County:
Classes for Baked Goods, 1890, 1900, and 1910

1890	1900	1910
Home-made Bread, 2 loaves, (Baker's excluded)	Special by R. Clarke, W.F., 2 loaves bread made from his flour	Bread, 2 loaves
Pie, Frosted	2 loaves bread society prize	Fruit cake not iced
Fruit Cake, plain, not iced	Fruit cake, not iced	Jelly cake, two layers
	Jelly cake, 3 layers	
	Best Pumpkin Pie	Pumpkin pie
		Apple pie
		Lemon pie
		Home-made biscuits
		Home-made buns

valued at $1.00.[191] In 1900, the classes increased to include a special prize for two loaves of bread baked from the flour milled by R. Clarke of Walters Falls, a three-layer jelly cake, and a pumpkin pie, the winner of which received a one-year subscription to the *Owen Sound Sun*.[192] By 1910, the classes increased to a total of eight because of the addition of buns, biscuits, and more varieties of pie (see Table 2.6).[193]

Like bread, other baked goods attracted donor-sponsored prizes. Sponsors donated prizes they believed female exhibitors would appreciate. For instance, at the 1925 Pinkerton Fair, the girl under eighteen years of age who won the "Best Loaf of White Bread and Best Apple Pie" won a pair of candlesticks valued at $3.50, and the winner of the "Best Fruitcake" won a parasol worth $5.00.[194] Donors awarded prizes they believed were coveted enough to encourage women to showcase their baking skills. The idea that young "ladies" needed and wanted candlesticks for their dining tables or attractive parasols for strolling down streets or attending garden parties was conveyed by awarding these items as prizes. Such prizes were symbols of middle-class respectability (see Figure 2.4).

Figure 2.4 Prize Cooking at Schomberg Fair
Women were proud of their exhibits and put on their Sunday best when attending the fair. "Schomberg Fair, Mary Potter, Schomberg, Prize Cooking," 1928. *Globe and Mail* Fonds, 1266, Item 15046, City of Toronto Archives. Image courtesy of the City of Toronto Archives.

Local mills and flour companies also offered special prizes to advertise their merchandise. At the 1925 Pinkerton Fair, local mill owners Wm. Knechtel and Son of Hanover offered fifty pounds of "O Canada" flour to the winner for the best bread made from their flour and twenty-five pounds of "Canadian Beauty" flour to the best pie made from their flour.[195] At the 1940 Collingwood Township Fair, a number of local millers offered special prizes, including the Collingwood Milling Company, which sponsored a ninety-eight-pound bag of North Star Bread Flour to the winner and a bag of Magic Pastry Flour to the second-place winner of "the best collection of Pies (to compare one Apple, one Pumpkin, one Raisin, one Lemon)."[196] At the 1940

Fair in Aberfoyle, contestants who used Ogilivie Flour to bake a winning entry in the bread class won a 12 ½-inch silver-plated cake platter in addition to the regular prize money, and if they used Ogilivie Flour to bake their fancy buns, tea biscuits, apple pie, or white layer cake, they won a 9-inch silver-plated platter for each winning entry.[197]

Large multinational flour companies sponsored sections of baking classes that required the use of their products. Robin Hood Flour sponsored baking classes across the province and placed advertisements in fair prize lists. For instance, at the 1940 Erin Fair, Robin Hood Flour doubled the cash prizes awarded in eleven of the regular baking classes if winners used their flour. Competing flour companies and dealers also enticed exhibitors to enter the classes they sponsored by offering generous prizes (see Table 2.7).[198] Paid advertisements in fair prize lists also served to inform exhibitors about why their products were the best. In 1940, Robin Hood Flour advertised that "At Fairs and Exhibitions right across Canada, in open competition with all other flours, Robin Hood won: More than 83% of all FIRST PRIZES, More than 76% of all SECOND PRIZES for BREAD, CAKES AND PASTRY" (see Figure 2.5).[199] In 1945, they advertised that it was the flour that "4 out of 5 Champion Home Bakers throughout Canada year after year use"[200] (see Figure 2.6). Robin Hood Flour's continued sponsorship and advertising campaigns at fairs in the postwar years reiterated the idea that "No other flour even APPROACHES this prizes-winning record!"[201]

Companies also sponsored classes to promote other food products, such as corn starch and corn syrup (see Table 2.7). At the 1954 Brussels Fair, Canada Packers Ltd. sponsored the "New Domestic Pie and Cake Competition," where exhibitors were required to use New Domestic Shortening in order to compete for a "5 lb. pail of New Domestic and box of Canada Packers assorted products."[202] Fairs offered classes with special merchandise prizes for items such as graham cracker or chocolate cakes so that companies such as McCormick and Neilson could promote their graham cracker crumbs or cocoa.[203] Baked goods were made with at least some, it not all, purchased ingredients, so national companies and local businesses took advantage of the opportunity to promote their merchandise.[204]

Agricultural societies were trusted institutions in rural communities, and fairs guaranteed a crowd, so companies tried to partner with them to convince consumers of the quality of their products. In the nineteenth century

DOUBLE YOUR PRIZE MONEY
BY BAKING YOUR EXHIBITS WITH

Robin Hood FLOUR

MILLED FROM WASHED WHEAT

1939 RECORD

At Fairs and Exhibitions right across Canada, in open competition with all other flours, Robin Hood won:

More than 83%
of all FIRST PRIZES

More than 76%
of all SECOND PRIZES

for

BREAD, CAKES AND PASTRY.

SEE PRIZE OFFER ON PAGE 40

Figure 2.5 Robin Hood Flour Advertisement, 1940
This advertisement for Robin Hood Flour informed baking exhibitors that their flour resulted in more winning entries than any other brand. Robin Hood Flour, "Double Your Prize Money," 1940 Erin Fall Fair Prize List. Erin Agricultural Society collection, 1862–1987, A1989.97, Wellington County Museum and Archives. Image courtesy of the Wellington County Museum and Archives.

4 OUT OF 5

CHAMPION HOME BAKERS

throughout Canada year after year use famous

Robin Hood FLOUR

ROBIN HOOD wins the favour of women who take pride in their baking because it mixes so readily—and so evenly—with the moist ingredients.
Use ROBIN HOOD Flour for your Fair baking this year and increase your winnings.

For Best Results In All Your Baking Use
Dependable—Economical

Robin Hood FLOUR

Figure 2.6 Robin Hood Flour Advertisement, 1945
In this paid advertisement published in the 1945 Erin Fall Fair prize list, competitors were told that "4 out of 5 Champion Home Bakers throughout Canada" used Robin Hood Flour. Erin Agricultural Society collection, 1862–1987, A1989.97, Wellington County Museum and Archives. Image courtesy of the Wellington County Museum and Archives.

and early twentieth centuries, rural consumers were still suspicious of food produced in factories and believed manufactured foodstuffs negatively affected their health and their pocketbooks.[205] Foods such as Campbell's Soup and Jell-O, however, intrigued them, because they offered to save time and expand their culinary choices.[206] Although most domestic sciences classes were for baked goods, by the postwar period more fairs had refrigeration and suitable display cases, so organizers expanded prize lists to include perishable food items such as jellied salads.[207] As with the classes that promoted flour and other baking ingredients, the promotion of ingredients such as Jell-O both reflected and encouraged changes in food culture. Fairs introduced people to new ideas and conveniences, and they were used by manufacturers and businesses to build public confidence in their products, demonstrate the quality of their goods, and showcase advancements for "enhanced rural living."[208]

In the 1950s and 1960s, baking competitions continued to flourish. Prizes, such as the $300 first prize for the light and dark layer cake contest at the 1951 Leamington Fair, stimulated competition. Some food exhibits had hundreds of entries, like the 171 entries made for the 1951 Harrow Fair pie classes.[209] New competitions, including the "Bake Queen Special" awarded at the 1960 Erin Fair for the top tray displaying bread, fruit cake, pie, light cake, *petit fours* and macaroons, inspired interest among exhibitors and displayed beautiful examples of "culinary art."[210] Other "Baking Queen" awards were given to women who excelled by earning the most points in the overall competition.[211] "Bake Queen" was a prestigious title that women earned by showcasing their talent. Winners illustrated their individual ability, but their success was sometimes also a matter of family pride. Multiple generations of families often competed, as younger women worked to uphold their mothers' and grandmothers' hard-won reputations. For example, at the 1968 Shedden Fair, twenty-year-old Deanna Bogart and her grandmother, Mrs Bogart, won most of the prizes in the domestic science competition (see Figure 2.7). Mrs Bogart was a previous winner of the fair's Silver Rose for baking, and Deanna had held the title of "Baking Queen."[212] These titles demonstrated a woman's success not just as a baker but as a wife and mother, or future wife and mother.

In the 1970s, more changes in home cooking and eating habits were demonstrated in domestic science classes. The inclusion of classes for "chop

Table 2.7
Erin Township Agricultural Society Fair, Erin, Wellington County:
Classes for Baked Goods, 1940

1940

Bread, 3 lb. loaf, to be made in pan 9 1/4 x 3 x 4 in. *†
Brown Bread, whole wheat
Fancy Buns, five *†
Plain Buns, half dozen *†
Plain Biscuits, half dozen *
Milk Rolls, half dozen *†
Plain Scones, half dozen
Best Apple Pie *
Best Pumpkin Pie
Best Raisin Pie
Best Layer Cake, dark, iced *†
Best Layer Cake, light, iced *†
Collection of Fancy Baking, 5 varieties only *
Collection of Plain Baking, 5 varieties only *†
1 Cake of Shortbread, made on a 9 in. pie plate *
Plate of Fancy Sandwiches, 16 sandwiches, 4 var. †

Baking Specials
For the best apple pie and half dozen biscuits made from Lily White Pastry Flour...
By P.J. Sinclair, Dealer in Flour...the best two loaves of bread, made from High Loaf Flour...
By Lloyd Lyons, Groceries [etc.]...the best load of bread made from Prairie Rose Bread Flour...
For the best Apple Pie made from Planet Flour...by P.J. Sinclair, dealer in Flour...
For a light layer cake, iced, made by a bride since the Fair of 1939...
By St. Lawrence Starch Co. Ltd...the best Lemon Pie...Durham starch to be used.
By St. Lawrence Starch Co. Ltd...the best Butter Tarts...Bee Hive Golden Corn Syrup to be used.
By St. Lawrence Starch Co. Ltd...the best quart of Canned Peaches...Bee Hive Golden Syrup to be used.
By St. Lawrence Starch Co. Ltd...the best pound of Divinity Fudge...Bee Hive Golden Corn Syrup to be used.
By St. Lawrence Starch Co. Ltd...the best Butterscotch Pie...Using Durham Corn Starch.
By the Canada Starch Sales Co. Ltd...for 1 nut and date loaf...[to use Crown Brand Corn Syrup]
By the Canada Starch Sales Co. Ltd...for Muffins, half dozen, whole wheat, flour or bran...[to use Crown Brand Corn Syrup]
By the Canada Starch Sales Co. Ltd...for 12 rolled cookies, 3 varieties, 4 of each...[to use Crown Brand Corn Syrup]
By the Canada Starch Sales Co. Ltd...for six syrup tarts...[to use Crown Brand Corn Syrup]

*Classes also sponsored by Robin Hood Flour
†Classes also sponsored by the McArthur Milling Company

Feeding the Family 101

Figure 2.7 A Family Practice at Shedden Fair
Deanna Bogart and her grandmother, "Mrs. Fred Bogart" of Southwold, "continued to dominate various domestic exhibitions" at the 114th annual Shedden Fair, according to the *St. Thomas Times-Journal*. Mrs. Bogart received the fair's Silver Rose for baking the previous year, while Deanna was crowned Baking Queen for 1967. The pair is pictured holding their winning entries. "A Family Practice," 1968. *St. Thomas Times-Journal* fonds, C8 Sh3 B1 F19 10i, Elgin County Archives. Image courtesy of Elgin County Archives.

suey loaf" and "Macaroni Salad" reflected agricultural societies' acknowledgement of rural women's interest in new food trends,[213] while classes such as "T.V. Evening Snack for a Teen-Ager" illustrated changes in dining etiquette and family dynamics. The staple classes also remained, and homemade bread continued to be exhibited, which highlighted bread's privileged

Figure 2.8 Canada's Flag Debate at Wallacetown Fair
The *St. Thomas Times-Journal* told politicians to "take note" that "Canada's flag issue was carried into the baking field" at the Wallacetown Fair. Contestants used their culinary talents to offer ideas for a flag design. The domestic science judge, Mrs William Beattie of Staples, is pictured here scrutinizing the entries. "Politicians, Take Note," 1964. *St. Thomas Times-Journal* fonds, C8 Sh3 B1 F18 15b, Elgin County Archives. Image courtesy of Elgin County Archives.

position as *the* home-baked good. Even though homemade bread may not have been a staple of all farm homes by this time, its traditional importance in the rural household and the skill with which a good loaf of bread was made ensured that the product continued to be exhibited at fairs and used as a symbol of domestic competence.[214]

In the postwar period, another noticeable feature of domestic science classes was the emphasis placed on showcasing "culinary art," rather than simply food. At many fairs, the "Domestic Science" department became the "Culinary Arts" department, and skill, artistry, and originality in crafting food was highlighted. Themed cake-decorating competitions became popular, providing women opportunities to illustrate their creativity by illustrating themes such as centennial celebrations, national symbols, public holidays, animals, and musical instruments (see Figure 2.8). These cakes illustrated women's ability to entertain fairgoers. Rural women wanted to display their creative side and show that their skills in the kitchen could be used for more than the prudent management of the household.

Conclusion

Women's food exhibits drew exhibitors, impressed organizers, and enticed fairgoers. Food exhibits were colourful and mouth-watering. They were a visual representation of the characteristics most valued in rural women: thrift, industry, domestic skill, and the nurturing of and responsibility to one's family and community. The continued cultivation and canning of garden vegetables and fruits, as well as the baking of bread and other foodstuff, reflected how despite the availability of mass-produced foods, rural women continued to take pride in making homemade goods that they believed offered superior taste and economy. The importance of rural women's homegrown and homemade foodstuff in the years before 1945 is supported by recent scholarship on farm families' consumption practices.[215] Women's desire to maintain these practices in the postwar period suggest that those practices continued to be appreciated economically, socially, and/or culturally. The desire to make one's own bread, for example, reflected the perceived quality of homemade food items, but also a desire to maintain valued rural traits. Other items, such as dairy products, were no longer worth the time or trouble, but the domestic skill women displayed by making canned goods and other foodstuffs awarded them recognition as careful consumers and useful producers. Despite the changes that occurred over the years, a great deal of continuity remained. Women engaged in market activities when it benefited them, and they gave up homemade production that no longer did. The continued dominance of food exhibits, especially prepared food exhibits, as women's work also served to perpetuate gender norms and traditional ideas about women's culinary abilities. Many women enjoyed this work and sought self-expression and personal meaning through the food they made, but they also understood that the appearance of their fair exhibits reflected not just on the items shown, but on their characters as well. Nevertheless, rural women believed that homemade foodstuffs benefited them and their households, and they also wanted to continue practices that helped to maintain elements of the rural traditions that they cherished.

3

Cultivating Beauty
The Flower Show

By the second half of the nineteenth century, agricultural societies began to include flower exhibits because they believed flowers would contribute to the fairgrounds' overall attractiveness and stimulate rural beautification projects. Generally, more women competed in flower shows than men. Society perceived women as having a "natural affinity for genteel culture,"[1] and fair organizers promoted ornamental gardening classes as a means of inspiring women to adopt principles of honest industry, moral fortitude, and refinement; all characteristics encouraged by other women's exhibits. Women were often responsible for taking care of their families' spiritual needs, and rural reformers argued that the cultivation of flowers involved a moral discipline that could be visually communicated. Women's application of labour and design principles resulted in a visual representation of their morality and taste. Improvers believed these displays would motivate others to adopt the same disciplined habits. In the twentieth century, advocates also associated flower gardening with an appreciation for scientific endeavours and modern aesthetics. Women could display these traits as evidence that they were progressive homemakers.[2] Still, an understanding of beauty remained central to these ideas. Women who demonstrated their knowledge of the concept, combined with discipline and good judgement, exemplified the respectable characteristics necessary for proper womanhood.

Nineteenth-Century Flower Shows

Trade catalogues that advertised nursery stock and newspaper advertisements for different garden vegetables, fruit trees and bushes, flowers, and other plants had existed early on in Upper Canada.[3] Nineteenth-century British North American newcomers, travellers, and long-time residents were both interested in the region's plants and flowers and also made great efforts to import ornamental plants that replicated their home countries' environment.[4] However, agricultural societies generally paid little attention to garden plants, especially ornamentals, when determining which exhibits to include at fairs. They were more concerned with offering premiums that benefitted farmers cultivating field crops or raising livestock, pursuits believed to be the agricultural economy's backbone.[5] Even the Toronto Horticultural Society, the first in the province, focused primarily on promoting fruit trees and vegetables rather than ornamental plants.[6] Horticultural societies were largely distinct from agricultural societies, although sometimes the organizations coordinated their exhibitions to ensure a stronger exhibit resulted. Most classes at horticultural shows during this period were for fruits and vegetables.[7]

Organizers of the first Provincial Exhibition in 1846 did not hold a flower competition. Still, they did show interest in ornamental gardening when they presented some prize winners with articles and books on the topic, such as *Gardening for Ladies*,[8] written in 1840 by Jane Loudon, an Englishwoman described as being "to Victorian gardening what Mrs. Beeton was to cookery."[9] The book sold over 200,000 copies and was soon reprinted in North America as the fuller version *Gardening for Ladies; and Companion to The Flower-Garden*. Loudon wrote the book because she found that many publications intended for professional gardeners were not suitable for amateurs.[10] She wanted to communicate the basic knowledge that female gardeners should know. *Gardening for Ladies* included the essentials of "kitchen gardening," but most of the text discussed proper flower-gardening methods and the plants one should cultivate. In awarding this book as a prize, the provincial association was acknowledging that flower gardening was a worthwhile pursuit for women.

Nineteenth-century rural reformers advocated ornamental gardening because they were critical of Ontario's lack of "cultivated" spaces. Although documentary and photographic evidence illustrates the existence of flower gardens during this period, reformers consistently lamented residents' inability to appreciate the "moral and spiritual welfare" that ornamental gardening afforded.[11] In 1847, the *British American Cultivator* published an article that criticized the state of horticulture. The author noted that Canadian farmers were "reproached with being sluggards in regard to their gardens."[12] He argued that farmers who had finished clearing their land and had "good health, good soil, and ample time" should be disgraced for not owning "a nicely cultivated and trimly kept garden and orchard."[13]

The author quoted at length from Dr Darlington, an American horticulturalist who emphasized the necessity of ornamental gardens for instilling higher principles. Darlington argued that food gardens sustained families' physical needs, but ornamental gardens were necessary for their spiritual needs. He argued that "the training of the most ornamental trees and shrubbery – the culture of the sweetest and most beautiful flowers – and the arrangement of the whole in accordance with the principles of a refined, disciplined, unsophisticated taste" was needed. He expressed that "all that is connected with comfort and beauty around our-dwellings [sic] – all that can gratify the palate, delight the eye, or regale the most fastidious of the sense" was essential if "the highest mental accomplishments" were to be achieved, and "the more sordid or grovelling passions" suppressed.[14] Darlington's association of ornamental gardening with beauty, intelligence, refinement, and purity of mind suggested that the pursuit created a more refined and civilized countryside. He proposed that even small improvements could affect significant change: "That the habitual association with interesting plants and flowers exerts a salutary influence on the human character, is a truth universally felt and understood ... Who ever anticipated boorish rudeness, or met with incivility, among the enthusiastic votaries of *Flora*? Was it ever known, that a rural residence, tastefully planned, and appropriately adorned with floral beauties, was not the abode of refinement and intelligence? Even the scanty display of blossoms in a window – or the careful trailing of honey suckle, round a cottage door – is an unmistakeable evidence of gentle spirits, and an improved humanity, within."[15] Failure to engage in ornamental gardening signified an absence of civility and a deficiency of character. Darlington lamented that many farmers did not adorn their homes with ornamental trees, shrubs, and flowers, and he asked, "What can be expected from a family, raised under circumstances so unpropitious to the formation of correct taste, or the cultivation of the finer feelings?"[16]

In the 1860s, reformers continued to criticize rural residents for not cultivating beautiful surroundings. In 1864, a person with the initials W.T.G wrote to the editor of the *Canada Farmer* to complain that farmers and townspeople only cared for "utility" when considering the landscape around them, rather than the "refreshing" and "improving" qualities that an ornamental garden instilled. He advised residents to make their homesteads more attractive by adding well-kept flower beds, ornamental trees, and shrubs.[17] Gardening was not a luxury, he argued, but a necessity if one wished to improve one's lot in life: "it is the home of the wealthy gentleman or farmer which proves generally most attractive. Not because the owners can better afford it than their poorer neighbours, but the industry and talent, which lie at the root of their prosperity, beget in them the love of improvement, and the garden affords the widest scope for it."[18] W.T.G believed a person's inability to cultivate a beautiful garden was a reflection of their indifference to improvement and a marker of coarseness and idleness. In 1868, the *Farmer's Advocate* urged farmers to adorn their homes, reasoning that a "love for trees and plants and flowers is natural to every refined and well developed mind ... Whatever makes a home pleasant and attractive, lessens the temptation to stray into paths of evil."[19]

In these examples, "farmers" were the target of criticism, but readers understood that farm wives were often responsible for maintaining home gardens. As the home's moral guardians, they were also responsible for enhancing their family's spiritual well-being. In 1874, a reader wrote to the *Farmer's Advocate's* columnist, Minnie May, that any woman who did not love flowers was "a mistake of nature" because the "delicate and the beautiful should have sympathy with all in nature that possess the same qualities."[20] Another concerned proponent of flower gardening wondered why flowers were not cultivated more extensively in Canada. He explained his surprise to "meet with ladies claiming to be passionately fond of flowers, who knew everything about dress, etc., yet cannot give the names of half a dozen different flowers."[21] The author happily reported, however, that these "careless florists" were disappearing and a more perseverant and successful "class of ladies" was emerging. He accompanied his description of the progress of women's work in ornamental gardening with two images, one that represented the fashionable but careless florist, and the other the more modest, attentive, and hardworking gardener (see Figure 3.1).[22] The careless florist's garden was in a state of disorder, while the attentive gardener's efforts resulted in a bounty of beautiful, healthy plants arranged in perfect symmetry. Each garden represented a women's character. The attentive gardener was able to control her environment through a disciplined application of labour and design principles. In the same way that women carefully presented foodstuffs at a fair, their presentation of ornamental plants created a vision of orderliness that reformers believed would inspire all who gazed upon them. Advocates of ornamental gardening suggested that women who said they did not have time to garden simply needed to reallocate the "waste time" they devoted to gossiping with friends to this elevating pursuit instead.[23]

Farm women were responsible for their families' happiness, and when rearing their children, they were expected to provide a suitable environment in which their children's spirits and characters could flourish. A writer for the *British American Cultivator* argued that it was a mother's job to imbue her children with desirable characteristics, such as "honesty, temperance, industry, benevolence, and morality," and eliminate contrary characteristics, such as "vice, fraud, drunkenness, idleness, and covetousness."[24] A mother was assigned the "daily, hourly task of weeding her little garden – of eradicating those odious productions, and plating the human heart with the lily, and the rose, and the amaranth, that fadeless flower, emblem of Truth."[25] The *Farmer's Advocate* also advertised that women purchase flower catalogues because they were "worth five times [their] price to any mother that wishes to have refinement, neatness, beauty, and adornment impressed on the minds of her family."[26] Rural reformers told women that flowers possessed a moral influence that pleased, cheered, and refined the mind and made the home more attractive, peaceful, and divine.[27] As a natural extension of woman's rule over the hearth, her participation in cultivating the literal and figurative family garden was necessary if her family was to thrive.

The Provincial Exhibition first offered a floral competition in 1849,[28] but it took much longer before many county and township fairs added flower classes to their prize lists. Most of the larger county fairs had at least some classes for table and hand bouquets and plant collections by the 1860s,[29] but many township fairs took longer. At the Erin Township Fair during this period, no official flower classes existed, although some discretionary awards

The woman whose flower seeds never come up unless they are scratched up.

Figure 3.1 *Above and opposite* Illustrated Florist
The *Farmer's Advocate* published an illustration that represented the fashionable but careless florist and the more modest, attentive, and hardworking gardener. The idea was that the state of one's garden embodied the state of one's character. "Illustrated Florist," *Farmer's Advocate* 10 (June 1875): 117.

were given for floral wreaths and arrangements in the 1850s and 1860s.[30] Even by 1889, the Erin Fair only had two classes: one for a collection of garden flowers and the other for a group of house plants.[31]

Most flower classes at this time focused on arrangements and collections of flowers, and women were the leading exhibitors. The *Guelph Mercury* remarked that the floral exhibit at the Hespeler Fair in 1889 "was large and fine," and described "A fancy piece by Miss Carrie Olaflin, containing a looking glass with a swan and waterlilies reflected in it, representing a lake, bordered with marigolds and evergreens and surmounted by the words, "Prosperity to Hespeler," was a work of art." Another attractive entry was the "fancy floral piece by Miss Whitmer" which presented "A sickle, on moss

The woman whose flower seeds all come up.

ground and everlasting flowers, with a bunch of wheat in the centre." It was noted that Whitmer's work was "was greatly admired, everything to make the display being gathered and made by the lady herself."[32] Newspaper descriptions of floral exhibits such as the entries above identified competitors by name, highlighting characteristics such as good morals, industry, and taste. Such descriptions allowed one to envision how each competitor artfully arranged their work. A reporter for the *London Advertiser* contended that tasteful bouquets and other flower exhibits represented the maker's skill and cultivated tastes and gave "pleasure to the thousands who study them."[33] Floral displays enriched the minds of both exhibitors and fairgoers by inspiring refinement. The visual knowledge imparted through a plant that reached perfect form or a tastefully arranged floral ornament based on classical elements of symmetry taught women how to understand these principles, cultivate them, and apply them.

Men also competed in early flower shows, typically those holding privileged status in rural society, such as judge or minister.[34] Their ability to illustrate taste did not emasculate them because their class position was associated with elevated sensibilities. By the 1880s, the other men who competed were typically owners of commercial nurseries and flower shops (this was often the case for larger provincial and regional fairs). For example, it was reported that "the largest exhibitors" at the 1890 Central Exhibition in Guelph were William Mann, James Gilchrist, Noah Sunley, John Marriott, and Robert Brooks. The newspaper report also noted that the plants displayed were "especially fine" and did "credit to our local men, all of whom have gained a provincial reputation."[35] These "local men" were nurserymen, such as Robert Brooks of Fergus, who had "swept everything before him in Toronto for begonias and English pansies."[36] Nursery owners competed at larger fairs to showcase their merchandise and win renown for the quality of their plants.

Smaller fairs, however, privileged amateur gardener, most of whom were women. Social reformers said women were naturally drawn to beautiful things, and therefore their femininity gave them a decided advantage in flower gardening. The local newspaper reported that flower exhibits at the 1868 Peel County Fair "elicited great praise and showed the good taste of the ladies who got them up."[37] Women, not men, were credited with having the ability to arrange flowers for competition tastefully.

Agricultural societies wanted to set an example for improving the countryside, so many societies made significant investments by planting flowers, trees, and shrubs to enhance the appearance of their grounds.[38] However, during fairs, the most cost-effective way to make agricultural halls attractive was to encourage large displays of plants and flowers. Newspaper reports testified that floral exhibits added a general brilliance to fair buildings. The local newspaper described the impressive floral display at the 1871 Peel County Fair and noted that, in addition to the flowers from "the carefully tended flower bed," there were also vases of wildflowers that exemplified "the oft repeated assertion that nature in her native wilderness can, for effect, successfully compete with art."[39]

Reporters noted that women who attended fairs regularly showed off "their best," whether that was a new fur coat or good looks.[40] At the 1857 Toronto Township Fair, the agricultural society's secretary remarked with pleasure that the showground became "brilliant with the gay dresses of the ladies."[41] A reporter pointed out how the "bevy of blooming, rosy-cheeked lasses, in their best bib and tucker" at the Markham Fair in 1869 were "closely followed by a band of bashful admirers."[42] Many men believed that women

Cultivating Beauty 113

made fairgrounds more attractive with their presence. Their food, fancywork, fine art, and flower exhibits emphasized the way "blooming, roseycheeked lasses" contributed to a more beautiful and civilized event.[43] In 1868, the *Farmer's Advocate* suggested that every township fair should offer a prize for the best bouquet, because it would draw people "to see the pretty flowers and pretty faces" and "the boys would love them no less because they admire a flower."[44]

Flower shows also provided a clear visual example of an area's agricultural potential. At the 1880 Western Fair, the fair's president, George Douglas, argued that in addition to the splendid display of livestock, grains, and agricultural implements that signalled the progress of the region to international visitors, the "flowers in the Horticultural Hall must tend to completely dissipate that prejudice with regard to the Canadian climate that has hitherto, more than anything else, prevented the better class of emigrants from seeking our shores."[45] Indeed, the extensive scholarship on the World's Fairs and other international exhibitions has shown how important a successful exhibit was to a region's or nation's reputation.[46] Many Canadians took pride in their status as an "Imperial breadbasket" for the British Empire.[47] Douglas contended that, along with crops, flowers showcased the province's agricultural possibilities, which Douglas believed attracted "the better class of emigrants."[48] His comment illustrates the contemporary belief that individuals who appreciated ornamental gardening were refined and intelligent, and therefore suited to contributing to the province's welfare.

In the 1870s and 1880s, rural reformers promoted the improvement of homes, farms, and landscapes. Wealthy owners built large country homes and hired professional gardeners to design and maintain their expansive grounds. Middling and well-to-do farmers also took pride in well-landscaped homesteads.[49] The Historical County Atlases published in Ontario in this period were idealized representations of homes, farms, and businesses, clearly illustrating the belief that residents should create orderly and attractive spaces. Figure 3.2 provides an example of a farm in Peel County that used trees as practical windbreaks and shade sources, as well as for purely ornamental purposes. Rural reformers believed a beautiful farm was an efficient farm, one where "flowers, gardens, trees and pleasant surroundings outside ... make children love the homestead," and therefore kept them "in love with the farm."[50]

Reformers were particularly concerned with rural depopulation in the late nineteenth and early twentieth centuries. They cited the need to keep young people from leaving the farm for the city's attractions as an argument for promoting beautification projects.[51] In the 1880s, the Department of Agriculture took on the task of beautifying farms and set the standard of what "an oasis of rural tranquility and beauty" should be.[52] Government personnel encouraged the adoption of ornamental plants to keep families happy working the land. Rural reformers warned that farmers' sons and daughters would become restless if their homes were not pleasant and enjoyable: "Children must be made to love their homes, else the attractions of cities and villages will surely lure them away."[53] An article in the *Guelph Weekly Mercury* urged farmers to consider the rights of their daughters, especially in terms of remuneration for their work, but it also encouraged them to make their homes attractive and "cultivate a taste in the girls for flowers" because these "features, with a moderate amount of work, should produce a happy contented home life on the farm."[54]

By the late nineteenth century, agricultural societies also held more flower classes because ornamental gardening had become a popular hobby in rural communities. The link between the mind and the soil had been well established and, by the latter decades of the nineteenth century, ornamental gardening was a respected recreational activity.[55] As floriculture's profile increased, rural communities became active in promoting ornamental gardens in private and public spaces.[56] The Victorian landscape style of carpet bedding (a patterned arrangement of foliage plants) and geometrical layout was popular in Canada by the 1860s, but many rural gardeners opted for an assortment of plants and flowers (depending on what grew best) rather than a manicured garden.[57] Fair displays also privileged individual garden items like flowers over garden designs as a way to encourage more recreational gardening. Some agricultural societies had extravagant "floral temples" set up during fair time,[58] but generally, recreating an ornamental garden or flower bed was impractical, and exhibits were typically limited to individual varieties of flowers or mixed arrangements.

Flower competitions also expanded by the late nineteenth century because horticulture was a growing industry in the province. Agricultural journals and newspapers regularly advertised flower seeds, and it was noted that "every little town has one or more persons who are making a living

Cultivating Beauty 115

PROSPECT HOUSE.-RESIDENCE OF JOSEPH GARDNER ESQ^R, LOT 5, I. CON. E. HURONTARIO S^T T.T.

Figure 3.2 Prospect House in Peel County
The pictures of individuals' homes, farms, and businesses in historical atlases published in the late nineteenth century were idealized. This illustration of Joseph Gardner's residence demonstrates that he carefully landscaped his farm to create an orderly and appealing design. "Prospect House-Residence of Joseph Gardner," *Historical Atlas of Peel County, Ontario*. Published by Walker and Miles, 1877.

profit on small seeds and plants for the garden, the window and the lawn; and in each city large establishments are profitably supported by the lovers of beauty."[59] The "landscape" lawnmower was also becoming a popular garden tool during this period among "all who desire to have a well-kept piece of grass."[60] Garden historian Edwinna von Baeyer contends that a distinctive shift took place in gardening practices in the 1880s. She argues that the rise of industrialism brought new wealth and amenities to communities. The increase in technology, the availability of markets, and greater population concentration created an atmosphere in which ornamental gardening became "a social marker, a status symbol."[61] In urban areas, the pursuit of ornamental gardens was closely linked with class, wealth, and leisure.[62] In the countryside, wealthier residents with hired help also tended to have more formal and elaborate gardens, but many rural households generally enjoyed flower gardening. For instance, the 1898 diary of Mary McCulloch reveals her passion for growing flowers. McCulloch was a single thirty-eight-year-old woman who lived with her mother and found paid employment as a housekeeper. She resided in Snelgrove, a small rural hamlet in Peel County. She regularly references her garden work, including how she cared for her vegetables, fruit bushes, and trees, and how much time she spent tending to her flowers. In April, she recorded that she sent the boys (children of a widower for whom she acted as housekeeper) "up to the Hutchinsons with some Rose bushes and some flower roots."[63] The next week she recorded that her brother George had brought her a white peony root from a Mrs Henderson, but she was disappointed when she found it "badly broken." She explained, "I am afraid it will not grow, but I hope it may, for they are so nice."[64] McCulloch recorded planting Sweet Williams (*dianthus barbatus*), Scarlet Lightning (*lychnis chalcedonica*), hollyhocks, roses, and Lily of the Valley. She planted seeds and bulbs, transplanted flowers, weeded her flower beds, exchanged flower bulbs with neighbours, and gifted flowers for special events.[65] She spent both time and money cultivating her flowers, and the diary details her efforts to stake out her flower garden or fix up hotbeds to protect early growth.[66]

Mary McCulloch's diary illustrates that flowers interested more than just wealthy residents in rural communities. We do not know if she participated in any flower shows, but we know that the women who won prizes came from many different backgrounds. For example, at the 1880 Peel County Fair, Miss McVean won the "Everlasting Bouquet" class. The McVean daughters lived on their family farm in the Toronto Gore district of Peel.[67] Elizabeth Rowden, who won the competition for a collection of annuals in bloom, lived with her widowed sixty-eight-year-old mother and was employed as a dressmaker in Brampton.[68] Mrs H. (Sarah) Baskerville won the Coxcombs class; she farmed with her husband and her young children in Chinguacousy Township.[69] These women illustrate how flowers, and flower shows, were enjoyed by women in varying circumstances and at different life stages.

The Peel County Fair's flower show expanded rapidly in the 1870s. At the show in 1868, only four classes existed, but the number increased to twenty-two by 1878. Peel County experienced a boom in the greenhouse industry

Cultivating Beauty

during these years that continued into the twentieth century. Brampton, the location of the Peel County Fair, came to be known as "the flower town of Canada" because of the success of horticulturalists, such as K. Chisholm and E. Dale & Son, in the late nineteenth century.[70] Women still competed in flower shows during this period, but most new classes showcased the plants of large-scale horticultural operations, and many of the local nurserymen dominated prize lists. At the 1878 fair, classes were offered for specific varieties of flowers such as dahlias, peonies, fuchsias, phlox drummondii, roses, verbenas, zinnias, gladioli, and marigolds, as well as for different types of table and hand bouquets, baskets, or larger collections of single or mixed varieties of plants and flowers, and specific flower designs or ornaments.[71] By 1880, the agricultural society had added six more classes, so that the flower exhibit had grown seven-fold since 1868 (see Table 3.1). By the 1880s, farm journals regularly featured articles that catered to the increasingly specialized horticultural industry in the province.[72] The changes at the Peel County Fair reflected a desire to satisfy the growing number of "nurserymen," and by the 1890s, most flower exhibitors at the Peel Fair were horticulturalists trying to market their merchandise.[73]

The Peel County Fair was located in a mature agricultural community, close to important urban centres, and therefore attracted more participation from large-scale horticulturalists than most township and county fairs. Smaller fairs continued to attract more female exhibitors. For example, an 1889 Nassagaweya Township Fair report boasted that the "display of house plants and flowers, cut flowers, would have done credit to more pretentious exhibitions," and four of the five winners listed were identified as women.[74] The less "pretentious" a show was, the more women entered as contenders.

Flower Exhibits in the Early Twentieth Century

The twentieth century continued to witness the growth of flower shows. The rising prosperity of rural Ontario led some women to seek ornamental gardening as a creative process.[75] By this time, some agricultural and horticultural societies had amalgamated, which increased the number of entries. Fair organizers became interested in how flower competitions could represent science and progress in a modern age. Scientists were researching flowers at

Table 3.1
Peel County Agricultural Society Fair, Brampton, Peel County:
Classes for Flowers and Plants, 1868 and 1880

1868	1880
Best Table Boquet Garden Flowers	Six standard dahlias
Best Hand Boquet of Garden Flowers	Six bouquet dahlias
Best Hand Boquet of Wild Flowers	Large vase bouquet
Best Collection Dried Plants	Table bouquet
Money-wort Plant*	Hand bouquet
	Bouquet of everlastings
	Bouquet of wild flowers
	Collection of 24 greenhouse plants
	Collection of peonies
	Pansies, 10 varieties
	Collection fuchsias
	Native Ferns, 4 varieties
	Foreign Ferns
	Collection of annuals in bloom
	Collection foliage plants
	Collection roses
	Single Geraniums, 4 named
	Doubled Geraniums, 4 named
	Collection Phlox Drummondii
	Collection Verbenas
	Collection double Zinnias
	Collection Gladiolas
	Best display of cut flowers
	Four Window plants in flower
	Two hanging baskets
	Collection ten-week stalks
	Collection Marigolds
	Two new or rare plants
	Coxcombs*

*Discretionary Prize

the provincial and national departments of agriculture and the Ontario Agricultural College. The new varieties of plants they developed demonstrated the way that science had "improved" plant life by increasing its suitability to Ontario's growing conditions.[76] The new types of flowers agricultural societies promoted, like their traditional favourites, showcased beauty and taste. Although the flowers shown at fairs represented all skill levels, every entry was thought to encourage others in the community to gratify their senses, enlighten their minds, and beautify their environment.

At the start of the twentieth century, many flower shows remained small, but they were growing. In 1910, the Erin Fair offered seven classes for geraniums, begonias, foliage plants, sunflowers, a bouquet, a collection of garden flowers, and a collection of house plants.[77] The same year, the Egremont Fair's flower show consisted of only five classes for a collection of flowers, cut flowers, a hand bouquet, a coleus plant, and a house plant of any other kind.[78] The Shedden Fair still did not have an official flower show, only a special prize for "The best and largest variety of flowers."[79] In contrast, the Arran-Tara Fair had grown significantly in size and offered twenty-two classes, including asters, pansies, verbenas, petunias, coxcombs, zinnias, dahlias, begonias, gladioli, geraniums, fuchsias, balsams, sweet peas ferns, and roses, as well as collections of house plants, foliage plants, large or table bouquets, small or hand bouquets, and displays of annuals cut and named.[80]

Government-sponsored horticultural research had begun in the nineteenth century, but in the twentieth century, researchers made significant strides in plant breeding. In 1908, the Ontario Agricultural College reorganized the Horticultural Department. J.W. Crow, initially a lecturer at the OAC but later a professor of Horticulture at the college, began experimenting with flowers in addition to developing new varieties of fruits and vegetables. His advances in the breeding of gladioli inspired Isabella Preston, who later became renowned for her hybrid plants.[81] Preston was not the first woman to breed flowers in Ontario, but her work left a substantial legacy for women in horticultural research. She was a star pupil of Crow and the first person to successfully crossbred two lily species, something previously thought impossible.[82] Preston later worked at the Central Experimental Farm in Ottawa and gained international acclaim for creating new ornamentals and publishing popular books and articles on floriculture.[83] She became one of the top hybridists in Canada, and her achievements were considered to be "all the more remarkable as she was a woman in a traditionally male occupation."[84] Preston and women like her encouraged other women to pursue horticultural pursuits and demand recognition for their work.[85]

Throughout the First World War, flower exhibits continued. In 1916, the *Farmer's Advocate and Home Magazine* noted that the war might cause some Canadian women "to turn their attention to more absolutely essential things in horticulture and agriculture."[86] However, most writers suggested that despite the demands on people's time during the war, they still should make some effort to plant flowers because there was "nothing more refreshing and encouraging when tired and blue than some blossoms to work among, or even a few minutes at the window gazing at them will cheer one up and give more courage to face the next duty."[87]

Though male horticulturalists came to dominate flower classes in places such as Peel, women continued to be the principal flower exhibitors at fairs in other regions. The 1919 Aylmer Fair offered twenty flower and plant classes for exhibitors. Of the thirty-two names recorded as winners in these classes, twenty-five were women and only seven men.[88] By 1925, the Erin Fair had increased the number of classes in the flower competition to twenty-three, and of the forty-two names listed as winners, twenty-eight were women. The judge of the competition, Mrs W. Fairbairn of Orangeville, was also female.[89] Culturally, flower gardening was feminized. Flowers evoked specific emotions: for example, moss signified maternal love, pansies represented thoughts and remembrance, and red roses expressed romantic love.[90] Women's association with the spiritual and emotional kept them firmly tied to the cultivation and exhibition of flowers.

The association between gardening and desirable personal traits continued in the 1920s. In 1926, Ada L. Potts published an article in the *Canadian Home and Gardens* magazine arguing that a person's garden revealed a good deal about their character. She also pointed out that gardening was "a reflector of the aesthetic side of mankind" and "as much an art as painting."[91] In another article, she referenced Alfred Austin's idea that one's garden was "interwoven with one's tastes, preferences, and character, and constitutes a sort of unwritten, but withal manifest, autobiograph[y]."[92] The *Canadian Home and Gardens* magazine catered to the leisure activities of the upper-middle and upper classes, so those reading Potts's articles were likely in a position to agree that one's aesthetic represented one's character. But it was not merely the wealthy who believed these ideas. The beautification and landscape projects promoted by farm journals agreed that a "flower garden is the ambition of many a good farmer's wife."[93] A good flower garden was inherently reflective of one being a good wife, mother, and woman.

Flower Exhibits during the Great Depression and the Second World War

Flower shows may seem superfluous during times of hardship, but they continued to expand during the Great Depression and the Second World War. 1934 was a notoriously difficult year for Ontarians; business failure rates and personal bankruptcy increased, while farm incomes had declined by over 40 per cent from five years prior.[94] Yet this was the same year that the *Erin Advocate* reported that the Erin Fair had the "finest exhibit and largest crowd ever."[95] American cultural historian Lawrence Levine notes that during the Great Depression, tens of millions of Americans attended movies weekly, despite the economic turmoil.[96] Levine argues that "even in the midst of disaster life goes on and human beings find ways not merely of adapting to the forces that buffet them but often rising above their circumstances and participating actively in the shaping of their lives."[97] Levine's point is not to minimize hardships and trauma, but to "assert that human beings are not wholly molded by their immediate experiences; they are the bearers of a culture which is not static and unbending but continually in a state of process, perennially the product of interaction between the past and the present."[98] The strength of flower shows at Ontario fairs during the Depression suggests that rural families valued their beauty, a relatively inexpensive luxury, and they worked to incorporate it into their lives, even during challenging times. Agricultural societies continued to promote flower exhibits and advised that flowers be "given a commanding position, preferably near the entrance where a pleasing impression may be created in the minds of all entering the hall."[99]

This period also experienced the expansion of women's roles and responsibilities in agricultural societies.[100] Women were beginning to gain authority over the management of flower shows, which helped them promote events that best represented their interests and talents. Flower arranging was a popular pastime among women and received increasing attention from horticultural writers. Florists and gardeners advocated for the seriousness of floral arranging, and they directed their advice to women because they believed they were the natural practitioners of this art. Women were also given credit for being more scientific and design-oriented – features typically associated with men – in their choice and presentation of flowers.

When women became "lady directors" and associate directors on most fair boards in the 1930s, they joined the committees for flower shows. As more women took over the Arran-Tara Fair flower show's management, for example, the competition grew. In the 1930s, an all-women committee managed the competition, to its largest size ever.[101] By 1950, the flower show included fifty-six regular classes and six "special prize" classes. Three women directed the flower department committee, and seventeen women served as lady directors on the fair board (see Table 3.2).[102]

Other fairs experienced similar expansion in their flower shows under women's direction. At the 1936 Seaforth Fall Fair in Huron County, the floral exhibit offered forty-five classes for various flower breeds and breed varieties as well as classes for collections of plants, bouquets, and arrangements.[103] The 1936 East Huron Fall Fair listed thirty-seven classes for different types of cut, potted, and arranged flowers and plants.[104] The growing strength of the flower show was evident in other counties across the province as well. In Lambton County in the 1930s, the Forest Agricultural Society and the Brooke and Alvinston Agricultural Society each offered more than thirty flower classes. In Oxford County, the Ingersoll Fair had thirty-two classes, and the Drumbo Fair had forty-four classes for breeds and varieties of cut flowers, potted plants, and decorative arrangements.[105]

Single variety plant and flower classes were common at these shows, but flower arrangements were increasingly popular, and here women excelled. Books on flower arranging started to become more common in the 1930s. A voluminous book on the topic published in 1935 and reprinted in 1938 was F.F. Rockwell and Ester C. Grayson's *Flower Arrangement*.[106] Rockwell and Grayson were prolific writers on gardens and flowers in North America, and they also addressed the art of flower arranging at home at exhibitions.[107] Their books sought to teach individuals how to master the principles of floral arranging because they believed that "Every man or woman who learns

Table 3.2
Arran-Tara Agricultural Society Fair, Tara, Bruce County:
Classes for Flowers and Plants, 1910, 1933, and 1950

1910	1933	1950
Collection Annuals, cut and named	House Plants, 6 varieties	Geraniums, in bloom, 3 different colors
Collection Asters, 6 varieties, 2 of each	Annuals, cut and named, 8 variety	Foliage Plants, 3
Collection Pansies, 12 varieties cut	Perennials, cut and named, 8	Collection of 3 Miniature Cacti, Separate containers
Collection Verbenas, 6 varieties cut	Dahlias, not less than 4, large decorative and cactus	Begonia, 2, Fibrous
Collection Double Petunias, not less than 6	Begonias, 3 tuberous	Fern, single sword
Collection Coxcombs, not less than 6	Foliage plants, 3	Baby's Tears
Collection Dahlias, not less than 4	Geraniums in bloom, 3	Begonia Tuberous
Collection Begonias	Dahlias, pompom or ball	Geranium, Ivy
Collection Geraniums	Asters	Gloxinia, in bloom
Collection Fuchsias, 3 in bloom	Pansies	Unusual or Rare House Plant
Collection Balsams, not less than 3 in pot	Petunias	Primula
Collect. House Plants, (professionals not to compete)	Zinnias	Collection of Dahlias in basket
Four German Stocks in Bloom	Gladiolis, 8 varieties	Bouquet in vase for living room table
Four Foliage Plants	Cosmos	Basket cut flowers, quality & arrangement
Large or Table Bouquet	Everlasting Bouquet	Floral Centre for dining table, 8" or under
Small or Hand Bouquet	Snapdragons	Floral Arrangement, suitable for Coffee Table
Hanging Basket	Helenium	Petunias, single, 3 or more varieties, long stems
House Rose in Pot	Window Box	Petunias, double, 3 or more varieties
Single Fern in Pot	Hanging Basket for veranda	Pansies, 8 blooms
Collection Sweet Peas	Helianthus or Sun Flower, 3 blooms	Zinnias, coll. giant
	Michaelmas Daisy	Zinnias, coll. Lilliput

Table 3.2 *continued*

1910	1933	1950
	Bunch Cut Salvia	Gladioli, 3 varieties, 2 of each
	Decorative Begonia, 1	Gladioli, 1 variety, 3 spikes (red)
	Amaryllis in bloom	Roses, collection
	Salpiglossis	Gladioli, 1 var., 3 spikes, white
	Marigold, African, 6 blooms	Dahlias, decorative, 1 var., 3 blooms
	Marigold, French, 6 blooms	Dahlias, cactus, 1 var., 3 blooms
	Bouquet in vase for living room table	Sweet Peas
	Basket cut flowers, quality and arrangements	Asters, double, mixed, 8 blooms
	Single Maiden Hair Fern	Chrysanthemum, 3 sprays
	Single Sword Fern	Salpiglossis, 8 sprays
	Gloxinia in bloom, 1 pot	Marigolds, African, 8 blooms
	Nasturtium, 6 blooms	Marigolds, French, 8 blooms
		Salvia, 6 stalks
	Phlox, 3 blooms, different varieties	Phlox, annual, 8 blooms
		Calendulas, collection
		Nasturtiums, 6 double
		Nasturtiums, 6 single
		Snapdragons, 6 spikes
		Boutonniere, 2
		Cosmos, 8 blooms
		Larkspur, double annual
		Dining Table Centre, suitable for Thanksgiving, not necessarily flowers
		Dining Table Decoration, suitable for a wedding anniversary

Table 3.2 *continued*

1910	1933	1950
		Special Section: Exhibitors to reside with five miles of Tara Bouquet, any other flower not on list Stocks, 5 blooms Dahlia, 1, decorative Rose, single bloom Scabiosa, 8 blooms Winter Table Decoration, mounted on base Cactus, any other variety Decorative Begonia Fuchsia, in bloom African Violet **SPECIALS: Limited to residents within 10 miles of Tara** Corsage Coll. African Violets, 3 var. Olive McDonald Flowers. Best dish garden… Tara Women's Institute - Basket Cut Flowers. Quality and Arrangement Floor Basket of Flowers, suitable for Church decoration… C.A. Abbott - Best basket of mixed gladioli

to arrange flowers beautifully makes his or her environment more attractive and becomes one or more enlisted in the army of workers striving to bring to the mass of mankind an appreciation of finer things."[108] They believed that when people appreciated fine things, society would take "a long step towards a higher civilization." Rockwell and Grayson contended that anyone who thought that flower arranging was trivial or inconsequential was "merely unintelligent." Flower arrangement, they argued, was "a true form of artistic expression" similar to "engraving, painting, or sculpture." They claimed that the "question of material is secondary; what counts is the assertion of an idea in terms of beauty."[109]

In the same way that women learned how to properly exhibit foodstuff at fairs by studying winning entries, those who advocated floral arranging emphasized visual learning. Rockwell and Grayson reminded exhibitors that by displaying their work, they were helping to foster improvement; "one need only visit any one of the innumerable local flower shows, view the first or second year entries in flower arrangement classes, and then go back a year or two later and note the improvement."[110] They argued that a woman should take pride in her reputation for beautifully arranging flowers, just as she should when she graces her living room with "rich and tempting tea cakes, her dainty china or chic gowns."[111] Rural and urban women alike were expected to know how to use flowers to decorate the home, yard, church, or fair hall. Fair flower classes represented the value placed on both the item itself and its cultivator.

Twentieth-century instructional books about flower arranging went on at length about principles of design, balance, unity, and colour, and they described various styles of arrangements from line arrangements, which included Japanese and Modern styles, to mass arrangements that focused on period identification such as French Empire, Victorian, Georgian, and Colonial. Rural and small-town women had a history of being intrigued by other cultures and historical periods, and this interest extended to their work with flowers.[112] In the 1930s, Rockwell and Grayson advocated for a mixture of principles they called "Conservative Modern."[113] Modern line arrangements required innovation, strength, severity, and contrast, while mass arrangements employed traditional flower breeds, the blending of period-specific colours, and fullness of composition. Conservative Modern designs were advocated as the most popular form in the 1930s because they were "said to blend line and mass principles in a happy combination which preserves the best of both."[114] Rockwell and Grayson expected gardeners to be innovative yet understand the principles of tradition; they had to apply robust design and express their femininity through their arrangements. This required ingenuity, especially if one was required to incorporate the more untraditional garden and field items, such as vegetables, fruits, or wild plants.[115]

By the 1930s, women were advocating for more female leadership in horticultural organizations. Loetitia Wilson, the only woman on the Ottawa Horticultural Society's board of directors for many years, contended that horticultural societies needed women leaders because they understood "the needs of the small gardener, have good taste in the choice of plants, [are] economical purchasers, and moreover, have time to devote to Society affairs."[116] Wilson also wanted more women to get involved because she knew that they formed the majority of home gardeners.[117] Wilson advocated for women's "scientific garden work" because she believed women were more adept at it, arguing that "hybridizing needs the neat fingers and patience of women, it calls for taste and perception of colours." She cited Isabella Preston as a woman whose work with lilies was an inspiration.[118] As with others who fought for more opportunities for women at this time, Wilson used the "separate but equal" argument that stereotyped female attributes by suggesting women were useful because they had unique characteristics. Wilson believed women's talent, patience, and taste were reasons they should occupy more significant roles in horticultural groups.

During the Second World War, women's involvement in fair management and flower exhibits continued to expand. At the 1940 Collingwood Township Fair, an all-female committee of five ran the "Plants and Flowers" department.[119] Most flower committees during this period were exclusively or mostly women.[120] The growth and recognition of women's contribution to agricultural societies that began in the 1930s continued in the postwar period. Women increasingly positioned themselves in leadership roles.

Flower Power: Flower Shows in the Postwar Period

Gardening's popularity as a recreational activity continued to grow in the postwar years, and so too did the number of flower classes at fairs. During this time, rural populations were declining in Ontario, and fairs sought additional ways to entice urban visitors to the fairgrounds. Agricultural societies embraced the opportunity to expand popular flower shows, and female organizers were also eager to expand their authority, so they worked hard to facilitate more extensive exhibits. Maya Holson argues in her study of private gardening in southern Ontario in the 1950s that it proliferated as a result of the right "economic conditions, advances in technology, increases in home ownership and the growth of new suburbs typified by single-family dwellings."[121] She also contends that the expansion of gardening during this time resulted from a return to traditional gender ideals, and a desire to use new tools such as chemical fertilizers as well as new cultivars.[122]

Rural women enjoyed gardening as a creative hobby. The association between flower gardening and improving oneself and one's community remained, but women also believed that flowers offered both a chance for personal expression and the physiological benefits of leisurely work outdoors. Ethel Chapman was a guest speaker at the 1953 OAAS Convention, and in her talk, "Fairs and Their Trends in Family Living," she noted that flower exhibits encouraged "another hobby that seems to mean a lot to the happiness of country women in their homes."[123] She explained that the Women's Institute had held an essay contest on the topic "A Country Woman's Day," and she "noticed that practically all the women said they hated washing the cream separator and they *loved* working in their flower gardens. It was recreation rather than work."[124]

Miss Margaret Dove of Toronto presented a paper at the 1954 OAAS Convention on flower arranging and argued that one should judge arrangements on four basic principles: design, scale, balance, and harmony.[125] Design signified the creator's ability to create a pleasing shape or outline for an arrangement; scale denoted the size and relationship of the plant material to the container; balance referred to the arranger's ability to create symmetry in their arrangement, and harmony meant that all parts of the arrangement blended well together. Colour was significant in creating harmony, and Dove suggested women refer to colour charts to create balance.[126] She recommended that exhibitors use containers that enhanced their flowers and suggested "neutral or harmonious colours and simple forms" as proper vessels.[127] In the same way that artists understood the fundamentals of painting or needleworkers understood patterns and styles, women who arranged flowers for competition were expected to follow basic principles before adding their own creative elements.

The increased variety of flowers offered in seed catalogues as well as journals reflected a growing interest in the kinds of flowers available and well-suited for Ontario gardens. In 1950, an article in *Your Garden and Home* reported that annual seed catalogues brought "a galaxy of new flower introductions that hold a thrill for every gardener."[128] It noted that "better and more interesting flowers" had been developed with x-ray treatment, colchicine (a poisonous alkaloid used to produce polyploidy in plants), and increased chromosome counts, and were a testament to modern science.[129] Flower gardening was big business in North America. Although amateur breeding still existed, plant breeders hired by large seed companies typically determined what new flowers went to market. Most rural women were less concerned about developing a new breed or variety of flower than they were about selecting plants that would grow well and beautify their properties. They showcased their gardening and taste by exhibiting plants and arrangements full of well-grown and attractively displayed flowers. New varieties of flowers, such as the larger Chrysanthemum cultivars shown in Figure 3.3, helped women achieve impressive exhibits.

In the 1950s and 1960s, flower classes continued to expand to include new single varieties and colour variants. For instance, at the 1959 Erin Fair, fifty flower classes were offered for a wide-ranging selection of flowers of different kinds, colours, and numbers of blooms.[130] The 1965 Arran-Tara Fair had fifty-eight regular flower classes and an additional ten special classes sponsored by local individuals and businesses and large corporations such as Simpson Sears (see Table 3.3).[131] In the 1960s, seed catalogues allowed women to purchase an increasing array of plant varieties.[132] Women could seek advice on what to grow and then see the actual flowers at fairs displayed as single blooms or in arrangements. Flower arranging remained popular, and judging standards continued to be based on design and plant selection. In an article she wrote for the *Farmer's Advocate* in 1964, Laura Chisholm advised potential exhibitors to follow the judging guidelines in the handbook published by the Ontario Horticultural Association in co-operation with the Agricultural and Horticultural Societies Branch of the Ontario Department of Agriculture. She also outlined some of the most popular themed

Figure 3.3 Blossoms Attract at Aylmer Fair
Flower exhibits were popular with all fairgoers, but the newspaper singled out women for having a particular interest in these displays. In this photograph, Joan Hobbs and Alice Matthews are pictured in front of the impressive flower display at the Aylmer Fair. "Blossoms Attract," 1968. *St. Thomas Times-Journal* fonds, C8 Sh2 B2 F3 29a, Elgin County Archives. Image courtesy of Elgin County Archives.

arrangements promoted, such as "Glamorous Hostess," "Barn Dance," "Tea Time," and "Housewife's Dream."[133] By this time, most flower shows had adopted floral arrangement classes that showcased greater creativity and more entertaining designs. Variety-specific classes remained, but flower show committees privileged themed arrangements and collections.

The 1970s was a rapidly urbanizing time, and therefore it is not surprising that fair organizers wished to reflect on the "good old days" as much as they wanted to encourage new and creative designs. At the 1974 Erin Fair, in addition to the use of flowers, arrangement classes incorporated weathered wood, dried material, moss, tree leaves, weeds, fruit, and fabric, along with specialized containers such as brass or copper tins, antique vases, old pitchers, or vegetables, such as pumpkins, to illustrate specific themes. Many of these themes recalled rural traditions and idylls, including "A Country Thanksgiving," "Autumn Gold," "Log Cabin Days," "Nature's Best," "Grannies Favourite," [sic] and "Autumn Bounty" (see Table 3.4).[134] Similarly, at the 1976 Erin Fair, more themes that harkened to the past (or the imagined past) were embraced, such as "Country Fall Fair," "Summer Memories," "Quilt Block," and "Country Kitchen."[135]

Other themes embraced more modern aesthetics or ideas. In 1974, classes for flower arrangements included themes for "Extremely Modern," "Patio Party," and "A Trip Around the World" (see Table 3.4).[136] In 1976, the growing multiculturalism of Ontario was represented by the theme "Italiano" at the Erin Fair, while current events and popular holidays were also recognized by the themes "Tribute to 76 Olympics" and "Hallowe'en" (see Figure 3.4).[137]

Women were eager for the creative opportunities these classes encouraged. Men were not excluded, of course, and organizers even tried to promote their participation in flower arrangement classes by offering classes for themes such as "Gone Fishing," which was open to male exhibitors only.[138] However, women dominated the organizing committees of flower shows. Traditional, single-stem flower entries that encouraged an individual plant's perfection still existed, but classes that showcased originality and fun received more attention. The women organizing flower competitions in the 1970s wanted to experiment with and express the new trends in floral arrangements. They also wanted to put on a successful fair and, more importantly, a successful flower competition. If that meant bringing more entertainment into the flower show, they were happy to do it.

Table 3.3
Arran-Tara Agricultural Society Fair, Tara, Bruce County:
Classes for Flowers and Plants, 1965

1965

African Violet, double, single crown
African Violet, single, single crown
African Violet, three varieties, single crown
Baby Tears
Begonia, 1 fibrous in bloom
Begonia, decorative
Begonia, tuberous
Cactus or Succulent
Coleus Plants (three)
Geranium, any variety house grown
Geranium, 3 different colours, house grown
Patient Plant (young plant preferred)
Asters, double, mixed 7 stems
Asters, single, 7 stems
Asters, Powder Puff, 7 stems
Cornflower, bouquet
Cosmos, 9 sprays
Chrysanthemum, 3 sprays
Dahlia, decorative, 1 bloom
Dahlia, 1 decorative, 1 variety, 3 blooms
Dahlia Cactus, 1 variety, 3 blooms
Dahlia, ball, 3 1/2 inches or over, 7 blooms
Dahlias, miniature, approximately 5" in diameter, 7 blooms
Dahlias, pom-pom, less than 2 inches in diameter, 7 blooms
Dahlias, any other variety
Calendulas, collection of 7
Gladioli, any variety, 1 spike
Gladioli, 1 variety, 3 spikes, pink
Gladioli, 1 var. 3 spikes, white
Gladioli, 1 var., 3 spikes, red
Gladioli, 6 spikes, 6 varieties
Gladioli, basket, mixed
Nasturtiums, 7 blooms
Pansies, collection of 7 blooms with foliage
Marigolds, African, 7 stems
Marigolds, French, 7 stems

Table 3.3 *continued*

Petunias, single, ruffled, 3 or more colours, 5 stems
Petunias, double, 3 or more colours, 5 stems
Annual Phlox, 7 stems
Stocks, 5 stems
Rose, single bloom, any colour
Rose, Peace, single bloom
Roses, collection
Roses, arrangement of all white, all pink, all yellow, or all cream, in low bowl
Salpiglossis, 7 sprays
Salvia, 7 stalks
Scarbiosa, 7 stems
Snapdragon, 7 spikes
Zinnias, 7 stems, giant
Zinnias, Lilliput, 7 stems
Dianthus, vase
Arrangement for a living room table
Floral Arrangement, 3 miniature containers
Centre for dining table, 10" or under
Arrangement, modern, roses predominating in low container
Bouquet, from our fields, natural
Bouquet for a sick child
Any flower, not listed
Specials
Miss Dent of Northern Nurseries - Corsage
Mrs. Sidney Rose - Bride's Bouquet
Mrs. George Morrison - Table Flower Arrangement to illustrate old song, with named displayed
Mrs. Fred Patterson - Tea Cup Arrangement
T.N. Duff - Collection of Floribunda Roses
Mrs. Earl Carson - Winter Coffee Table Arrangement
Miss Emily Grant - Collection of Cactus Dahlias (not less than 6 blooms)
Geo. Keith & Son, Toronto - Best basket of Annuals (no gladioli or dahlias)
Horticultural Society - Best basket of Gladioli and Dahlias, arrangement considered
Simpson Sears Special - Best Basket of Gladioli (not less than 12 spikes)

Table 3.4
Erin Township Agricultural Society Fair, Erin, Wellington County:
Classes for Flowers and Plants, 1974

1974

Vase of cut flowers, annuals
Vase of cut flowers, biennials or perennials
Bowl of cut flowers
Annuals, 6 varieties, exhibited in 6 qt. basket
Best specimen of a cut flower
1974 introduction - named
Coleus, 3 plants potted separately in 4" pots
Climbing vine - not over 4 feet
Terrarium - must be established growing plants
House plant, in bloom
House plant, decorative
Dish garden, 3 cacti and 3 succulents
Table plant, in bloom, under 10 inches
Begonia, with bloom
Asparagus fern, sprengeri
Most unusual plant, named
"A County Thanksgiving" horizontal arrangement for dining table
"Autumn Gold", tones of yellow, in brass or copper container
"Log Cabin Days", wild flowers in an old pitcher
"Nature's Best", weathered wood, dried material
Wall plaque, driftwood natural materials may be formed, tinting permitted
"Ice Storm" - white branches and foliage
"Grannies Favourite", cut geraniums, own foliage
Christmas Wreath - moss, evergreen, dried materials
"For the Birds", featuring sunflowers
"Oriental Mood"
"Autumn Bounty", Ontario fruit and flowers in a pedestal
"Extremely Modern", for a living-room corner
"Evening Glo", floating candles, for a coffee table
"Patio Party," using ground or pumpkin etc., as container
Handmade planter with growing plants
"A trip around the world", featuring a rock
"Gay Pari", a tower of bright flowers
"Ontario", one-sided arrangement featuring maple leaves
"Mexico", featuring cacti
"Scotland", one-sided arrangement using tartan and Scotch Thistle

Figure 3.4 Depicting Halloween at Erin Fall Fair
This Halloween-themed flower arrangement at the Erin Fall Fair illustrates that organizers were embracing more creative and entertaining displays in the postwar years. "Depicting Halloween, 1976." Erin Agricultural Society Private Photo Collection, Erin, Ontario. Image courtesy of the Erin Agricultural Society.

Conclusion

Nineteenth- and twentieth-century agricultural fair flower shows provide a window into rural society's values and ideas. The ideas of discipline and taste were central components of promoting flower exhibits early on. The legacy of a disciplined application of labour and design principles to produce visual representations of character continued into the twentieth century. Horticultural enthusiasts and rural reformers promoted orderliness and aesthetic judgement as effective tools for inspiring organized habits and improved minds.

Although women were the main exhibitors at flower shows over the years, some agricultural societies in the late nineteenth century privileged men in an effort to enhance the "seriousness" of their shows and also to showcase the best commercial floriculture. Undeterred, women used the opportunity to compete in more flower classes, not withdraw from them. In the 1930s and 1940s, most official organizers of flower shows were women, and they showcased gardening skill and creativity of design, which found full expression in the postwar period.

Many people saw the cultivation of flowers as a civilizing act that uplifted rural families and beautified the countryside. Fair organizers believed that the "best class" of people enjoyed flowers. However, the range of rural women who exhibited flowers shows that individuals of all socio-economic statuses were eligible for such a classification. Yet, although women's desire to organize, compete in, and visit flower exhibits continued to be connected to their willingness to portray traditionally "feminine" characteristics such as thrift, beauty, taste, and artistry, they also sought opportunities to provide leadership in flower competitions and determine which elements of the show were most important to them.

4

Providing Comfort, Refinement, and Respectability
Ladies' Domestic Manufactures, Fancywork, and Fine Art

The previous chapters illustrate that rural communities valued women's food and flower exhibits because they respected women's ability to contribute to the physical, psychological, and spiritual well-beings of their families. Domestic manufactures, fancywork, and fine arts were also central to women's capacity to provide for, comfort, and uplift their families. Domestic manufactures represented their ability to make useful household goods such as cloth and clothing, while fancywork included more decorative skills such as lace work and embroidery. Fine arts typically designated drawn and painted artwork, and later included photography. Domestic manufactures were among the first exhibits included to showcase women's work in the countryside, while fancywork and fine art classes were added to most township and county fairs by the 1870s and 1880s. Whereas both men and women exhibited fine art, domestic manufactures and fancywork in the form of knit, sewn, crocheted, and embroidered goods were female affairs. An important element of these handcrafted items was the legacy they provided for their makers. Unlike butter, bread, jams, flowers, and other perishables, these exhibits could outlast their creators. The objects women made for fair competition often remained as lasting tributes to their makers and were passed down in the family for generations. These artifacts contribute to our understanding of the cultural framework of women's lives.

Women who exhibited home-woven textiles, homespun wool, hand-sewn quilts and clothing, and knitted apparel represented themselves as important contributors to household economy and beacons of thrift and industriousness. Such items, and the traits associated with them, evoked the independent and self-supporting rural household, a symbol of the prosperity and opportunity that boosters claimed awaited immigrants to North America throughout the nineteenth century. By the later part of the century, the focus on domestic manufacture gave way to more emphasis on fancywork and fine arts. The increased availability and affordability of mass-produced textiles and clothing signaled a shift in rural households' production. Some items were no longer worth producing at home. At the same time, middle-class standards of refinement permeated the countryside and created a growing demand for craft and artwork that showcased women's talents.

Into the twentieth century, agricultural societies promoted the creation of things both beautiful and useful, but even more emphasis was given to exhibits that allowed women to express artistic initiative and individuality. What becomes clear is that the story of women's participation in these exhibits is one of continuity rather than change, not necessarily because the exhibits themselves did not transform with the times, but because the values associated with the work had lasting currency in rural communities. Women's "thriftiness" remained valued due to cyclical periods of scarcity and because the meaning of thrift transformed over time. In the twentieth century, a growing sense of thrift as needed for intelligent consumption was layered on top of the Victorian sense of thrift as a moral indicator of self-discipline and industry.[1] Thrift was not contradictory to consumer society; it represented smart consumer choices and the ability to save in one area of life to spend in another.

Domestic Manufactures, Fancywork, and Fine Arts in the Nineteenth Century

Women's skill in manufacturing cloth, clothing, and other household furnishings was a significant part of their assigned value in the nineteenth and twentieth centuries.[2] Early fair classes for women's work included homemade cloth, flannel, and woolen socks – all items that showcased women's "Domestic Industry."[3] These items represented women's contributions to the household economy,[4] and the list of classes later expanded to include homespun yarn, handmade socks and stockings, woollen mitts and gloves, counterpanes, quilts, mats, clothes, and hats: the things women made for their families or for sale or barter.[5]

Provisioning for the household by creating woven, sewn, and knitted items, however, was not just about saving money. Both women and men associated desirable characteristics such as being "thrifty" with these activities. An article published in the *Farmer's Advocate* in 1868 argued that a "good housewife" kept regular accounts of household income and expenditures and was cognizant of the family budget.[6] A women who was an efficient housekeeper was considered "a pearl among women" and "one of the prizes in the great lottery of life."[7] Women's ability to spin yarn, manufacture rugs, sew dresses, or knit dozens of warm mittens illustrated personal attributes of industry and broader ideals of independence and self-sufficiency. Agricultural societies sought to reward women who had these attributes and encourage others to adopt them.[8]

In the late 1850s and 1860s, agricultural societies expanded classes for domestic manufactures, and provided clearer classifications between exhibits. For example, at the 1857 Toronto Township Fall Fair, the "Manufactured Goods" division included fulled cloth (woven or knitted wool subjected to heat, moisture, and friction to increase its thickness and compactness) and flannel, while "Ladies' Work" consisted of embroidery, needlework, crotchet work, sewing and knitting.[9] By the 1870s, most fairs classified domestic manufactures as separate from items that were more decorative. Domestic manufactures were practical household goods that rural families needed and could create for themselves.

Agricultural societies promoted homemade items because they believed material expressions of thrift and self-sufficiency were part of a moral order that also emphasized "self-discipline, hard work, sobriety, honesty, diligence, and industry," all central to contemporary of middle-class respectability.[10] The rhetoric of improvement celebrated family solidarity, household industry, and images of domesticated women who were "industrious, self-sacrificing, and patriotic."[11] Agricultural improvers generally saw their role as to advance the interests of the nation through the expansion of agricultural productivity.[12] While they promoted a modern capitalist agriculture that meant interdependence with the rest of the society, the ethos of self-sufficiency and the independent farm family was also powerful. Agricultural improvers encouraged the advancement of domestic manufactures to promote thrift as a strategy of personal and communal well-being.[13] Local reporters praised the work of women exhibiting domestic manufactures at the 1866 Puslinch Independent Fair and emphasized that the exhibits were proof that the wives and daughters of Puslinch were "thrifty" and efficient homemakers.[14]

Agricultural societies promoted classes for homespun cloth, but it is worth noting that women in nineteenth-century Ontario were not limited to homespun in their daily lives. Rural general store accounts show that rural families did buy ready-made fabrics such as cotton or silk, or fabrics with more sophisticated patterns or of superior quality than they could or desired to produce at home. Other families spun yarn but had a weaver produce their blankets and cloth with it. Families also produced their own textiles and fashioned their own goods, but few relied solely on homespun cloth.[15]

Nevertheless, homespun constituted an important part of farm production for much of the nineteenth century, especially in low-income areas, and was a useful source of income for rural women whose labour outside of the home was limited.[16] Self-sufficiency in cloth production was not the goal of rural households; women produced cloth as a response to the opportunities made available by the market.[17] Agricultural societies promoted homespun cloth because it represented women's prudence and industry. Historian Beatrice Craig argues in her work *Backwoods Consumers and Homespun Capitalists*, that rural women, like men, were "utility not profit maximizers"; they understood the opportunities offered by the market and took advantage of them, but they also tried to balance the need for material goods with their need for time for other tasks, and for leisure.[18]

The homespun cloth production in Ontario waned in the late nineteenth century, and so did classes for these items at fairs. At the Erin Fair, the amount of homespun cloth required for competition decreased from ten yards in 1864 to five yards in 1889, and to an unspecified amount by 1901.[19] Agricultural societies initially supported local home- and factory-made cloth and praised the work shown at fairs as "little inferior in appearance to the product of the English looms."[20] After the Canadian production of handwoven cloth peaked around 1870 and was followed by a significant decline by the twentieth century,[21] it was not long before only homespun stocking yarn remained (see Table 4.1).[22] By the early twentieth century, the industry had been transformed, and agricultural societies no longer saw the future in homemade manufactures.

In her study of nineteenth-century New England agricultural fairs, Catherine Kelly has suggested that declines in textile production were unsettling for rural improvers because they conflicted with their perception of the self-sufficient farm family. She argues that fair organizers had to confront this change while they tried to maintain "a ritual aimed at creating tradition."[23] Fairs did create a sense of continuity with the past, but Kelly's argument overlooks their main impetus, which was to improve agricultural life and evolve. Agricultural societies updated their domestic manufactures classes to reflect the reality that fewer homespun fabrics were being made, but increased the amount of sewn, knitted, crocheted, etc., items (see Table 4.1). While homespun cloth waned, homemade clothing and decor did not.

The transformation of readymade fabrics into homemade goods still represented thrift and hard work. Women's homemade clothing, quilts, and household décor demonstrated that they embraced the advances of a market society that provided an increasingly wide variety of fabrics and sewing notions, but still encouraged traditional modes of economy. Unlike weaving, which had significant input costs and decreasing market demand, sewing continued to be important for the creation and maintenance of family clothing.[24] The purchase of cheap yarn allowed women to knit relatively inexpensive socks and sweaters, and readymade fabrics allowed them to sew low-cost clothing. Sewing and knitting could be done at home, even by those who were sick, weak, or supervising children.[25] The sewing machine also became an affordable household purchase that alleviated "toiling with the needle," which was said to be more destructive to women's health than other household labours.[26] The types of classes offered at fairs represented items that still held value for rural residents as marketable commodities, cost-saving articles, or treasured items.

Women's domestic manufactures and fancywork were part of a message about how women should use their industry to enhance the respectability of their families, inspire their neighbours, and glorify their country. Laurel Ulrich Thatcher notes in *The Age of Homespun: Objects and Stories in the*

Table 4.1
Erin Township Agricultural Society Fair, Erin, Wellington County:
Classes for Domestic Manufactures, 1864, 1889, and 1910

1864	1889	1910
	1)	1)
Quilt	Woollen counterpane	Woollen quilt
Pair blankets made by hand	Cotton counterpane	Woollen blankets
Ten yards fulled cloth	Woolen quilt	Sheets
Ten yards flannel	Cotton quilt, made by hand	Woollen socks
Specimen of knitting	Cotton quilt, machine made	Woollen stockings
Spun stocking yarn	Log cabin quilt	Homespun stocking yarn
Straw hat	Tufted quilt	Pair cotton stockings, fancy
Best made shirt	Pair blankets, made by hand	Rag carpet
Specimen of fancy needle work	Pair sheets, made by hand	Carpet, any other kind
	Blankets, factory made	Hooked mat
DISCRETIONARY	Flannel, 5 yards	Mat, any other kind
Cone basket plain	Winsey, 5 yards	Pair woollen mitts, gent's
Scarf	Carpe, 5 yards	Woollen mitts, gent's fancy
Braiding	Hooked mat	Pair gloves gents
Embroidery	Homespun stocking yarn	Ladies' mitts, fancy
Fancy wreath	Pair cotton socks	Gent's flannel shirt
	Pair woollen socks	Gent's flannelette night shirt
	Pair woollen mitts	Ladies' flannelette night dress
	Pair woollen stockings	Sofa afghan
	Woollen jacket	Ladies' woollen jacket
	Straw hat	Baby's dress
	2)	Silk quilt
	Woollen counterpane	Ladies' plain shirt waist
	Cotton counterpane	White bed spread
	Woolen quilt	Ladies' white night dress
	Cotton quilt, made by hand	Knitted counterpane
	Cotton quilt, machine made	Crochet counterpane
	Tuffed quilt	Cotton quilt, machine made
	Pair blankets, made by hand	Cotton quilt, hand made
	Pair sheets, made by hand	Log cabin quilt
	Blankets, factory made	Tufted quilt
	Fulled cloth, 5 yards	Fancy quilt, any other kind

Table 4.1 *continued*

1864	1889	1910
	Flannel, 5 yards	Woollen counterpane 2)
	Winsey, 5 yards	Woollen quilt
	Gentlemen's plaid	Woollen blankets
	Hooked mat	Sheets
	Homespun stocking yarn	Woollen socks
	Pair cotton socks	Woollen stockings
	Pair woollen socks	Homespun stocking yarn
	Pair woollen mitts	Rag carpet
	Pair cotton stockings	Carpet, any other kind
	Pair woollen stockings	Hooked mat
	EXTRAS	Mat, any other kind
	Knitted drawers	Pair woollen mitts, gent's plain
	Gloves	Woollen mitts, gent's fancy
	Ladies Mitts	Woollen mitts, driving
	Blankets	Ladies' mitts, fancy
	Blankets	Gent's flannel shirt
		Gent's flannelette night shirt
		Ladies' flannelette night dress
		Sofa afghan
		Ladies' woollen jacket
		patching on woollen garment
		Silk quilt
		Ladies' plain shirt waist
		Ladies' white night dress
		Darning an old sock or stocking
		Knitted counterpane
		Crochet counterpane
		Cotton quilt, hand made
		Log cabin quilt
		Gent's white shirt machine made
		Button holes

Creation of an American Myth that a desire for refinement grew among rural families in the nineteenth century, and, while they responded to rhetoric for agricultural improvement, they also understood that "women might be both manufacturers and ladies."[27] Domestic manufactures provided function and comfort, fancywork conveyed a family's taste and refinement, and both displayed beauty and utility. When these items were brought together and displayed at fairs, reporters emphasized how they did credit to women's character. The *Brampton Times* reported that the "display of ladies' work" at the 1868 Albion Fair "was very creditable to the good taste and industry of the blooming daughters and hearty matrons of Ontario's Albion, and the other productions of their facile and industrious hands displayed on the tables, gave good evidence of their ability to furnish forth good cheer to all friends and visitors at their hospitable homesteads."[28] In 1865, the Eramosa Agricultural Society proclaimed the domestic manufactures exhibits showcased "some exceedingly clever work in coverlets."[29] Before the 1870 Peel County Fair, the agricultural society expressed its full confidence that "thanks to the nimble fingers of the ladies," which allowed them to excel in their work, the exhibits under their domain would be a success.[30] Journalists praised women who used their feminine skill to artistically arrange and present home-crafted goods that showcased the "supremely comfortable and neatly decorated homes which are the pride and glory of Canada, and credit to their inmates."[31] The *Daily Globe* proudly reported that the women's domestic manufactures and fancywork, including the "comfortable looking lamb's wool socks and stockings and tastefully made shirts," at the 1881 Etobicoke Township Fair found many admirers and "gave evidence that the farmers' daughters of Etobicoke are clever needlewomen, and possessed of much taste.[32]

Commentators on women's work emphasized women's tastefulness and craftsmanship, but the utility of homecrafts to create comfortable spaces remained essential. At the 1899 West Elgin Fair, the ladies were praised for exhibiting articles that displayed elegant artistry and were also useful rather than simply ornamental.[33] Men in agricultural societies appreciated the function and form of women's exhibits because they too tried to achieve the same combination of traits in their show stock. They understood that in order to receive the best prices for their breeding animals they needed to combine "beauty and utility,"[34] so they supported the idea that women's work should be similarly balanced.

"Fancywork" was the term given to elaborately decorated items. This included things such as ornamental needle and lace work, crocheting, embroidering, beading and braiding work, as well as wire and wax flowers and arrangements made from moss or hair. Some items were strictly ornamental, while others were useful as well. Most types of fancywork did not appear in Ontario fair prize lists until the 1860s, and their presence signalled changes in the countryside. By Confederation in 1867, Ontario's economy was best described as agricultural and artisanal rather than commercial, but that was changing.[35] From the 1870s onward, the industrial sector of the province increased significantly, especially in industries linked to consumer goods.[36] Despite cyclical downturns in the mid-1870s and early 1890s, Ontario's economic growth significantly accelerated by the 1900s.[37] Rural families were enticed by a greater number of goods and services, and at the same time they were influenced by a Victorian middle-class culture of consumption.[38] The things one owned had increasingly become important markers of identity, and the popularity of fancywork was a reflection of this.

Fancywork was often used to adorn household furniture and decorate rooms. As more home furnishings became affordable in the late nineteenth century, rural women could take advantage of investing in store-bought sofas and other furniture that increased the comfort of their homes, while still exhibiting thrift by decorating those pieces with homemade fancywork. David Handlin explains in his study of the American home that nineteenth-century women justified the purchase of home furnishings because a "handmade bedspread and pillowcase rendered a store-bought bed suitable to the values of economy. A display of flowers from the garden justified a table or even a piano."[39] Fancywork signalled both an increased capacity to pursue more leisurely activities, especially among middling farm families and town and village residents, and tangible evidence of their middle-class status while still allowing them to maintain their identities as efficient housewives. By the Victorian period, the attainment, or at least the understanding, of beauty was a sign of respectability and progress that was important to anyone – rural or urban – who desired upward mobility.[40] Women found pleasure and economy in making knitted, crocheted, and other items, but they also found social recognition.[41] Being both industrious and a lady was not a contradiction in the minds of rural people.

Not everyone was happy about rural women's consumer choices or leisurely pursuits. During the mid-nineteenth century, concerns had already been expressed in Upper Canada that farmer's daughters were becoming "unproductive hands" in the household. In 1846, the *British American Cultivator* reprinted a lengthy cautionary tale that advised farmers to prepare their daughters for labour on the farm, not a leisured life that would limit both his and her prosperity. The author described a conversation he had with "one of the most respectable farmers in the country" who explained how he had done his utmost to educate his daughter – including in the creation of "fine needlework, embroidery, and drawings" – but was distressed because:

> In the loss of her mother she is my whole dependence, but instead of waiting upon me, I am obliged to hire a servant to wait upon her ... I told her a few days since that my stockings were worn out, and that I had a good deal of wool in the chamber, which I wished she would card and spin. Her reply was, in a tone of unaffected surprise, "Why father, no young lady does that; and besides it is so much easier to send it to the mill and have it carded there." – Well, I continued, you will knit the stockings if I get the wool spun? "Why, ne, father! Mother never taught me how to knit, because she said it would interfere with my lessons; and then, if I knew how, it would take a great deal of time, and be much cheaper to buy the stockings at the store."[42]

The author warned farmers not to create a distaste for labour among their daughters. He argued that even if domestic manufactures took time and could be purchased cheaply, "they who might produce it must be sustained at an equal expense, whether they work or are idle."[43] Rural newspapers often featured articles that suggested fathers should educate their girls to be "wives, mothers, heads of families, and useful members of society," not "vain, thoughtless, dressy slattern[s]."[44] Women were to embrace their duties in the countryside and not be "ashamed of honest industry," or "temped by the appearance of a better dress, a broach or a bonnet."[45]

What is missed by such rhetoric is the importance of fancywork as an artistic outlet. Decorative items were popular in rural households.[46] The skilled creation of fancywork gave evidence of a woman's "special refinement" and her respectability. Rural women who valued their identities "as inhabitants of a particularly female sphere"[47] created fancywork as a way to distinguish themselves from men and other women who had neither the ability, nor perhaps the leisure time and desire, to develop such skills. Furthermore, unlike cleaning or mending, chores that were relegated to servants or children whenever possible, the fashioning of clothing and fine sewing were high-status tasks and not generally viewed as drudgery. Typically needlework, especially fancy needlework, required a level of skill and experience that meant female heads of household performed the task.[48] This work was high in the hierarchy of the household, therefore young women who exhibited their skill at fairs could be especially boastful.[49]

Refined rural women appreciated the elegancies of life but recognized that they did not guarantee happiness; it was the "cheerful heart" that could "arrange the most discordant material into harmony and beauty."[50] In the same way that women organized their environment through ornamental gardening or displays of flowers and food, fancywork gave women a measure of control. The same discipline that allowed women to make straight and even stitches, maintain a uniform tension when knitting, or implement the symmetrical design of a quilt brought order to their immediate environment. A woman's ability to create, not just to purchase, items of beauty gave her options when fashioning her domestic realm. Furthermore, just as she could alter her landscape with flower borders, she could alter her home with a crazy patch cushion.

In the 1870s and 1880s, interior design crusaders encouraged women to embellish their homes with their own handiwork so as to advance the moral growth and gentility of their families.[51] In the 1880s, rural journals encouraged this movement by publishing instructions for fancywork such as crocheted cotton "tidies" (antimacassars), and knitted afghans.[52] The home, however, had its limits as a medium for public expression. Fair exhibits, on the other hand, presented women with the opportunity to receive broader communal acknowledgement. The introduction of this type of work at fairs signalled agricultural societies' acceptance that items showcasing women's taste, refinement, respectability, and skill were worthy of display and something all rural women should aspire to create. In fact, when women failed to display enough of their work at fairs, as was the case at the Harriston Fair in 1901, the local newspaper chastised them for shirking their duty: "We trust that our lady exhibitors will see that this does not occur again."[53] The individual merit of each item may have held little interest for male reporters who sometimes glossed over "the galaxy of human femininity,"[54] but fancywork's contribution to the fair display was still important to them. As previously discussed, agricultural societies pragmatically promoted women's exhibits as a way to increase attendance at fairs, but they also recognized that fancywork exhibits and other attractive displays elevated these events.

By the late nineteenth century, most fairs had classes devoted to fancywork for a variety of sewn, knit, embroidered, crocheted, tatted, appliqued, or braided items. Even stuffed birds and animals, wax sculptures, and human hairwork can be found in prize lists. The most popular items were embroidered fabrics and lace work. Classes for embroidery generally did not specify the type of embroidery, whether low, raised, or laid work, but did indicate the type of material used, such as cotton or muslin, silk or satin, or cloth. At the 1889 Erin Fair, three embroidery classes were distinguished by the fabric used, and although the two lace work classes were specified, point lace and Honiton lace, most other classes were vague in description (see Table 4.2).[55] Specific categories of lace work allowed for creativity in design and appearance. "Honiton lace" was a very fine hand-made bobbin lace ranked first among the English laces because of Queen Victoria's Honiton lace wedding dress.[56] It was also considered simpler to make than French and Belgian lace designs, but the designs could vary considerably.[57] Agricultural societies likely kept classes broad to encourage maximum participation and allow more individuality, even if it made judging more difficult.

The popularity of fancy needlework among female exhibitors illustrates that personalized items were important to rural women, but it also shows a commitment by some women to making traditional handicrafts despite the availability of machine-made and imported items. Advocates of reviving the art of making lace by hand argued that design was central. They claimed that the "charm of variety and the beauty of novelty can only be found in the work of skilled hands, guided by fanciful minds, and not in the productions of iron wheels set a-going by steam."[58] It was hoped that women who

Table 4.2
Erin Township Agricultural Society Fair, Erin, Wellington County:
Classes for Fancywork, 1889

1889	
1) Shirt men's fine, unwashed machine made Silver wire wreath Wax fruit or flowers Hair work Embroidery on cotton or muslin Embroidery on silk or satin Embroidery on cloth Lace point Lace honiton Etching Ornamental beadwork Pillow shams Tidy, woollen Tidy, cotton Bead work Braiding Tatting Crochet work Worsted work Fancy wool work Sofa cushion Scarf, woollen Motto Collection of stuffed birds and animals Single stuffed bird or animal 2) Shirt men's fine, unwashed machine made Flower wreath Silver wire wreath Wax fruit or flowers Hair work Embroidery on cotton or muslin Embroidery on silk or satin Embroidery on cloth	Lace point Lace honiton Etching Ornamental beadwork Pillow shams Tidy, woollen Tidy, cotton Bead work Braiding Tatting Crochet work Worsted work Fancy wool work Sofa cushion Scarf, woollen Motto Collection of stuffed birds and animals Single stuffed bird or animal EXTRAS Wall bouquet Paper picture Painted plaque Ottoman Foot rest Macrame work Table mats Painting on muslin Woollen blankets and drapes Card bracket and drape Slippers and cap Arassne work Shaving sheet Thermometer holder Ribbon on plush Fancy table drape

1) Items made previous to the last fair
2) Items made since the last fair
Extras: Items that received discretionary prizes

engaged in needlework had learned from tradition, but they were also induced to "aim still higher, so that by exerting the fanciful and imaginative faculties so largely possessed by the refined of the fair sex, they may attain the same perfection in diversity and beauty of design."[59] Women were encouraged to take up needlework and other forms of art and craft as a means of preserving the past, but also, more importantly, to enlighten their spirit, refine their tastes, and innovate for future generations.

Fundamental to nineteenth-century femininity was the concept of taste. Some social reformers questioned if rural women, who had "a severe limit on their culture and accomplishments," were capable of the delicacy and elegance of manners demanded of the feminine sex.[60] Rural women, however, proved that taste was possible in the countryside, even if pretentious displays of wealth were rare. Rozsika Parker argues in her history of embroidery that taste became more about morality and spirituality than class standing by the eighteenth century, and that good taste was a regulating factor of women's lives, conveyed by their actions, words, and creations.[61] She contends that femininity was a social marker, and that "every footstool and screen and pair of slippers made a statement about the family's social aspirations."[62] Historian Richard L. Bushman also notes that refinement was less about an artificial imposition than "a personal quality like courage or kindness, ingrained in one's character and among the most admirable of the virtues."[63] He explains that social reformers promoted the idea that taste was not a property of class, but was possible for everyone.[64] The pursuit of beauty, not wealth, allowed people of different backgrounds and stations in life to exhibit taste. Therefore, rural women's labour did not preclude them from also displaying refinement. Rural women's fancywork was a way to acknowledge their taste as much as it was to display their aspiration to or arrival at middle-class respectability.

Another important consideration is how the things women made reflected their roles as custodians. One example of women's creations as "devices for building relationships and lineages over time"[65] was the hairwork shown at fairs across Ontario in the late Victorian period. Hairwork was a common class at fairs in the 1890s, but even in the 1860s some fairs promoted this work, and reporters commented on the beautiful specimens shown.[66] It was usually composed of the hair of loved ones, both living and deceased, woven into wreaths, bouquets, wall decorations, and other keepsakes. Wreaths and bouquets fashioned into lifelike twigs and flowers were especially popular. While we might view these items as strange, at the time they were treasured both as remembrances of loved ones and mediums for personal expression.[67] Some women collected hair from many family members and wove them into one wreath serving as "a lasting symbol of family unity."[68] Hairwork and other forms of decorative craft were most popular at fairs at the "height of the middle-class parlor as memory palace" during the 1890s, when objects were used to create a ritual setting packed with meaning and symbolic associations.[69] Parlour objects acted as shrines to a host of personal sentiments and connections. Ulrich argues that when individuals inscribed their names on these objects, they "assured some sort of immortality."[70] A hair wreath went beyond an inscription because it left behind a physical remnant of a person. Such an object was as inalienable and sentimental as one could make, so women used hairwork to express solidarity, memorialize family, and create unity in their relationships.

Fine art exhibits were added later in the nineteenth century at many fairs in the province. Agricultural journals such as the *Farmer's Advocate and Home Magazine* instructed readers on the art of painting, and images of female painters were used to represent the target audience of such instruction.[71] Like fancywork, fine arts allowed rural women to illustrate their creativity and style through respectable and refined activity. The importance of fine arts at World's Fairs and other large national and international exhibitions has received significant attention by historians and art historians,[72] but virtually no attempt has been made to consider this work at local agricultural fairs. Part of the reason for this might be that the fine art displayed at such fairs was often considered of lesser quality and not worth serious attention. Indeed, even the art shown at large provincial exhibitions was not highly regarded. The *Ontario Farmer* reported that, at the 1869 provincial show in London, Ontario, "In the fine arts, there were specimens enough, such as they were, but many were mere daubs and blotches."[73] The following year the reporter gave a more generous assessment of the work, explaining that "progress and improvement were distinctly perceptible." However, a general lowering of expectations was also expressed. The *Ontario Farmer* suggested that, because Canada was a new country, the quality of fine arts should not be judged too harshly: "Even in the fine arts, where from year to year in the past there have been those deficiencies which might be reasonably looked for in a new country, and in regard to which perhaps too severe a style of criticism has been indulged in, there was this year very fewer daubs and gaudily coloured pieces, fewer old acquaintances familiar to the eye by successive appearances on such occasions; and a larger proportion of really meritous new things, proving that we have a race of native artists who already do us no discredit."[74]

The reporter covering fine arts at the Guelph Central Fair weighed the successes and failures of both male and female artists, but he was generally more dismissive of female painters, simply noting that their work had been either "fairly" or "tolerably" painted.[75] The most attention he paid a female painter was to Nancy Strickland, a regular exhibitor who showed her work at fairs across the province, for her copy of a Myles Birkett Foster painting. Foster was an artist renowned for his picturesque depictions of English landscapes and rural life.[76] He noted that she had "rendered the colouring well, but in which, apparently from a too servile imitation of detail, there are evidences of a cramped execution."[77] These comments are illustrative of why women's art was dismissed more broadly. Unlike the "professional" male artists that displayed their work, women's art was considered "amateur," an idea that was reinforced by women who exhibited copies of famous paintings. Still, even women who exhibited original artwork were not believed to have the skill of their male counterparts. Most of the male artists who exhibited at fairs had received some form of formal training and their art was how they made a living, but women generally did not have the advantage of formal training, and because their work was created in leisure, it was considered of little or no value.[78] Furthermore, along with male hobbyists, female artists were assumed to apply "a studious adherence to established conventions, a presumed lack of seriousness and an absence of innovation."[79] Women were also believed to engage in "art-making" because they saw it as a social accomplishment that ensured or enhanced their social position.[80] It was therefore assumed that women who exhibited artwork did so to showcase good taste, respectable skills, and genteel sensibility, rather than out of any real desire to achieve artistic merit.

Like art and fancywork, domestic manufactures could display originality and artistry that was praiseworthy. At the 1853 Erin Fair, women were admired for the "ingenuity" and "workmanship" necessary to craft beautiful quilts.[81] Although quilts were a staple of home manufactures at the time, they still allowed women to show their inventiveness and resourcefulness by constructing intricate patterns and selecting appropriate colour schemes. The women who exhibited quilts at the 1865 Arthur Fair were praised for the nine handsome quilts shown, which "proved beyond question that the ladies of Arthur are possessed of considerable taste."[82] General patterns were common in many forms of handicraft, but women still had the opportunity to create unique pieces of work.[83]

The size and scale of women's exhibits grew during the nineteenth century, so that by the twentieth century, domestic manufactures, fancywork, and fine arts promoted an array of handcrafted items. Despite some rhetoric about which items were "useful" and which were "ornamental," generally the things women exhibited at fairs were said to demonstrate their thrift and industry while conveying beauty and taste, honouring the past, and fostering solidarity.

"Ladies' Work" and Arts and Crafts in the Twentieth Century

The number of classes at fairs for "Ladies' Work" and other arts and crafts continued to increase in the twentieth century and the expansion of domestic manufactures was especially evident. Although women exhibited fewer homespun items, the importance of sewn items grew.[84] Despite the increased affordability and availability of ready-made clothing, most rural women did not turn whole-heartedly to department stores, mail-order catalogues, or even corner stores to buy ready-to-wear garments. The early twentieth century saw the expansion of rural women's organizations, such as the WI, which encouraged domestic science instruction. Many women were relieved of the burden of making their husbands' clothing, but cheaper fabric and the sewing machine were still used to make their own and their children's clothing.[85] Even in the postwar period, rural women boosted family income or saved money by producing value-added products.[86] More than simply economizing, however, women in the twentieth century were also expected to be up to date on current fashions and consider new styles and fabrics.[87]

During the First World War, agricultural societies and local community groups altered the decor in exhibit halls by adding patriotic flags and displays, but little changed in terms of the types of classes offered. The same sewn and knit goods exhibited before the war continued to be shown, although often more items were added to prize lists. Class changes usually reflected changes in fashion rather than shifts caused by wartime measures. The only detectable effect the war had on "Ladies' Work" at the Egremont Fair was the retraction of a rule restricting exhibitors from showing anything other than items made in the previous two years. The agricultural society explained that "On account of so many ladies working for patriotic purposes, Ladies' Work of any year may be shown."[88] Canadian women were engaged in supporting the war effort by raising funds, producing goods and foodstuffs, and taking on more unconventional work roles.[89] Thousands of women mobilized and concentrated their efforts in groups such as the IODE, the Canadian Red Cross, Women's Institutes, local women's councils, and the YWCA to organize and coordinate war relief projects to raise funds for patriotic purposes. The WIS in Ontario raised an estimated four million dollars for the Red Cross, which included money to finance medical relief and transport, field kitchens, and knitted and sewn materials and apparel.[90] Mrs Walter Buchanan of Ravenna described women's knitting contributions in a poem published in the *Farmer's Advocate*.[91] Her poem concluded by encouraging soldiers to push on and think about the women supporting them back home:

On, Canadian soldier,
Bravely take your knocks;
Think of nimble fingers
Busy knitting socks.[92]

By the 1920s and 1930s, women's and children's clothing classes were the most popular. For example, at the 1925 Seaforth Fair in Huron County, the only class in the "Domestic Needle Craft" section specifically for men's apparel was for "Men's sleeping garment, machine made," while classes for women's clothing included aprons, house dresses, and slips, in addition to the larger category of classes for "Ladies' Work," which included six classes for "Ladies' Wear" and six classes for "Infants' Wear," but no men's wear (see Table 4.3).[93] The availability of cheaper ready-made men's apparel meant that many women no longer manufactured most of their husbands' clothing. Constructing garments was a laborious task, and even though affordable

Table 4.3
Seaforth Agricultural Society Fair, Seaforth, Huron County:
Classes for Ladies' Work, 1925

1925

Domestic Needle Craft
Quilt, fancy
Quilt, pieced or patchwork, cotton
Comforter, silk or cotton
Bed spread
Apron, unlaundered, most practical
House dress, most practical
Men's sleeping garment, machine made
Ladies' slip
Pair knitted sox, hand made
Pair mitts, hand made
Floor Mat, braided
Floor Mat, hooked

Ladies' Work
Lace, Irish crochet
Lace, filet, cotton
Lace, knitted, cotton
Tatting, display
Drawn Work, display, 3 pieces
Drawn Work, Italian
Embroidery, cross stitch
Embroidery, display, 3 pieces
Embroidery, modern conventional, colored
Fancy Work Bag
Curtains, hand made

Dining Room Furnishings (sub-section)
Luncheon set, 5 pieces
Tea Cloth
Set of table mats
Set of table doyleys
Tray cloth
Centre piece, embroidered in silk
Centre piece, embroidered in cotton
Centre piece, linen with colored border
Serviettes, four, hand trimmed
Table napkins, 4, initialed or monogramed
Buffet set, three piece
Table cloth and 2 napkins, embroidered, most practical

Bed Room Furnishings (sub-section)
Day slips
Pillow slips, pair, embroidered
Pillow slips, pair, other hand work
Pair of towels, other hand work
Pair of guest towels
Fancy sheet and pillow slips to match
Dresser set, three piece
Dresser set, washable

Ladies' Wear (sub-section)
Night robe
Camisole, hand made
Fancy handkerchiefs, 3 samples handwork
Fair Bedroom slippers
Fancy collars and cuffs
Ladies' scarf, fancy

Infants' Wear (sub-section)
Short dress, washable
Fancy dress, handmade
Bonnet
Wool Jacket
Bathrobe or kimona
Set of underwear

Living Room Furnishings (sub-section)
Table runner, colored
Centre piece, colored
Sofa pillow, hand made
Lamp shade, hand made

Miscellaneous (sub-section)
Single piece, fancy work, not listed
Collection of fancy needlework, 6 pieces

fabric and better sewing machines encouraged women to make some clothing for themselves and their children, they tended to buy men's clothes.[94]

Domestic manufactures remained central to rural women's organizations such as the Women's Institutes, which were led by women who believed instruction in domestic science was necessary if rural families were to succeed. Social reformers such as Adelaide Hoodless argued that women's proper management of the farm home and their ability to produce household goods would reduce their need to purchase ready-made items and benefit their pocketbooks.[95] Farm journals regularly advertised patterns for women's and girls' dresses, nightgowns, and aprons, and even toys.[96] Although some cried out against a "pleasure-mad, luxury-chasing people" on "the path of wanton extravagance," and insisted society needed a nation-wide campaign to return to "the highway of thrift and sensible living" after the rush of post-war indulgence,[97] the reality was that many families continued to value economy.

The ability to fashion, alter, and mend one's clothing was especially important during times of scarcity.[98] During the Great Depression, many fair classes remained the same, but new classes also materialized that promoted thrift, such as the Erin Fair's 1934 class for "Darning on worn sock."[99] At the 1934 West Elgin Fair in Wallacetown, classes included a "Child's Dress, made from old garment," and an "Apron, made from flour sacks."[100] Also in 1934, the North Bruce and Saugeen Agricultural Society offered a "Thrift Section" of domestic manufactures which included "Best Apron, made from flour or sugar sacks," "Best Child's Pantie Suit, made from sacks," "Table cover, made from sacks," "Best Quilt, made from waste material," and "Collection of Six Articles for Kitchen Use, made from sacks."[101] Making do or making over was an important way to economize. Men proudly reported on their wives' ability to reduce household expenditures,[102] and research has shown that farm women's pattern of self-sufficiency prevailed at most family income levels,[103] illustrating a shared commitment to these practices.

By the 1930s, "Ladies' Work" committees emphasized that they wanted new articles, not previously shown items. They amalgamated the two sections of classes that had allowed for new and old work into one section that required work to be made within the last year or two. For example, at the 1930 Collingwood Township Fair, a rule specified that "Works of Art and Ladies Works ... may be exhibited for two years, after which time, the exhibit will be ruled out. This rule will be strictly enforced."[104] Organizers also warned women that judges were "authorized by the Directors of this Society to discard all soiled, defaced or old work and instructed to award the prizes to new work. In case of no competition, unless the exhibit is worthy, the prize will be withheld." Fair organizers discouraged exhibitors from entering old or damaged entries because they believed they had a responsibility to promote the best and latest in domestic industry. The slogan, "New Styles; New Ideas, New Handwork" reflected the value placed on up-to-date exhibits that illustrated current trends and practices.[105]

Women were expected to be knowledgeable about current fashions and consider new styles, fabrics, and "in vogue" materials if they wanted to receive first prize. It was no longer enough to exhibit skillful items; women had to also illustrate that they were aware of the latest in products and design. For example, at the 1930 Collingwood Township Fair, a class for housedresses specified that entries should showcase a "smart new idea," while those for boudoir jackets and caps were expected to present a "dainty new style." Colour was also important. Women had to know what fashionable and appropriate colours were, especially in men's wear classes, where work was to be "strictly new and appropriate Colors."[106] When making items for their husbands and sons, women were to respect the gendered styles and colours of clothing considered appropriate.

In the 1940s, fair exhibits drew attention to women's wartime efforts. The Women's War Work Committee of the Canadian Red Cross organized to supply military forces and civilians with both the necessities and comforts of life.[107] Red Cross branches throughout Ontario coordinated their efforts and created a quota system to ensure that women were producing needed articles.[108] Traditional gender roles were upset in many cases during these years, but the renewed effort to supply men and civilians overseas with home-made goods emphasized women's maternal and household duties, helping to "patch-up," if not restore, traditional notions of womanhood.[109]

Fairs served to promote and display the wartime supplies and clothing that women were making. In 1940, new classes emerged that represented women's war efforts and patriotism. Convalescent jackets, knitted helmets, and sleeveless sweaters were all items needed for soldiers and created to display at fairs.[110] At the Collingwood Township Fair, a special class was offered for the "Best Collection of Red Cross Work by any Society,"[111] and a whole category of classes was created at Sombra Agricultural Society's Fall Exhibition in Lambton County for the "The Red Cross Division." Items included knitted goods and sewing for soldiers, airmen, or sailors, including socks, mitts, scarves, sweaters, helmets, jackets, hospital gowns, pajamas, and bed jackets, to be made according to Red Cross specifications and donated to the cause.[112]

By 1945, fairs across Ontario were committed to supplying items to the Red Cross and other government relief agencies. At the Burford Fair in Brant County, the local Red Cross Society supplied exhibitors with the material needed to make items for their exhibits. The prize money as well as the items went to the Red Cross as a special donation from the exhibitors. This included socks (navy, service; grey, wheeling; and sea boot socks), scarves (Navy and Amy), sweaters (khaki, wheeling wool; heavy service wool), men's pyjamas, navy gloves, mitts, a quilt, and refugee clothing (three articles: girl's pantie dress, boy's suit, and boy's pyjamas).[113] By 1945, the Collingwood Township Fair created nine classes for the "Patriotic Division of Domestic Needlecraft." Again, items such as socks, scarves, mitts, turtleneck sweaters, helmets or caps, and gloves were promoted, as well as classes for a "box suitable for sending to member of the Armed Services Overseas" and a collection of articles made by any "Patriotic Society," to include "Sewing for civilians, British, Russian, etc.; Quilts, Children's dresses and bloomers; Dresses for girls; Knitted Clothing for Armed Services and Civilians; Layettes." Additionally, the same special prize of a pair of blankets offered by Collingwood resident Max Faith since 1940 was offered to the winning entry for the "Best collection of Red Cross Work by any Society."[114] As had the First World War, the Second World War required thousands of women to make knitted and sewn goods for soldiers and refugees. Fairs highlighted the work women did and promoted their campaigns.

Thrift remained a virtue for rural women even during the period of prosperity that followed the Second World War.[115] At the 1950 Arran-Tara Fair, sixty-seven classes in "Ladies' Work" displayed a range of sewn and knit items in categories such as "Children's Wear," "Ladies' Wear," "Men's Wear," "Dining Room Furnishings," "Bedroom Furnishings," "Living Room Furnishings," "Kitchen Accessories," and "Miscellaneous."[116] Nine additional donor-sponsored special classes awarded prizes for items such as a child's play dress or a woman's sun dress. New categories and classes focused on women's ability to decorate the home with a greater variety of needle and craftwork items, while more classes for leisure apparel signaled families' increased ability to participate in recreational activities. Traditional classes also remained and continued to be promoted as reflecting rural skills and values. For example, Fleming Furniture in Owen Sound sponsored a homemaker's special at the 1950 Arran-Tara Fair that consisted of an appliqué quilt, hooked rug, crocheted doily, pint of preserved peaches, and a basket of mixed flowers. This class emphasized how a resourceful and talented rural homemaker drew on a variety of skills to create a comfortable home.

By the 1960s, most fairs had established a fairly consistent "Ladies' Work" prize list that included classes for home accessories and manufactures, infants and children's wear, ladies' wear, and some men's apparel. Even though some agricultural societies reduced domestic manufactures classes slightly over the course of the decade, most of the classes remained the same and represented a broad range of useful things for the home and family (see Table 4.4).[117] The items rural women exhibited showed that they still took pride in fashioning clothing and items of household décor, an idea supported by studies of their home production. The report of a 1979 study by the Council on Rural Development Canada indicated that 60 per cent of rural women surveyed produced household clothing for their families.[118] Agricultural magazines also featured articles on homemade clothing, including trendier items such as knit and crocheted ponchos.[119] Whether women were making traditional items such as men's shirts, or new ones such as a barbecue apron embellished with liquid embroidery trim, they used their skill and industry to make useful items that they showcased at fairs.[120] Gail Bartlett, a long-time member of the Binbrook Agricultural Society and past president of the homecrafts division, recalled that she used to make all of her daughters' dresses and entered them in the local fair competition. Most of her entries won first prize, and when her daughters "went to school the next day after the fair they wore the dresses and they were so proud because they had the ribbon. They took their ribbon to school to show everybody."[121]

Fancywork and fine arts remained popular in the twentieth century, but they underwent changes that reflected new technology and aesthetic standards. The arts and crafts movement in Canada that began in the late 1890s was primarily led by women.[122] The promotion of craft across Canada by

Table 4.4
Arran-Tara Agricultural Society Fair, Tara, Bruce County: Classes for Ladies' Work, 1960 and 1970

1960	1970
Children's Wear	**Children's Wear**
Cot Quilt	Crib Cover (liquid embroidery)
Denim Overalls, 2 to 5 years	Child's Slipper Socks
Child's Dress, smocked, 5 yrs. or under, Mercerized Cotton	Baby's Sweater, Bonnet and Bootees (hand crocheted)
1 Pair Children's Sox	Baby's Sweater, Bonnet and Bootees (hand knitted)
Baby's Sweater, Bonnet and Bootees (knitted)	Child's Pyjamas (flannelette) Size 6 to 10 years
Baby''s Sweater, Bonnet and Bootees (crocheted)	Child's Sweater (bulky knit)
Infant's Flannelette Nightie	Boy's Shirt (cotton)
Child's Pyjamas (flannelette) Size 6 to 10 years	Child's Skating Sweater (pullover)
Girl's Plaid Pleated Skirt, 8 to 10 years	Child's Wool Cap or Hat & Mittens to match
Girl's Short-sleeved Pullover, size 8 to 10 years	**Ladies' Wear**
Boy's Shirt, cotton flannel, size 8 to 10 years	Overblouse, cotton (sample of material attached)
Girl's Jumper, 6 to 10 years	Skirt, wool, Map Lead Tartan (sample of material attached)
Child's Wool Cap or Hat and Mitten Set	Pyjamas, cotton, (sample of material attached)
Ladies' Wear	Shorty Nightgown (sample of material attached)
Lady's Slacks, other than cotton, sample attached	Duster, cotton (sample of material attached)
Overblouse, cotton	Sun-Dress, cotton (sample of material attached)
Skirt, cotton	Day-Time Dress (sample of material attached)
Pyjamas, broadcloth	Gloves, fancy (hand knit)
Nightdress, cotton crepe	Hat (wool yarn)
Slip	Lady's Sweater, cardigan (Mohair)
Sun-Dress, cotton	
House Dress	
Gloves, fancy, hand-knit	

Table 4.4 *continued*

1960	1970
Yarn Hat Handkerchiefs, 2 hand-trimmed (mounted) **Men's Wear** Knitted Socks, coarse, factory yarn Knitted Socks, fine, fancy Knitted Gloves Knitted Sweater, sleeveless Man's Sport Shirt Man's Plaid Work Shirt Pair Pyjamas, flannel **Dining Room Furnishings** Luncheon Set, five pieces Set of three Hot Place Mats 4 Place Mats, any work, white or coloured Crochet Table Cloth, 48" or larger Buffet Set, Crocheted **Bedroom Furnishings** Quilt, showing fancy quilting, plain color Quilt, pieced Quilt, pieced and applique, cotton Quilt fancy applique 1 Pair Pillow Cases, embroidered 1 Pair Pillow Cases, crochet trim 1 Pair Pillow Cases, knitted lace trim **Living Room Furnishings** Living Room Furnishings Cushion, wool or wool trim Cushion, Corduroy Cushion, any other kind	**Men's Wear** Socks, hand knit (in Fingering or Double knitting yarn) Socks, hand-knit, fine Vest, Bruce Tartan (sample of material attached) Men's Sport Shirt, Paisley (sample of material attached) Men's Sweater, bulky knit **Dining Room Furnishings** Lace Table Cloth Luncheon Set, 5 pieces (liquid embroidery) Luncheon Set, 5 pieces (hand-embroidered) One Centre and 4 Place Mats (machine embroidered) Breakfast Cloth, gay color (not plastic) **Bedroom Furnishings** Quilt, cotton, showing fancy quilting (plain colors) Quilt, pieced (cotton) Quilt, pieced and applique Quilt, applique and embroidered 1 Pillow Slip (colored, crochet trim) 1 Pillow Slip, made from flour sacks (liquid embroidery trim) 1 Pillow Slip, hand embroidered (not cross stitch) **Living Room Furnishings** Cushion, Crewel Embroidered Cushion, smocked

Table 4.4 *continued*

1960	1970
Chair Set, 3 pieces Collection Crochet Centre Pieces… Knitted Doily Tatted Doily	Cushion, toss, one pair Ruffled Centre Piece Three Doilies, mounted (3 patterns, under seven inches)
Kitchen Accessories Half Apron, cotton Half Apron, fancy Large Kitchen Apron Cobbler's Apron, sample mat. Attached 3 Pot Holders, crocheted 1 Pair Tea Towels, hand-made, embroidered	**Kitchen Accessories** Tea Apron Large Kitchen Apron, with bib, pot-holders to match (sample of material attached) Cobbler Apron (with sample of material attached) Shower Gift (made from 5 kitchen articles)
Miscellaneous Wrapped package for a Bride's Gift Article made from flour sack, over 10" x 12" Sugar-Starched Crocheted Novelty Article of Needlework by a New Canadian	

various organizations meant that more women created embroidered goods, lace, crocheted dollies, and other decorative work than ever before.[123] Beginning in the 1900s, changes to home design called for a refashioning of the formal, Victorian parlour into a family room that was more functional for everyday use, but many forms of fancywork continued to be popular because they were compatible with modern design.[124] Some items, such as hair wreaths, wax fruit, stuffed birds and animals, and paintings on muslin, no longer reflected the aesthetic of the times and were dropped from most prize lists by the 1910s, although stuffed birds managed to remain until the 1920s.[125] Item such as crocheted tablecloths, lace-trimmed pillow cases, and embroidered towels continued to be made because they had beauty and served modern functions (see Figure 4.1).[123] Even during the Second World War, when the 1943 OAAS Women's President, Miss Lillian M. Rutherford, advised

Providing Comfort, Refinement, and Respectability 163

Figure 4.1 Prize Tatting at Schomberg Fair
The stylish, modern attire worn by Mrs Leslie Holmes suggests that women who cared about contemporary fashion considered fancywork to be compatible with chic attire. "Schomberg Fair, Mrs. Leslie Holmes, Schomberg, prize tatting," 1928. *Globe and Mail* Fonds, 1266, Item 15036, City of Toronto Archives. Image courtesy of the City of Toronto Archives.

female delegates, "Let us be patriotic and leave fine needlework off lists for the duration,"[127] women continued to show fancywork.

Fancywork was a way for rural women to maintain traditional skills and create connections with their past. Figure 4.2 is a photograph taken at the 1955 Waterdown Fair of four women surveying fancywork. The caption read: "Looking at the needlework display at the Waterdown Fair this week these ladies decide that women still have time to express their creative urge in intricate stitches and patterns as their grandmothers did."[128] This description reveals two ideas: first, that women continued to value fancywork as an outlet for their creativity and artistry and second, that the creation of fancywork was an activity that connected women with their mothers, grandmothers, and other female relations. Various forms of fancywork continued to be made, and some that had gone out of fashion saw a revival in the postwar years.[129]

As in the nineteenth century, most art classes in the first half of the twentieth century were for oil and watercolour paintings, as well as pencil, crayon, and ink sketches. The subject matter also continued to be mostly landscapes, portraits, flowers, fruit and vegetables, animals, and local landmarks or scenery (see Figure 4.3). In the post-war period, attempts to engage with more popular styles of painting were also evidenced. Abstract painting was more common in fair prize lists by the 1970s. At the 1974 Erin Fair, a class was added for this style (but so too were pictures depicting pioneer scenes). The judges of artwork tended to be the same women judging other crafts and handiwork, and the art shown in the postwar period was not perceived as having great artistic merit because it was generally the work of amateurs rather than professionals. Reporters expressed appreciation of local talent, but few detailed accounts of the artwork were given.

One significant change in the art department was the expansion of amateur photography. Photography classes had existed in the nineteenth century, but usually only professional photographers competed.[130] By the 1930s, however, most fairs had added some classes for Kodak pictures to their prize lists, and the number of amateur photography classes expanded significantly in the postwar years. The popularity of photography illustrates the accessibility of the technology. The Kodak film camera and its subsequent mass-production allowed more amateurs to participate. Kodak cameras were advertised to farmers as a way to keep a record of the improvements they made to their farms, and they were told that "In every phase of farm work and farm life there are pictures that are interesting and valuable. And *you* can make them."[131] It was also reported that more women were becoming successful professional photographers, while others "had a fancy for the Kodak as an amusement." Some women purportedly confined their talents to photographing children and weddings because these scenes were suitable to domesticity.[132]

Another important shift in the postwar period was the merging of fine art, fancywork, and other crafts into an "Arts and Crafts" department (see

Figure 4.2 Judges Inspecting Needlework at Waterdown Fair
The *Hamilton Spectator* pictured the judges, Mrs Lewis Binkley of Carlisle, Mrs Gordon Ofield of Dundas, Mrs Charles Cook of Waterdown, and Mrs Charles Harper of Carlisle at the Waterdown Fair inspecting needlework entries, and noted that "women still have time to express their creative urge in intricate stitches and patterns as their grandmothers did." *Hamilton Spectator*, "Waterdown Fair," 1955. Local History & Archives, 32022191097548, Hamilton Public Library. Image courtesy of the Hamilton Public Library.

Table 4.5). This title change is significant, especially when individual classes are considered. Quilt classes at the Erin Fair, for example, were once part of the domestic manufactures' competition. By the 1950s, quilts were elevated from home manufactures and exhibited in the fancywork competition, and by the 1970s they were deemed art and craftwork. Once women had considered quilts beautiful, but routine and functional, products of the rural household. Now they were art: a traditional craft made by a select number of women with specialized skills. Quilts had always represented the combination of usefulness and ornamentation that a lot of women's other exhibits

Figure 4.3 Top Art Prize at Rodney Fair
Mrs Eva Bitterman of Rodney, supervisor of art at the Rodney Fair, points to the winning oil painting on the opening day of the fair. The oil painting was by seventy-two-year-old Mrs Myrtle Babcock of Wardsville. "Top Art Prize," 1976. *St Thomas Times-Journal* fonds, C8 Sh2 B1 F11 21a, Elgin County Archives. Image courtesy of Elgin County Archives.

symbolized. When crazy quilts were popular in the 1890s, newspaper reports commented on how they made a particularly brilliant display at fairs.[133] Women used fabric scraps to make quilts, which highlighted their thrift, but this fabric was beautifully pieced together in a way that evidenced their taste.[134] The artistry required for quilting was recognized, as always, but now that fewer women were taking up the practice it had become a specialized skill rather than a domestic necessity. Antique quilts were also beginning to be considered valuable artifacts in Ontario, and their display in art galleries

Providing Comfort, Refinement, and Respectability

Table 4.5
Erin Township Agricultural Society Fair, Erin, Wellington County:
Classes for Arts and Crafts, 1974

1974

(Originals) Drawing or sketch, pencil or charcoal	(Miscellaneous hobbies) half apron, cross-stitch
Local scene, oils with title	Place mats, set of 4, any media
Portrait, any medium	Batik, any article
Street scene, named	Candle, hand-made
Bird life, any medium	Christmas table arrangement, stationary, not to exceed 12" x 14"
Modern art, other than paint	
Abstract painting	Costume Jewellery
(Copy Work Only, Oils) flowers winter landscape	Crochet article - novelty
	macrame - any article
(Pastels) Pioneer scene	Copper enamelling, any article
Flowers and/or fruit	Block printing, any article
(Crafts and Hobbies) Quilt, crib, nursery design	An article made from a square yard of material
Quilt, child's	Something new from something old, STATING ORIGINAL ARTICLE
Quilt, applique	
Quilt, pieced	Article made from terry cloth
Quilt, plain material, quilting only to be judged	Stuffed toy, hand-made
	Child's Toy, hand-made, using any other media
Quilt blocks, pieced and/or applique, 4 different, named and mounted	
	Slipper socks
Pillow cases, one pair, embroidered	Picture, framed, embroidered or crewel
Pillow cases, one pair, cross-stitch	
(Living Room) Hooked Rug, latchet	Wall hanging - not framed, any media
Hooked rug, wool strips	Amateur Photography - 3 coloured prints of local interest, mounted and named
Article of needlepoint, finished	
Afghan, crochet or knit	
Tablecloth, crochet	Any other hand-made article
Cushion, smocked	

and museums spoke both to their meaningful history, and to the artistic value they held.[135] In 1967, agricultural societies offered classes for heirloom quilts as a way to commemorate Canada's Centennial.[136] Women exhibited quilts at the fairs not as common home manufactures, but to preserve an art form they loved. Although fewer women were making quilts by the 1970s, they were still a cherished part of rural society's identity, and they continued to hold significant meaning.

Craftwork allowed women to personalize their homes in a way that displayed positive ideas about thrift and skill. *Country Guide: The Farm Magazine* provided a section for "Country Living" that catered to rural women. In 1979, it included recipes, craftwork instructions, fashion notes, and profiles of interesting women. A "Country Crafts" section was a regular feature that provided instructions for how to make things such as knit, crocheted, sewn, and hooked cushion covers and table mats, or embroidered pictures.[137] Most of the fancywork was traditional forms of household goods, but some women tried to replicate other art forms. For instance, Mrs George Davis was congratulated for her "outstanding" entry at the 1960 Erin Fair when she won first prize for a filet-crocheted wall panel of the Last Supper.[138]

The Legacy of Women's Work

Perhaps there is no greater testament to the meaning of women's handiwork than the way in which it was cherished long after the women themselves were gone. The examples of women's exhibits that have been meticulously cared for, not only by the women themselves, but by their family members, reveal their significance. Two such items include a pair of finely knitted fancy children's socks and a woman's cream-coloured flannel nightgown with embroidered trim. Both were made around the turn of the twentieth century by Sarah Wheeler Jackson and donated to the Wellington County Museum by her granddaughter (see Figures 4.4 and 4.5].[139] Sarah Jackson was born to an established family in Erin Township in 1856, and she married her husband David Jackson in 1872 at the age of sixteen.[140] The Jacksons were farmers and active participants at the Erin Fair; Sarah herself is said to have shown her handiwork and cookery there for over fifty-seven years.[141] The fine knitting of the socks and the homemade eyelet embroidery in satin stich with scalloped edge on the nightgown illustrate the skill she possessed. The nightgown was stitched with a sewing machine, while the embroidery is believed to be handmade.[142] Both items show no sign of wear, and the care with which they were crafted and maintained suggests both pride in their creation and strong sentiment. Indeed, as Kathleen Cairns and Eliane Leslau Silverman suggest in *Treasures: The Stories Women Tell about the Things They Keep*, material possessions are often kept as a reflection of personal identity, hold rich biographical meanings, and help women to tell the stories of their lives.[143] For Jackson, socks and nightgown symbolized her economy and skill, but the fine knitting she used to construct the socks and the use of embroidery on the trim of the nightgown illustrate her understanding of what beautiful and useful objects could be.

The ability to fashion clothing was passed down from Sarah Jackson to her daughter Clara, who, like her mother, left evidence of exhibiting at fairs in the form of a blue cotton men's work shirt with white stitching (see Figure 4.6). The shirt was made in 1925 and shown at the Erin and Acton Fairs that year. Clara Jackson Robertson was married in 1896 at the age of nineteen to George Gray Robertson, a twenty-five-year-old blacksmith.[144] A regular exhibitor, Robertson was an accomplished seamstress and quilter who won many prizes at fairs for her quilts and clothing, as well as for baking and fancywork.[145] This particular shirt was exhibited but never worn, and kept as a treasured item. The care taken is evident; donated to the Wellington County Museum and Archives by her daughter-in-law, the ninety-one-year-old machine-stitched shirt looks as if it was fashioned yesterday. For Robertson, the shirt represented many things. Sewing was a marketable skill, and the shirt illustrated her proficiency in this line of work. By exhibiting the shirt at fairs, Robertson demonstrated both the value she placed on her work and her desire to maintain the family tradition of participating at these events. The treasuring of this everyday item suggests pride of ownership and recognition of personal skill and ability.

Fanny Colwill Calvert was also proud of the fancywork she kept safe over the years. Calvert immigrated to Canada from England in 1890 and settled in Guelph two years later. After a second marriage that failed, Calvert pursued her own career, becoming an active businesswoman in the community by managing rental homes and owning a shop that sold drapery and milling supplies.[146] Despite her business success, Calvert still made time for artistic endeavours, and the oil paintings, wood carvings, sculptures, and craftwork she left behind are proof of her commitment and ability. Calvert was especially interested in ornamental needlework. She exhibited Battenberg lace, tatting, crotchet edgings, and Irish crotchet at fairs such as Galt, Fergus, Elmira, Barrie, and many others. It was reported that she received numerous first prizes for her work, and that she used the prize money she earned to

Figure 4.4 Sarah Jane (Wheeler) Jackson's Prize-Winning Socks
Sarah Jane (Wheeler) Jackson, "Sock," circa 1900. 2004.14.5.02, Wellington County Museum and Archives. Photograph by author, taken courtesy of the Wellington County Museum and Archives.

pay her taxes.[147] In 1914, she entered thirty items of art and fancywork at the Caledonia Fair alone that year.[148] The more than fifty samples of crochet and lace work donated to the University of Guelph Archives and Special Collections display a strong attention to detail, intricacy, and precision (see Figure 4.7). While the monetary rewards Calvert received were clearly incentives to exhibit, the significant time Calvert took to make and exhibit these items

Figure 4.5 Sarah Jane (Wheeler) Jackson's Prize-Winning Nightgown
Sarah Jane (Wheeler) Jackson, "Nightgown," circa 1900. 1954.97.1, Wellington County Museum and Archives. Photograph by author, taken courtesy of the Wellington County Museum and Archives.

and their preservation over time suggests both personal and subsequently familial affection for these objects.

Other material objects passed down in families that connected women to fairs were not items they had entered for competition, but rather things they created with the awards they and their families received. For example, women sometimes made quilts out of fair ribbons. Beyond prize money,

Figure 4.6 Clara (Jackson) Robertson's Prize-Winning Work Shirt
Clara (Jackson) Robertson, "Shirt, Work," circa 1925. 1994.34.1, Wellington County Museum and Archives. Photograph by author, taken courtesy of the Wellington County Museum and Archives.

agricultural societies awarded other items that encouraged exhibitors to participate, including useful material goods or trophies and plaques. Ribbons, however, were the most common objects awarded in addition to premiums. For agricultural societies, ribbons were relatively inexpensive to buy, but they were symbols of accomplishment. The problem with ribbons, however, was that they had little utility beyond their representation of achievement. Sometimes, therefore, women repurposed fair ribbons into quilt pieces that, when sewn together, created a family heirloom that honoured a family's achievements, history, identity, and community ties.

Scholars have argued that quilts created bonds among female relatives because quilting involved a transfer of skill from mothers to daughters or other female relatives that reaffirmed female connectedness.[149] Fair ribbon

Providing Comfort, Refinement, and Respectability　　　　173

Figure 4.7 Fanny Colwell Calvert's Irish Crochet
Fanny Colwell Calvert, "Irish Crochet," circa 1910. Regional History Collection, XR1 MS A 743, University of Guelph Archives & Special Collections. Photograph by author, taken courtesy of the University of Guelph Archives & Special Collections.

quilts also, however, demonstrate that women could use quilts to connect with their male kin. For example, Bernice, Irene, and Edna Rudd used the ribbons won by their father, William James Rudd, to construct a unique quilt that showcased Rudd's success at livestock competitions at regional, national, and international fairs between 1891 and 1901 (see Figure 4.8).

The Rudds farmed in Eramosa Township, Wellington County, and bred prize-winning North Devon cattle.[150] William Rudd's youngest daughters, Bernice, Irene, and Edna, selected from the many ribbons he had won at prestigious cattle shows to create a quilt that represented his achievements.[151] The unique design of the quilt reflected the various shapes, lengths, and colours of the silk ribbons. Rather than the smaller ribbons that were typically distributed at township and county fall fairs, the larger regional and

Figure 4.8 Bernice, Irene, and Edna Rudd's Commemorative Quilt
Bernice, Irene, and Edna Rudd, "Quilt, Commemorative," circa 1901. 2000.26.1,
Wellington County Museum and Archives. Photograph by author, taken courtesy
of the Wellington County Museum and Archives.

international fairs offered large silk ribbons or banners that were impressive because of their scale and quality. Their visibility, and the high level of competition they represented, undoubtedly amplified pride of ownership.[152] The number of ribbons they won and top finishes they achieved show the success of the Rudd livestock. At the time the quilt is thought to have been made, Bernice was twenty years old, Irene was fourteen years old, and Edna was twelve years old. Their father was fifty-six.[153]

Like other women who constructed quilts, the Rudd sisters likely worked together. But this quilt meant more than a chance to forge sisterhood and learn valuable skills; it also signified the girls' interest and pride in their father's and family's achievements. While it is difficult to know if the young women were involved in the cattle business, they cared enough about it to sew an object that memorialized their family's success in that industry.

This family's quilt was made from prize ribbons, but not all ribbons distributed at fairs denoted winning exhibitors. Fair ribbons were also used to distinguish agricultural society members, fair directors, and judges. Designatory ribbons conveyed an individual's position or contribution to a community organization or event, and thus symbolized their civic engagement and public influence. Elizabeth Ann Donaldson used designatory ribbons from a variety of organizations and events in her region to create a beautiful crazy quilt in the early twentieth century (see Figure 4.9). Like the Rudd family quilt, Donaldson's honoured past accomplishments, displayed values and identity, and connected the skill and interests of male and female family members. Her crazy quilt consisted of a variety of ribbons from community events and organizations, as well as scraps of silks, satins, and velvets of various textures, colours, and patterns. Donaldson stitched hundreds of patches of fabric by hand into eight twenty-six-centimeter squared blocks, machine sewn together in four rows of four and backed with fabric printed with chrysanthemums and forsythia on a light green background. She embroidered each block by hand in a variety of stitches and added embroidered motifs, many of which depicted flowers.[154]

Donaldson's quilt reflected her situation in life. At thirty-two years of age she married William Donaldson, who was ten years her senior and kept a general store and post office in Hillsburg. She and William never had any children.[155] Elizabeth's circumstances likely gave her the time for leisure activities and access to an array of sewing supplies, fabrics, and patterns.

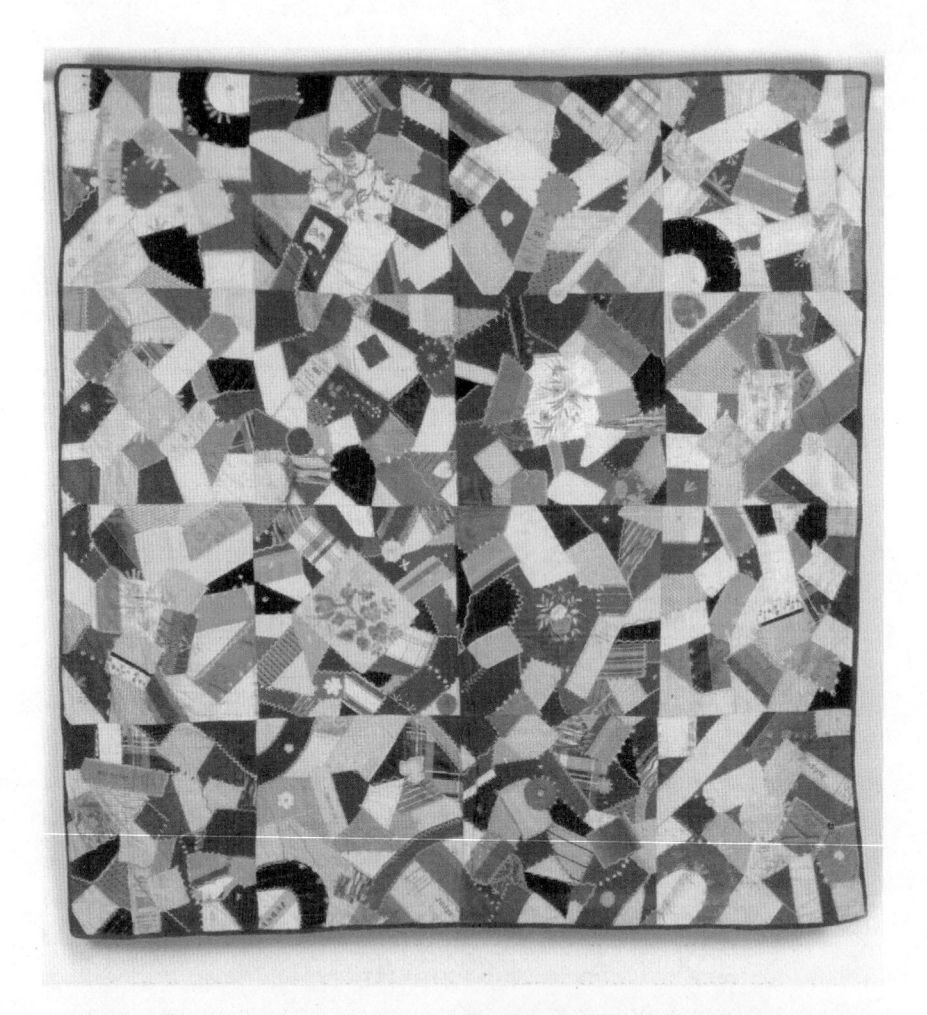

Figure 4.9 Elizabeth Ann (Huxley) Donaldson's Crazy Quilt
Elizabeth Ann (Huxley) Donaldson, "Quilt, Crazy," 1910–1920. 1977.37.1,
Wellington County Museum and Archives. Photograph by author, taken
courtesy of the Wellington County Museum and Archives.

The quilt itself was a token of her creativity, talent, and interests, but it also honoured her husband by incorporating seven designatory ribbons used to signify his civic involvement, including his position as judge at the Luther Agricultural Society Fair (a township fair that no longer exists, but was originally located in Wellington North) and his directorship in the Erin Agricultural Society for 1896 and 1897 (see Figure 4.10).[156]

Providing Comfort, Refinement, and Respectability 177

Figure 4.10 Erin Agricultural Society Directorship Ribbon Sewn onto Crazy Quilt
Elizabeth Ann (Huxley) Donaldson, "Quilt, Crazy," 1910–1920. 1977.37.1, Wellington County Museum and Archives. Photograph by author, taken courtesy of the Wellington County Museum and Archives.

Crazy quilts that used silk and other fine materials were often made to be displayed and treasured, rather than for everyday use. Donaldson's is a perfect example of how quilts communicated more than just a pleasant image. Anyone who saw this quilt would have admired its design and creativity, but would also have observed that it conveyed the Donaldsons' dedication to local agricultural societies and therefore commitment to and standing in their community. Janet Floyd argues that in the nineteenth century, "quilts began to be used as important props in the increasingly elaborate staging of particular life events and friendship groups."[157] Elizabeth's crazy quilt positioned her and her husband among the rural reformers who believed that agricultural societies and fairs were necessary for the betterment of the community. The additional designatory ribbons she used from local voluntary groups and civic events further highlighted the Donaldsons' conviction that residents should support their communities.

Fair ribbon quilts could also be very personal, especially when they were made from the ribbons women won themselves. One such quilt was made by Christena Nelena (McMillan) Neal, known as Lena, from ribbons she reportedly won at the Wellington County Fair in Fergus (see Figure 4.11). She was a life member of the Northgate Women's Institute, which may have influenced her desire to compete and display her domestic skill. She was born in 1898, and married her husband, Leslie Howard Neal, in 1923. They lived on a farm in Arthur Township and had eight children between 1925 and 1938.[158]

As with Donaldson's crazy quilt, Neal's displays a variety of colours, textures, and fabrics, but her quilt is less skillfully constructed. The design is pleasing, but on closer inspection the stitching is uneven and the fabric haphazardly cut. No attempt at embroidery was made. Neal used seven third-place ribbons, four fourth-place ribbons, and one unspecified sweepstake ribbon in her design; it appears first- and second-place prizes eluded her (see Figure 4.12). It is unclear what classes the ribbons represent and when they were won, and the quilt's construction suggests Neal was only an occasional seamstress. Still, the quilt remains in good condition, and the time and care taken to fashion the quilt, and the pride she revealed in the awards she had won, should not be ignored.

In her material cultural study of clothing and the aesthetics of self, Sophie Woodward argues that by understanding how women assemble an outfit, one can better understand "the complex construction of surface" that takes place. Clothing, she argues, is "part of the multifaceted surface, being the site where the self is constituted through both its internal and external relationships."[159] Quilts, like clothing, mediate the relationship between an individual and the outside world. Individual expressions of creativity, taste, and skill are combined with external considerations such as the patterns, technology, and materials available. So, in addition to representing material goods, Neal's quilt expresses the broader ideas she had about worthwhile pursuits and markers of personal success. Although she did not receive first prize, the ribbons she won at the Wellington County Fair still had significant enough meaning for her that she took the time to create an object that would forever honour and bear witness to her accomplishments and interests.

Figure 4.11 Christena Nelena (McMillan) Neal's Quilt
Christena Nelena (McMillan) Neal, "Quilt," circa 1950–60. 2010.36.1, Wellington County Museum and Archives. Photograph by author, taken courtesy of the Wellington County Museum and Archives.

Figure 4.12 Wellington County Fair Ribbons Sewn onto Quilt Christena Nelena (McMillan) Neal, "Quilt," circa 1950–60. 2010.36.1, Wellington County Museum and Archives. Photograph by author, taken courtesy of the Wellington County Museum and Archives.

Perhaps there is no greater example of the importance women's exhibits had for their creators than the recognition this work received in their obituaries. The obituary for Nancy Strickland, the painter who was criticized for the "too servile imitation of detail" in her work,[160] illustrates how the recognition she obtained at fairs clearly outweighed any criticism she endured. Strickland was a long-time exhibitor of ladies' work and fine art at township, county, provincial, and world fairs, and her obituary in 1886 revealed the significant place her art had in her life. After describing her years as a schoolteacher, the notice went on to note that Strickland "had a rare gift for fancy needlework, for painting and drawing and in these she acquired great proficiency. Her skill to these things [sic] was well known throughout this country, as many diplomas and medals testify."[161] The notice reported that Strickland had received diplomas and medals from international competitions as well, including the Centennial Exhibition at Philadelphia in 1876 and the Paris Exposition in 1878. The author of the obituary acknowledged

Strickland as a strong Christian, a clever and good woman who led a beautiful life, and a woman of "excellent judgment combined with force of character," but it is still clear that her identity as an artist and exhibitor was central to her and those around her.[162] Strickland's death was reported across the province, and the notices emphasized this identity. The *Huron Expositor* announced that "Miss Nancy Strickland, of Oshawa, who has for years secured the lion's share of prizes for the ladies' and other work at the Provincial, Western and other fairs, died recently from the effects of a tumour."[163] The daughter of an immigrant farmer who had come to Ontario from England with his wife and four daughters,[164] Strickland would remain single throughout her life, becoming the primary caregiver for her parents. For a woman who clearly devoted so much of her life to others, the achievement she found at fairs was undoubtedly a source of pride and self-expression.

Conclusion

By displaying their domestic manufactures, fancywork, and fine arts at fairs, women also displayed characteristics such as economy, hard work, and refinement. Moreover, their exhibits' ability to blur "the distinction between self and object"[165] meant that each creation was a deeply personal one. The women I interviewed who exhibited at fairs explained that they became "hooked" on showing their work because of the joy it instilled. Gail Bartlett recalled how she was introduced to fair competition and the effect it had on her: "I thought [fairs were] just selling. Everybody selling their goods and their vegetables, and maybe like a bazaar. It was a shock to see all these prize ribbons and people entering and winning. And the next year the neighbour said, "Well you sew, come on, you enter," so I did enter and I had several firsts ... I've been hooked ever since to the fair."[166] Bartlett went on to explain that she believed women who exhibited did so because "it's just in our blood." Bartlett was proud of her work and the many awards she won over the years. She explained she sewed every day, and it was important to her that she showed this work to others: "I really like showing what I can make. I'm not out there to say I'm out to win this year, but I like to exhibit ... I just love to show and see what other people put in the fair."[167] For fair women, the items they made and won were important to them, and many times these objects provided legacies for their families as well.

Rural women's objects are critical to our understanding of their lives because they help us reconstruct what women did, the values and ideas they held dear, and how they wanted to be remembered. Home manufactures, fancywork, and fine arts include similar commitment to household provisioning, thrift, and industry, as well as middle-class respectability, taste, and refinement as that highlighted in previous chapters, but they also illustrate that rural women cherished individual expression and artistry and worked hard to create legacies that their families and communities would remember them by.

5

Getting Outside the Inside Show
Female Horse and Livestock Exhibitors

The three previous chapters focused on women's presence inside the exhibition hall. Whether they were admiring the work of others or exhibiting canned peaches and doilies, where women belonged on the fairgrounds seemed clear. But this chapter illustrates how female participants went beyond the exhibition hall and contested traditionally masculine categories of competition. Horse and livestock shows were initially seen as male preserves, but over the course of the nineteenth and twentieth centuries women challenged that gendered space and created a place for themselves in the show ring.

Women first began participating in the "outdoor show" by exhibiting horses. Their horsemanship was put on display once fair organizers realized audiences wanted to see the "spectacle" of women's driving and riding classes. Women's participation in cattle, pig, sheep, and poultry showing was slower to materialize. Initially, the only recorded female exhibitors in cattle shows were widowed women who could claim title to their livestock. By the mid-twentieth century, however, youth organizations such as 4-H made livestock showing an acceptable activity for both boys and girls, which encouraged more women to enter the show ring. Still, while women participated, reports of their involvement emphasized their feminine characteristics. By highlighting women's beauty rather than skill, their delicacy rather than strength, and their uniqueness rather than prevalence in the show ring, their participation was highly gendered. Over time this improved, and many women found both validation and acclaim in livestock and horse showing. This transition, although incomplete, gave women the confidence to expand their influence in agricultural societies and undertake larger leadership roles in agriculture in the late twentieth century.

An Overview of Women and Animals in the Countryside

Historians writing about women's work in nineteenth-century Ontario have recognized their importance in caring for and profiting from livestock.[1] Farm women and girls were often responsible for the daily upkeep of and products derived from farm animals, especially chickens and dairy cattle (see Figure 5.1). But even though they often had a great deal of influence over livestock care and by-product production, they very rarely owned the resources of production – the animals themselves.[2] In her discussion of farm women's labour, Marjorie Griffin Cohen explains that the "concentration of ownership of the means of production by males meant that women's labour throughout their lives would be subject, either directly or indirectly, to male power and authority."[3] Women were able to exercise more control over the money they earned from poultry and dairying, but generally major decisions about livestock and farm management were undertaken by men.

Livestock breeding improved in Ontario during the nineteenth and early twentieth centuries because quality animals were necessary for efficient production.[4] Very few women, however, had a significant effect on horse and livestock breeding in Canada during this period. Perhaps this is not surprising, considering women's general exclusion from horse sports and livestock competitions. Examples of women involved in race horse breeding in Britain at the end of the nineteenth and beginning of the twentieth centuries exist, but these women were usually members of the upper classes.[5] Harness racing was another equestrian sport that originated "as a sport centered on male camaraderie, based on shared maleness and class status."[6] Despite the popularity of harness racing in North America, few women participated in the sport or bred harness horses.[7] Even in the early twentieth century, the record of Canada's great horsemen is just that: a record of men.[8]

Livestock breeding was considered a masculine pursuit. It was generally argued that women's "natural" characteristics of tenderness, kindness, com-

Figure 5.1 Woman Feeding a Calf
Reuben Sallows, "Woman Feeding a Calf, 1910." C 223-2-0-0-5, I0002255, Archives of Ontario. Image courtesy of the Archives of Ontario.

passion, and their nurturing instincts made them well suited to care for animals, but they were thought to lack the scientific minds, business acumen, and know-how necessary to breed livestock.[9] Only men were deemed to have the qualifications necessary "to attain the highest success in the art."[10] These traits included strong analytical powers, according to which a "man judges as a whole instead of in detail," persistence and perseverance in adhering to a plan, an educated eye to detect "form and quality," a mind free from prejudice and bias, an understanding of cause and effect, a measure of caution so as not to be "prone to jump to conclusions from insufficient data," and the ability to be "an artist, capable of forming an ideal model of perfection."[11] To be a successful breeder, a person had to have the capacity to be a judge. As Cecilia Morgan notes in *Public Men and Virtuous Women*, conservative writings "juxtaposed feminine foolishness and flightiness with solid, reasoned male judgement," positioning logical thinking as a masculine trait, and ignorance and excessive emotion as feminine.[12] Furthermore, the sexual nature of breeding animals itself may have been seen as reason enough to exclude "the delicate sex."

Although rare, some examples of famous female breeders exist. One such was Mrs Eliza Maria Jones of Brockville, Ontario. Jones was born Eliza Maria Harvey on 24 December 1838, in Maitland, Upper Canada. She came from a prosperous Upper-Canadian family and was educated in Montreal and Scotland but returned to the family farm to care for her five younger siblings when her mother died. When she married and settled in Brockville, she started her own small dairy farm. Jones began her herd with grade cattle (typically, animals with no recorded pedigree and used for commercial rather than improved breeding purposes), but after purchasing an outstanding Jersey-cross cow in 1873, she was convinced that purebred dairy cattle were necessary for a successful dairy operation. Jones was a proponent of scientific agriculture and utilized the latest in farm technology and nutrition to produce superior butter, which she sold to prestigious clients across North America, shipping over 7,000 pounds a year. She personally oversaw all aspects of her farm and hired three men to help with the labour.[13] Her husband, Chilion Jones, was largely absent and had little to do with her dairy operation,[14] so the hired men dealt with her Jersey bulls, animals considered "often treacherous and difficult to handle."[15]

Jones developed her purebred herd from cattle she purchased from Romeo H. Stephens of Montreal, whose own herd was developed from cattle purchased from the Royal Herd at Windsor, in addition to cattle she imported from New England and the Isle of Jersey.[16] She met with immediate success when she exhibited her cattle at agricultural fairs in the 1880s. Although she complained that her domestic and family affairs did not allow her to do more showing, as they "must always come first," she still exhibited across Ontario, Quebec, and New York State, and sold cattle throughout North America. Jones was a member of the Canadian Jersey Cattle Breeders' Association, an author of popular books on dairying and dairy animals, a frequent contributor to agricultural journals, and an international butter judge. In 1897, she was declared "the best dairywomen on the continent" by the Toronto periodical *Farming*.[17]

In 1899, the *Farmer's Advocate and Home Magazine* published a series of biographies of successful Canadian livestock breeders and Jones was the only female breeder featured. Her farm near Brockville, named Belvedere, was described as beautiful with a "commodious dwelling" that gave "evidences of culture and refinement, and the absence of any indications of extravagance or unnecessary display," but it was from "the reputation of her far-famed herd of high-class Jersey cattle, and their signally successful career in scoring records in the show ring and in practical tests for butter production, that Mrs. Jones' name has become so widely known." The article went on to chronicle the various purchases and sales of cattle Jones had made, the awards she had won, the production records she had broken, and the impressive success she had breeding quality dairy cattle. Most importantly, Jones was congratulated for her ability to "make dairying pay": "The reason we have dwelt at length upon Belvedere is because the establishment there had been of more real practical use than any place of the kind before. Here are no costly appliances, no fancy fixtures, no artificial care, but good, plain and PAYING business management. As a fine old farmer said to his wife, when on a visit there: 'Golly, Maria, Mrs. Jones ain't got a single thing that we can't have too.' This it is that has made Belvedere an object lesson for the whole Dominion, and proved that it is within the power of anyone to make dairying pay. All her life, Mrs. Jones labored to teach people how to make the *most* butter, the *best* butter, and, above all, butter produced *at the least cost*."[18]

Jones' superior managerial skills, as well as her ability to breed quality cattle, were praised by the author as an example of how others (men included) should conduct their farming operations. When asked if she still planned to exhibit whatever cattle she had left after downsizing her herd in 1896, Jones explained that due to her increasing age and "family cares," and the fact that "she could not add to her reputation," she would likely never exhibit again, but her "interest in stock and dairy matters [was] keener than ever."[19]

Still, although Jones clearly demonstrated women's ability to be successful farm managers and livestock breeders, and she was praised for the instruction she provided on making "dairying pay," she was viewed as the exception rather than the rule. This was the case in most professions for women in nineteenth-century Canada, especially ones that involved large capital expenditure and direct competition with men.[20] As the Ontario dairy industry began to move from the farm to the factory around the turn of the century, male labourers began to dominate dairy processing.[21] Some women continued to process dairy at home, where butter was more commonly made than cheese.

Another area of farming traditionally under female control was the poultry industry. Farm journalists contended that women were "better adapted" to poultry-raising than men because they could "more faithfully attend to the many little details that go to make the sum total of success" in this line of work.[22] They advised women to make a name for themselves by supplying markets with superior poultry and eggs,[23] but generally women only raised stock – breeders were traditionally men.[24] By the mid-twentieth century, more large-sized flocks existed and a greater emphasis was placed on meat production. Women might still work with chickens on the farm after 1960, but if these were large flocks, they typically did not manage the operation.[25] In many livestock industries men and women needed to work together, but this collaboration did not mean equality.[26] The difference between working with livestock and making decisions about breeding, management, and marketing were important, and women's exclusion from these responsibilities limited their ability to claim full partnership on the farm and in the show ring.

Female Horse and Livestock Exhibitors in the Nineteenth and Early Twentieth Centuries

Early horse and livestock owners and fair exhibitors were almost always men. It is rare to find female exhibitors' names in winners' lists, even for items specifically defined as "Ladies' Work," and finding women's names in horse and cattle competitions is especially unusual.[27] The examples are few, but they do exist. Mrs Gapper was a regular horse exhibitor at the Home District Agricultural Society's fairs before 1850. At the 1836 Fair, she won second and third prizes in the mare class. Out of all the livestock exhibitors, she was the only woman.[28] At the 1840 Spring Fair, she was again the only female exhibitor, and this time she won the mare class for draft horses.[29] The next year, Gapper won a first and second prize in a particularly strong horse competition.[30] Some other women who exhibited livestock and horses at fairs included Mrs McCormick, who won second prize for the "Best Sow" class at the 1847 Eramosa Fair in Wellington County,[31] Janet McRobbie, the second prize winner of the yearling filly class at the 1853 Puslinch Agricultural Society Show,[32] and Mrs Gordon, who was the only female exhibitor listed in any livestock or horse class at the 1889 Central Exhibition in Guelph. She won second prize in the two-year-old gelding class for heavy horses.[33]

These women were all widows. "Mrs Gapper" appears to have been Mrs Gapper Sr, the mother of Mary Sophia Gapper O'Brien, whose journals describing her time in Upper Canada between 1828 and 1838 have become popular primary sources for historians investigating pioneer women in Ontario. Gapper was a wealthy woman whose husband, the Reverend Edmund Gapper, was both a squire and rector (positions of high social standing) of Charlton Adam in England before he died in 1809. She came to Canada with her daughter Mary to help look after her son William's household while his wife Fanny expected their second child.[34] When Mary met Edward O'Brien and married, Gapper decided that she would remain in Upper Canada and live principally with her daughter and occasionally with her other children.[35] Gapper's sons and son-in-law were all active members of the Home Agricultural Society and they likely encouraged her to enter her horses at the fair.[36]

Mrs Abigail McCarter McCormick continued to exhibit at the Eramosa Fair decades after her husband Robert died in 1828. She had three daughters at the time of her husband's death, the youngest being only one year old.[37] Her family was established in the community, and she may have believed that participating at the fair was an important way to market her livestock, and other farm and domestic produce, while maintaining her family's social position despite the loss of her husband.

Janet McPherson McRobbie was born in Scotland in 1830 and immigrated to Canada with her family in 1841. She married Lodwich McRobbie, a farmer and one of four brothers who moved from Perthshire, Scotland, to Canada around 1833. Lodwich died in 1852 at the age of thirty-four, leaving twenty-two-year-old Janet with a son not yet one year old. She later remarried and had six more children,[38] but her decision to exhibit at the Puslinch Fair the year after her first husband's death must have been difficult. The McRobbie family were active exhibitors and members of the Puslinch Agricultural Society, however, and Janet herself seemed to enjoy competing, as her many awards attested.[39]

The "Mrs Gordon" who exhibited at the Central Exhibition in Guelph was likely Mary Gordon of Elora, the widow of Andrew Gordon, who died in 1883 of tuberculous.[40] When she entered her horse at the District Fair in 1889, she would have been sixty-eight years old. Gordon was listed as the head of her household in the 1891 Census, living with her daughter Isabella as well as three young lodgers – two twenty-five-year-old males and one female.[41] Gordon also had two adult sons, one of whom, George, was a harness maker like his father. Perhaps he helped his mother exhibit horses to showcase his own work.[42]

These examples illustrate the responsibility that many women took for their families, property, and chattel after their husbands passed away. In her work on women's legal history, Lori Chambers has argued that, despite being meant to safeguard wives when their husbands died, in practice "the dower portion often failed to provide widows with adequate support because it gave the wife a life interest in realty, a right to enjoy but not to alienate such land; it did not give her a claim on any personal property or the money or chattels owned by the husband."[43] The concern of contemporaries, including the editors of the *Upper Canada Law Journal*, was that the law did not provide widows with a guarantee that they would recover money after the death of a spouse, and thus failed to provide relief for those who needed it.[44] The women who entered their animals at fairs, however, demonstrated their positions as heads of households, and maintained ownership of their land and chattel, even when they had adult sons. Widowed women with either a particular interest in breeding stock or a desire to market their animals and farm products could continue to exhibit, perhaps with the help of their children and relatives. Whether they wanted to maintain the legacy of their late husbands, create their own reputations, or simply exhibit as they did before their husbands died, we cannot tell. What we do know, however, is that women sometimes exhibited their animals, and although it is unlikely that they were "at the halter" in the show ring, their participation nevertheless created a foundation for future change.

Nineteenth-century women were generally discouraged from participating in public events,[45] and men were critical of women who "appropriated" male occupations. They believed men alone should be breadwinners, and that women had neither the "intellectual initiative" nor "disciplined rationality" that was required for most professions.[46] By the mid nineteenth century women were participating at fairs, but their participation was mainly limited to competing in home manufactures and domestic produce.[47] For example, an 1880 article in the *Brampton Conservator* encouraged parents to take their children to the fair to stimulate "a spirit of emulation" for their respective tasks: "Let the boys have some poultry or animal to show; the girls, some flowers or fruit they can watch the season through, and then exhibit as the result of their care." The newspaper noted that, in respect to girls, "If one shows a talent for housekeeping, let her exhibit canned fruit or a loaf of bread of her own making," while "if the boys have any mechanical skill or genius in any direction, let it be encouraged and made the most of."[48] Girls were expected to showcase domestic skills, while boys could pursue agricultural and mechanical undertakings and "genius in any direction" – other than domestic skills, of course.

While gendered assumptions of female traits limited women's opportunities, women were still proving that they had the skill and drive to work with farm animals, and the talent and intelligence to manage farming operations. In her study of female English cheesemakers, Sally McMurray argues that many dairywomen participated in all matters of decision-making, and occasionally some men turned the entire operation over to their wives so they could focus on other aspects of the farm or go out for day labour.[49] In British North America, the Hudson's Bay Company hired an experienced English dairywoman, Mrs Capendal, to run an operation at Fort Vancouver in 1835. When Capendal and her husband arrived, however, she was disappointed with both the cattle and conditions of the fort, and soon returned to England, having found things at Fort Vancouver "different to what she expected."[50]

Capendal's departure illustrated her managerial expectations and ability to protest when unsatisfied. Other women who moved to western Canada appeared more eager to embrace a new environment, learn new skills, and develop their talents outside of the domestic sphere. For instance, some woman who came to Alberta during the late nineteenth and early twentieth centuries had to develop their riding skills to visit friends and participate in everyday ranch activities. Violet Pearl Skypes's horsemanship allowed her to herd cattle and train horses, and Mary Inderwick cited her superior riding ability as the reason male employees accepted her authority.[51] Women's ability to "communicate knowledgeably about the tasks, the men, the horses, and the dreams for the future of the ranch" were vital for the success of their operations, but also to their personal happiness on the range.[52] Women of various socio-economic and geographic backgrounds had different experiences with livestock, but most rural women in Ontario had some relationship with and responsibility for the animals that formed the basis of the competitions at fairs.

Still, Ontario agricultural societies remained wary of introducing competitions that highlighted women's skill outside the domestic sphere. In the 1860s, the *Farmer's Advocate* published articles that criticized the women's equestrian events as fair attractions in the United States. In one satirical article published in the *Farmer's Advocate*, reprinted from an American newspaper, *Moore's Rural New Yorker*, the author, tongue-in-cheek, "congratulated" agricultural societies for hosting women's equestrian races, which it argued illustrated "how a modest woman, with her blood up, may be most skillfully thrown from her horse, heels over head, into the soil of a race track, mount again and win the applause of the refined throng who admire the performance, and the premium offered by the Agricultural Society to encourage and develop such skill."[53] Other reports of American women's horse racing were also published in the *Farmer's Advocate* and their "unnaturalness" was emphasized. In Decatur, Illinois, twelve women raced on horseback, "every horse under full run, the ladies were applying the whip, and the air was filled with hats, ribbons, laces, and "fixins," which were said to "have no place on the race track."[54] When one woman riding bareback was thrown from her horse, she was still able to remount and cross the finish line in third place. She received loud applause from the audience, despite being "covered with dirt by this fall, and her clothes torn almost in shreds."[55] Examples such as these served to warn against women's horse racing by illustrating both their danger and the unfeminine behaviour they occasioned.

Despite such criticism, by the late 1860s female riding and driving classes came to Ontario. These classes judged women's ability to perform functional tasks such as having a horse walk, trot, reverse, etc. Women's riding and driving classes were very different from the races that also took place at fairs. Both women and men could feel comfortable about a class that emphasized control, rather than a chaotic free-for-all to the finish line. Women's equestrian classes were created to evaluate their ability to handle a horse under saddle or behind a cart in much the same way they were expected to ride or drive during regular travel. People in the province usually drove rather than rode horses to make daily trips and visits, and this is likely the reason that "ladies' driving" classes were initially more popular than riding classes.[56] Still, both types of classes were promoted, and in 1868 the Beeton Agricultural Society allowed women to compete in "the special prize for riding" free of charge, to encourage their participation.[57] Although some women hesitated to showcase their horsemanship, driving and riding classes necessitated that women be seated unlike "line" classes wherein a leadsperson was required to walk and run their horse using a lead line and halter, the purpose of which was to better showcase the confirmation of the animal. Driving and riding classes also allowed women to maintain feminine decorum by wearing dresses and skirts without hindrance.

Women's riding and driving competitions became popular. In 1879, four women competed in the ladies' driving class in front of a packed crowd at the Cooksville fair. The winner, Mrs Ireland of Meadowvale, showed she was "truly an expert with the 'ribbons,' and managed her horse as few men could do."[58] The "ladies riding race" was reported as being among the "best of the sports" at the 1881 Peel County Fair and Miss Annie Lyons, who won the class, received a premium of $6 for her effort. The popular "lady driver class" earned the winner, again Ireland, "a pair of cabinet photo frames valued at four dollars."[59] At the 1881 Centre Wellington Agricultural Exhibition, "the great attraction of the day was the lady riders," and although only four women competed, the local newspaper reported that "the judges had more difficulty in selecting the best rider than in any other part of their judicial duties."[60]

Women's driving and riding classes offered organizers a way to entice people to the fair without succumbing to more contentious forms of entertainment such as vaudeville shows or carnival acts.[61] Although some reports of women's riding and driving classes emphasized their sensationalist quality, many local fairs offered such classes by the late nineteenth century.[62] Women who rode or drove horses were not necessarily novel, but their doing so for money in front of a grandstand of people was. The potential drama that resulted from a difficult horse with an unskilled rider provided additional entertainment. Working with horses could be dangerous, and although some people questioned women's suitability for the ring, others were fascinated by it. The *Brampton Conservator* reported that at the 1880 Burwick Fair, "One

Unlucky Accident occurred. Miss Lizzie Chafor's horse fell and Chafor's leg was broken just above the ankle."[63] The risks of competing were clear, but often this simply helped to further sensationalize the event.

Ladies' driver classes were featured in advertisements for the 1884 Central Exhibition in Guelph alongside "Indian Club Swinging and Bar Bell exercise," demonstrations, bicycle races, and music performances.[64] The Wallacetown Fair in 1889 featured a "lady drivers" class with a hefty purse of $9 alongside men's racing.[65] A report of the 1910 West Elgin Fair celebrated the fact that the fair was "a purely agricultural exhibition, devoid of the amusement element that is so necessary to many other fairs to maintain their existence,"[66] but the reporter failed to note that the fair's success also relied on the popular "Speeding Events" and classes such as the "best lady driver" in order to draw crowds.[67] Even in the twentieth century, women's driving classes maintained an element of novelty that set them apart from other horse show events. The *Aylmer Express* reported that at the 1921 Springfield Fair, various attractions featured in front of the grandstand included music by the local band, horse racing, and a "contest for lady drivers" in which a local woman, Mrs Anson Chambers, won first prize. A similar contest for women under sixteen years of age took place in which Miss Mary Winter won first and Miss Viola Chambers second prize.[68] Along with band performances, speed races, and other novelty contests, women's driving classes were singled out as being special fair features. The "best lady driver" was a popular class because it provided an unusual spectacle for fairgoers.[69]

Reporters rarely failed to describe women's appearance in these competitions. Female exhibitors, like men, were expected to create an attractive image in the ring by showing a well- turned-out cart or buggy, horse, harness, and becoming personal attire. Unlike men, however, women were expected to bring grace and beauty to the show. The local newspaper described the female drivers at the 1910 Wallacetown Fair who "lined up before the judges and spectators" as making "a picture that would be difficult to excel."[70] When J. Lockie Wilson, the superintendent of Ontario Agricultural and Horticultural Societies, wrote to the province's agricultural societies outlining rules to follow when judging lady drivers, he explained that, in addition to sitting in the carriage and holding the reins and whip appropriately, a lady driver should also sit in "a comfortable, yet graceful position." He suggested that the competition should require the lady to receive a passenger, and also make an exit from and entry into the carriage. Lockie Wilson described the precise actions judges should watch for, including where the lady's hands should be placed on the reins or her feet on the carriage, noting that judges "should see that this is done correctly and gracefully."[71]

Organizers and fairgoers emphasized women's appearance and demeanour as a way to promote acceptance of their place in the ring.[72] Fair organizers' desire for attractive participants perhaps prompted the 1938 Hanover Fair Horse Show Committee to award the first-place lady driver with a "Permanent Wave, value $5.00."[73] Similarly, the 1899 Arran-Tara Fair organizers awarded the winner of the "Best Lady Rider" class with a pair of embroidery scissors.[74] Horsewomen could not distance themselves too far from such feminine characteristics as grace and domesticity. Although Ireland's impressive driving skills won her numerous prizes and a reputation, recognition of her ability was tempered by descriptions of other women's driving classes that focused on the appearance of the ladies or the novelty of their competition.

By the 1920s and 1930s, horse and livestock competitions were becoming less gendered. Women now competed alongside men in light horse and pony driving, riding, and line classes. Exhibitors who entered driving and riding competitions were usually judged on their skill as well as their horses' conformation and ability. Line classes, however, were focused solely on conformation. Leadline showmanship classes had emerged by this time,[75] but generally line classes were used to judge the physical attributes of a horse (height, width, strength, sounds feet and legs, etc.). Women who exhibited in these classes were emphasizing both their contribution to the horse's breeding and their own ability to show horses to their best advantage, otherwise they would not have exhibited the animal themselves.

Women who competed in the general horse show demonstrated that they were not afraid to compete alongside men. The horse show prize list from the 1924 Erin Fair recorded that Mrs J.W. Snow won second for the colt or filly foal in the carriage horse competition, and Mrs R. Sloan won third for the brood mare class and third for the colt or filly foal in the roadster competition.[76] At the 1938 Erin Fair, Miss Mary Ansley of Brampton won first place for her half-breed brood mare and Miss Vivian Clark of Norval for her two-year-old gelding or filly in the line classes.[77] All these women were competing against men to win these awards. The horse show directors and judges continued to be men, but more women were exhibiting. Some women showed with their husbands, promoting the stock raised on their family farms, while others showed individually, highlighting women's ability to stake their own claim in the industry.

Rural School Fairs, Boys' and Girls' Clubs, and 4-H: Training Women to Show in the Twentieth Century

By mid-twentieth century, more women were showing horses and other types of livestock, but the progress continued to be slow. When the first rural school fair was introduced in 1909, it influenced agricultural societies to consider offering more horse and livestock competitions for both sexes.[78] Rural school fairs sponsored by the Ontario Department of Agriculture taught children how to show horses and livestock, and young women were therefore accustomed to the show ring when the agricultural societies held their fairs.[79] At the 1920 rural school fair in Vaughan Township, York County, livestock classes included Barred Rock chickens and roasters (hatched from eggs supplied by the Ontario Agricultural Department), heavy horse foals, light horse foals, purebred or grade dairy calves, purebred or grade beef calves, and a lamb class.[80] Generally the animals shown were under one year of age. Students had to train their animals and ensure they were halter-broken before show day. Some classes were judged solely on the animal's conformation, while others were meant to evaluate students' handling skills. Showmanship classes emphasized a child's ability, rather than the appearance of the stock owned by their parents. Also, unlike competitions for "Manual Training" and "Baking and Sewing," the first being for boys only and the latter for girls only, the livestock and horse competitions were open to both sexes.[81] Boys and girls competed against one another for top prize.

At the same time that rural school fairs were becoming popular, Boys' and Girls' Clubs, the forerunners of 4-H Clubs, which were also government and community-sponsored, became widespread across the province. Often their "Achievement Day" shows were held at fairs sponsored by agricultural societies.[82] Boys' and Girls' Clubs' tended to segregate the sexes in the first half of the twentieth century; girls participated in homemaking and garden clubs and boys participated in crop, machinery, and livestock clubs. This arrangement was typical, but not universal.[83] For example, at the 1917 Erin Fair, a special class was offered for "Calf, pure bred or grade, to be fed and cared for by a boy or girl under seventeen years of age for at least six weeks before the Fair."[84] There was also a class for "Two pigs, Bacon Type, pure bred or grade, to be fed and cared for by a boy or girl under seventeen years of age for at least six weeks before the Fair."[85] In 1919, a Boys' and Girls' Calf Club was established in Grenville County, and for the project both boys and girls purchased young Holstein heifers and recorded their milk production.[86] In 1921 in Elgin County, it was reported that there was a St Thomas District Boys' and Girls' Purebred Livestock Club,[87] and at the 1931 Elgin County Fair, a junior section in the dairy cattle competition offered classes for both boys and girls to show their heifers. The major prizes that year for the "Best showman of dairy cattle under twenty-one years," "best dairy type calf over six months, and under one year," "best purebred Holstein calf," and the "best calf raised on Royal Purple calf meal" all went to female exhibitors.[88] The first Boys' and Girls' Club in Leeds County was the Brockville District Pig Club, which began in 1923 with eleven male and three female members between the ages of fifteen and eighteen.[89]

At other local Ontario fairs sponsored by agricultural societies, however, competitions remained gendered. There was still only a Boys' Calf Club at the 1938 Aylmer Fair,[90] and at the 1940 Belmont Fair it was a "Boys' Calf Club and the Boys' Foal Club" that competed.[91] In Victoria County, it was not until 1939 that the first co-ed club was formed when the Poultry Club accepted female members.[92] From photographic evidence of beef and dairy calf clubs in the 1930s, it appears that most, if not all members, were boys.[93] Agnes Foster's account of her club work in the 1930s confirms the division that existed. She explained that garden and homemaking clubs, as well as public speaking contests were acceptable for girls, but livestock clubs were not, a fact she regretted deeply. "I had five brothers, all of whom were involved with agricultural clubs. Since I knew how to show a calf, lead a colt, choose show potatoes, and tie a sheaf of grain for competition, I worked with them on their projects. I watched with envy while the boys brought home shields, trophies, silver trays, won trips to Chicago or the Royal 500, and participated on winning judging teams in county, provincial, and Canadian competitions. Oh how often I wished I had been born a boy."[94]

Foster's experience was likely similar to many young women across Ontario. They may have done the same work as their brothers on the farm, but they were not supposed to display their skills publicly. Although some young women participated in livestock shows at this time, it was still seen as curious. In 1937, a newspaper reported that "Miss Bompas of Bells Corners represented alone the fairer sex amongst more than a score of men and youths as she smartly showed a fine Ayrshire heifer at the Ottawa Fair." The report noted that Bompas did not "mind being the lone girl in the ring, with the hundreds of pairs of eyes watching" and that "Naturally there would be the odd feminine touch. Miss Bompas had a brush handily set in a dress pocket and now and then she would whip it out and carefully brush the coat of her Ayrshire. She kept it prettied up as she walked around."[95] The article neglected to report on Bompas's final placing in the class, and even though it notes how "smartly" she exhibited her heifer, the fact that she was the sole female exhibitor and the "cynosure of all eyes" suggests that her participation was very unusual. The reporter tried to dissipate any anxiety caused by this unusual presence by describing her feminine attributes, such as her focus on keeping her heifer "prettied up" while they walked around the show ring.[96]

As Agnes Foster noted, and photographic evidence indicates, it was usually boys who were afforded the opportunity to travel to fairs to compete and develop their skills, as well as to meet others in the industry, which benefited them later in their careers.[97] Boys participated in these competitions because, as future farmers, they were expected to use what they had learned in these clubs to be successful in their future husbandry. Draft horse clubs were especially male-oriented. Although the junior section of the 1935 Erin Fair beef cattle competition was open to both boys and girls between the ages of seven and twenty years, in the heavy horse division, the Wellington Foal Club required that members be "boys between the ages of twelve and twenty years at time of entry."[98] At the 1938 Manitoulin Fair, the Kagawong Boys' Foal Club was also strictly male. Why girls could show calves but not draft horse foals is significant. Girls were not supposed to be active participants in field work on the farm, which often involved the use of heavy horses for plowing fields, planting or harvesting crops, or pulling logs from the woods. Furthermore, the size and strength of draft animals were considered reasons enough to discourage women from showing them (see Figure 5.2). In her work on plowing matches, Catharine Anne Wilson examines how men viewed plowing as a "manly art," where "rugged, independent work and

Figure 5.2 Heavy Horse Stake Class at Erin Fair
This photography shows a line-up of heavy horses led by men. It was extremely rare to find women exhibiting draught animals for most of the nineteenth and twentieth centuries. "Heavy Horse Stake Class," 1949. Erin Agricultural Society Private Photo Collection. Image courtesy of the Erin Agricultural Society.

mastery over nature were the core tenets of rural masculinity."[99] Heavy horses were symbols of men's work and strength, and therefore they were not suitable for women to exhibit.

Some people struggled with the idea of including girls in activities meant to prepare boys for their responsibilities as farmers. From the beginning, youth agricultural clubs were meant to teach boys about agricultural technology, crop innovation, and animal husbandry. Domestic science associations and clubs for girls were premised on women's duties in the home and community. Boys' and Girls' Clubs were modeled after earlier American youth organizations,[100] as well as rural Ontario organizations such as the Farmer's Institutes and Women's Institutes that separated the sexes into agricultural and homemaking activities.[101] In her study of the Women's Institute in Ontario, Margaret Kechnie argues that even though women themselves believed that "farm women's work should not be limited to the home" and that they "should learn about crops and soil and everything else around the farm," nonetheless "providing farm women with agricultural education was never a part of the extension education offered through the WI."[102] Agricultural youth organizations initially had trouble deciding how inclusive they should be. While it was fine to promote a girls' "Homemaking" or "Gardening" club, there seemed to be more apprehension about livestock husbandry training for girls.

Boys' and Girls' Clubs were changing, however, and by the 1940s girls started to join livestock clubs (see Figure 5.3).[103] During the Second World War, women, especially young women, were tasked with playing a larger role in agricultural production at home or in organized groups, such as the Farmerettes or the Women's Land Brigades, while men were away.[104] In addition to the other rights women were gaining in the mid-twentieth century, they also achieved more opportunities to showcase their farming work.[105]

In 1952, Boy's and Girl's Clubs became 4-H Clubs, although the work that continued under the name "4-H" was split into two divisions, the 4-H Girls' Homemaking Clubs and the 4-H Agricultural Clubs.[106] Maintaining the designation of "Girl's Homemaking Clubs" was important to Miss Florence P. Eadie, who served with the Ontario Department of Agriculture for thirty-six years and was credited with establishing the format of the Homemaking Clubs. She was adamant that to maintain "the integrity of the Homemaking Clubs," the word "girls" had to remain in the title.[107] This decision, along with society's general reluctance to see men perform domestic tasks, ensured hat membership in homemaking clubs remained almost entirely female. 4-H livestock clubs, however, encouraged more girls to exhibit at the same time that they began serving in leadership positions for other organizations such as Junior Farmers. In 1952, Mae Todd was elected president of the Bruce County Junior Farmers, the top executive position in the county organization. She used her position to emphasize that the "aim or purpose of the Junior Farmer movement is to bring together rural youth ... At club meetings, boys and girls learn to take part in discussion groups, to learn proper business procedure in conducting meetings, and the most important factor is that they learn to co-operate and to work willingly with others."[108] Todd highlighted the shared interests and goals of girls and boys in Junior Farmers, explaining that "Many Juniors take an active part in seed and livestock judging competitions, curling bonspiels, hockey and softball series, debating and

Figure 5.3 Metcalf Calf Club
Young girls were now showing calves, although their feminine dress was not compromised by taking up the lead line. "Carleton County, Metcalf Calf Club–1942." 4-H Collection, RG 16-275-2, Container 1, Archives of Ontario. Image courtesy of the Archives of Ontario.

public speaking events, choir concerts, as well as our Annual County Barn Dance held at Paisley in October."[109]

The inclusion of girls in livestock judging competitions was important because it signified that women were capable of accurate observation, sound judgement, courage, honesty, and clear communication – characteristics that had long been attributed solely to men.[110] In the 1950s, a film captured the inter-county judging competition at the Ontario Agricultural College in Guelph. Most participants were male, but one segment of the film showing participants judging poultry focuses on a young woman handling and inspecting a chicken. The film's commentator announces, "if you've already noticed, boys aren't the only ones who win places on competition teams, and this girl is not going to be outdone by any boy."[111] 4-H livestock club leaders – who were all male during this period – increasingly encouraged their female members to compete in judging competitions (see Figure 5.4).[112]

Figure 5.4 4-H Members Judging Livestock at Shedden Fairgrounds
Three young 4-H members are pictured judging Suffolk ram lambs at the Shedden Fairgrounds. In the postwar period, many young women participated in livestock showing and other competitions at fairs. "Annual Livestock Judging in Shedden," 1960. *St. Thomas Times-Journal* fonds, C9 Sh2 B1 F46 13a, Elgin County Archives. Image courtesy of Elgin County Archives.

By the 1950s, female members of 4-H clubs were regularly winning cattle competitions and receiving accolades for doing so. In 1954, Katherine Merry received significant attention when she beat 180 other 4-H members to win the prestigious Queen's Guineas at the Royal Winter Fair. The Milton newspaper, *The Canadian Champion*, reported that Merry's win "was indicative of the high regard and esteem in which [she] is held by her fellow 4H Club members throughout Ontario." The reporter described Merry as "a deserving winner – for six years she has been a top club member – once before she won the reserve award. Our readers may be interested to learn that nearly every night for the past six months, Katherine has walked her steer two to three miles. Halton may well be proud of Katherine."[113] Sandra Duffy won the coveted top showmanship award for the Elgin 4-H Dairy Calf Club at the 1958 Aylmer Fair in what was reported to be "an exceptionally outstand-

Figure 5.5 Elgin 4-H Dairy Calf Club Showmanship Winner Sandra Dufty Sandra Dufty of St Thomas won the showmanship award in an "exceptionally outstanding class." "Event Winners," 1958. *St. Thomas Times-Journal* fonds, C8 Sh2 B2 F3 6c, Elgin County Archives. Image courtesy of Elgin County Archives.

ing class" (see Figure 5.5).[114] At the 1959 Rodney Fair, Aldborough 4-H Junior Calf Club member Patricia McLean won both the showmanship and the Shorthorn yearling steer classes. The report of the show acknowledged that McLean and her steer "Moondoggie" had beaten a large number of excellent entries, and that the pair went on to compete at the Royal Winter Fair that year (see Figure 5.6).[115]

Figure 5.6 4-H Aldborough Junior Calf Club Showmanship Winner
Patricia McLean

At the 105th Rodney Fair, Patricia McLean took first prize in showmanship and first place in the steer class with her one-year-old Shorthorn, Moondoggie. Following her success, McLean planned to enter her prize-winning steer to the Royal Winter Fair. "Rodney Fair - Thousands Attend," 1959. *St. Thomas Times-Journal* fonds, C8 Sh3 B1 F11 2a, Elgin County Archives. Image courtesy of Elgin County Archives.

Female "Showmen" Show Men What They Can Do: The Postwar Period

Adult women, not just young women, were also becoming more active at fair cattle shows during this period. Mrs McKay could often be found in the beef cattle ring along with her husband James McKay at the Erin Fair, and many other local fairs. Mr McKay was the herdsmen of the successful Aberdeen Angus farm, Malden Farms, owned by Dr W.F. James, but both he and his wife managed the cattle and the farming operation. When the bull owned by Dr James won the Royal Winter Fair in 1953, it was Mr McKay, with Mrs McKay at his side, who first accepted the award.[116] The passion the McKays had for cattle was transferred to their daughter Evelyn, who also participated regularly in the cattle show ring.

Women were also more active in the horse ring in the postwar period. The exhibition of heavy horses remained a masculine pursuit,[117] but women thrived in the pony and light horse classes. Gender-specific classes for lady riders and drivers still existed, with prizes that were equally gender-specific. The first prize for the "Lady Rider" class at the 1949 Erin Fair was a Danby Toaster, and the winner of the "Lady Driver" class won Weston Bakery merchandise.[118] Generally, female exhibitors competed alongside men in regular light horse classes. Mrs Mary Welsh, Mrs Irene Wheeler, Mrs Mary Woods, Miss Joyce McMillan, and Mrs Edna Day all took home prizes in the general light horse division classes at the 1949 Erin Fair (see Figure 5.7).[119] Women also owned, and sometimes bred, the animals they exhibited. At the 1950 Ancaster Fair, Mary E. Welsh of Brantford proudly posed for a picture with her winning saddle pony, Jimmy Fleetfoot.[120] Women were now actively competing with men in many horse show divisions.

Female riders had begun to compete at elite equestrian competitions around the mid-twentieth century, and their success contributed to the acceptance of horsewomen on the fairgrounds. At the 1949 Royal Horse Show, Mrs Dorinda Hall-Holland of London, Ontario, beat professional riders to win the open jumping stakes class. Her win provided a thrilling evening that left "No yawning, squirming spectators that night!"[121] The following year, Eva Valdes competed for Mexico, becoming the first woman on an international jumping team.[122] It was not long before women in other countries

Figure 5.7 Women Competing in Carriage Class at Erin Fair
In this class for the best carriage horse under 15.2 hands, women and men competed side by side. In this picture, women outnumber men. "Carriage under 15.2," 1947. Erin Agricultural Society Private Photo Collection, Erin, Ontario. Image courtesy of the Erin Agricultural Society.

also won spots on their national teams. In 1952, Mrs George Jacobson of Montreal was selected for the Canadian Show Jumping team, becoming the first woman to represent Canada internationally.[123] The next year, seventeen-year-old Shirley Thomas of Ottawa was chosen to join the Canadian team and competed successfully at Washington, Toronto, and Harrisburg, New York. It was recorded that the "excellent performances of the very young, good-looking Ottawa girl were given extra applause."[124] Women's appearance remained a point of interest for reporters, illustrating the attention placed on women's looks, but their talent in the show ring could not be denied, and encouraged other women to follow suit.

Gender Identity Persists in the Show Ring

Despite women's success in the show ring, stereotypes remained that emphasized women's appearances. In the 4-H literature and handbooks distributed to members, farmers were typically depicted as male, especially when performing tasks such as driving tractors, working with machinery, and selecting and judging breeding stock.[125] Although girls and young women were now participating in work previously thought to belong to men, such as working with cattle and learning farm equipment safety, agriculture as a profession remained a masculine domain and gendered ideas about men's and women's roles continued. Even when young women found success showing cattle, gender distinctions were obvious. For example, it is unlikely that had a boy won the reserve champion junior female calf prize at the Erin Fair, the newspaper headline and caption would have read as it did when Judy Merry won that honour in 1952. The newspaper article, titled "Beauty and the Beast at Erin," commented that "Pretty girls were only one of the features of the Regional Shorthorn Show held at Erin on Monday. There were many fine-looking cattle, too, in the opinion of the judges." The report noted that Judy Merry, the daughter of a well-known Shorthorn breeder in Oakville, W.H. Merry, had "been showing cattle for many years" and that she "pleased connoisseurs both ways."[126]

Despite the differences that existed and the barriers that remained, the young women who participated in livestock clubs and showcased their talent at fairs helped pave the way for women's further involvement in horse and livestock showing, as well as in the industries more broadly. Girls who grew up being members of 4-H cattle, tractor, or crop clubs pursued their passions in these fields for many years. Switzer recalled how her participation in 4-H helped her believe that she could do anything she set her mind to. Born in 1950, Switzer described how she was treated no differently than the boys in her 4-H clubs, which included livestock and tractor clubs. Switzer had been raised on what she termed "an equal opportunity farm." Her mother was the one most interested in livestock, and while she was responsible for the turkeys and laying hens, her passion was the dairy herd. She milked, kept records, and made the breeding decisions. Switzer recognized that her parents were "really ahead of their time," and that she was fortunate to have grown up with a mother who modelled this behaviour and father who encouraged her "to do whatever [she] wanted to do." She also credited the 4-H clubs she belonged to for creating an atmosphere where boys and girls competed on equal footing. Switzer described how "4-H opened up so many avenues for me." She stated that she could do anything she wanted in 4-H, and that "it wasn't until I got out into the other world that I found out that that's not the way it is in the real world ... I found out men weren't all like this ... that they had no idea a woman could do this." For Switzer, her ability to show a calf or drive a tractor was a source of pride. Her family and her participation in 4-H taught her that there were no limits to what she could achieve, on the farm or otherwise, so it was even more startling for her when, as a young professional woman and later as a married woman, she met people who questioned her abilities and discouraged her from performing tasks that were supposedly not women's work. She recognized, however, that there were changes taking place. Switzer recalled, while "it would have been unheard of to have women on the cattle committee in 1960," in time women became important committee members and participants in most aspects of the fair, especially in livestock competitions.[127]

Another woman whose experience showing cattle led to a lifelong commitment to agriculture was Phyllis MacMaster. MacMaster was born in Hawkesbury, Ontario, the first of seven children in her farm family. Her mother was responsible for domestic tasks but she also worked in the dairy barn, and when MacMaster was asked about her childhood recollections, she explained that in her family, "you went with your parents to the barn. I think that's how I developed my appreciation for agriculture and why I worked in that industry."[128] MacMaster noted that from a young age she helped in the barn by feeding calves and pushing in hay to the cattle. But it was when, in her teenage years, MacMaster joined 4-H and experienced the excitement of showing cattle that she really developed a passion for the dairy industry. She also participated in homemaking clubs, but 4-H Dairy Club competitions were especially exciting and she enjoyed competing in the show ring. MacMaster later went on to become one of a few well-recognized female dairy-cattle judges, and she was proud that she was often asked to provide her placings and reasons at dairy-judging schools and competitions. She explained: "That gives me recognition that I'm doing a good job, that [your peers] value your opinion." After she received her bachelor of science degree in Agriculture at the University of Guelph, she became one of the first women to work in the Ontario Ministry of Agriculture and Food (OMAF) and went on to become the first female agricultural representative in the province in 1984. MacMaster credited her career choices to her early love of farming, and believed she benefited from a shift in public opinion in the late 1970s, when women became increasingly recognized for the work they did on the farm. When she stepped into her position as agricultural representative, she felt her employers "had a lot of confidence that [she] would be able to work and develop relationships with farm families."[129]

Conclusion

Despite the positive experiences remembered by June Switzer, and the significant advances for female horse and livestock exhibitors, women who exhibited farm animals remained a minority. Although youth clubs encouraged more girls to take up the lead line at livestock shows, the idea that men were breeders, managers, and farmers, and women were caretakers and helpmates, did not disappear. Young women gained access to the show ring, and even older women took up the lead line, but the power relations remained uneven – men were still the ones organizing, officiating, and judging these events, as well as making up the majority of exhibitors. Furthermore, the stereotype of the hardy, masculine farmer continued, and despite significant strides in women's sense of belonging and authority in livestock industries, the gendered dimension of food-animal production limited women's ability to be full stakeholders.[130]

Over time, however, rural communities accepted that at least some women were talented show people and knowledgeable exhibitors in the horse and livestock show rings. Once seen as novel participants whose femininity limited their ability and credibility, women were starting to become more accepted as farm partners and showpersons.[131] The transition to gender equality in agricultural industries is not complete yet, but in the second half of the twentieth century, women built on earlier progress to significantly advance their participation in equine sport and livestock competitions. Like women's experience on agricultural society fair boards, women who exhibited farm animals found that gaining recognition and respect was often slow, but they nevertheless embraced opportunities to move into arenas previously closed to them. By the 1970s, a masculine agricultural system and separate gender expectations remained,[132] but women were making advances and building on the achievements of the women who came before them – including acquiring increasing numbers of first prize ribbons in the show ring.

6

From Church Booths to Baby Shows
Women on Display

Whether women were fairgoers or fundraisers, performed feats of acrobatics and skydiving, or simply lived long enough to win awards for being the oldest people on the fairgrounds, their participation at fairs nourished, amused, and entertained. This chapter examines how women contributed to fairs beyond their participation as agricultural society members, directors, and executive officers or as exhibitors in agricultural and domestic competitions. They also used fairs to meet with friends, neighbours, and potential suitors, fundraise for community organizations and causes, make money as professional entertainers, perform athletic, musical, and dramatic talents, or display their abilities as successful mothers, Dairy Princesses, or Fair Queens.

This chapter also illustrates how women's range of activity on fairgrounds increased. In the last chapter, women moved beyond the exhibition hall to enter the livestock and horse barns and show rings. This chapter shows them moving beyond the dining, theatre, and exhibit halls and into the midway at the turn of the century and in front of the grandstand by the interwar years, becoming visible in every space of the fairgrounds.

Women Managing Fair Banquets and Fundraising with Food and Fun

In the nineteenth century, women typically supplied food to fairgoers by serving lunches and dinners at agricultural society banquets, and agricultural societies paid for their service. By the twentieth century, they volunteered to sell food on the fairgrounds to fundraise for women's organizations and the causes they supported. Groups such as the Ontario Women's Institutes, Imperial Order Daughters of the Empire, Christian Women's associations, Women's War Work Committees for the Canadian Red Cross, and many other church and community groups benefited. Selling food was a popular way for women to fundraise because, as any fairgoer would have told you, a trip to the fair was incomplete without some sort of delicious treat. Cheryl MacDonald notes in her history of the Norfolk County Fair that "Food has been so much a part of the fair that it has been mostly taken for granted,"[1] and she suggests that few records specify precisely what was served. However, local newspaper accounts, fair board minutes, Women's Institute minutes, and oral histories provide details. Dessert foods like pies, cakes, ice cream, as well as sandwiches or more hearty meals like roast beef dinners were popular fare. No fair was complete without a selection of delectable foodstuff. Indeed, rural Canadian communities have a long history of celebrating with food. Harvest festivals and banquets were held during times of celebration and thanksgiving,[2] while individual families expressed appreciation of their neighbours for helping during work bees by hosting hearty meals.[3] At fairs, women utilized those same culinary skills to feed fairgoers.

In the nineteenth century, most reports about fair food focused on agricultural society banquets held at the end of fairs at local hotels. Initially, women were not guests but workers who served items such as "roast beef and plum pudding, with their various accompaniments."[4] The proprietors' wives or female staff did the cooking, not farmers' wives or women on the fairground. Guests made speeches to the "gentlemen" in the room, and published lists indicate an all-male attendance. Agricultural societies hosted banquets for male members, directors, and guests to enjoy one another's company, congratulate each other on jobs well done at fairs, and reflect upon agricultural improvements made in the community and province. At the 1880 Exhibition dinner of the West Riding of York and Township of Vaughan Agricultural Societies held at the Woodbridge Hotel, John Abell, the president of the society and a local implement manufacturer, congratulated society members for the agricultural improvements they had made. The group then toasted the Queen and the royal family, and sang the national anthem as well as a song entitled "Jolly Good Fellow" and a "Scotch song which was much enjoyed." The fair's judges also gave speeches that elicited "peals of laughter," followed by a pledge to the Toronto Industrial Exhibition and an address from Captain McMaster advising farmers to "make their boys farmers instead of clerks."[5] Although loyalty and service were equally familiar to women, they had neither the political power nor public authority to speak about future agricultural progress. Furthermore, the dinners' raucous nature, the boisterous singing, and much-noted drinking were not seen as respectable for women.

At banquets, men asserted their right to lead by reflecting on past activities and charting the course for the future without women's interference. Women's efforts were sometimes acknowledged, as demonstrated by "the enthusiastic and vociferous manner in which they were toasted during the *banquet*"[6] of the 1857 Toronto Township Fall Fair. Still, when men sat down to a fair dinner, discussion focused on improved cattle breeding and farmland, not how to better educate women on the production of domestic goods or advance their skills in lacework or beading.[7] Women's work was considered worthy of display but not given serious consideration by agricultural societies.

By the end of the nineteenth century, agricultural societies began to invite women to fair banquets. At the 1892 Woodbridge Fair, the wives and daughters of agricultural society members and prominent guests attended "an elaborate banquet" in the Agricultural Hall on the fair's last evening. Guests made speeches and drank to "her Majesty's health,"[8] but the boisterous atmosphere of earlier dinners was absent. By this time, fairs had become much more than livestock shows; they were now symbolic of rural families' interests rather than being exclusively focused on men's agricultural and industrial pursuits.

In the twentieth century, women's groups, rather than independent innkeepers or hoteliers, often organized fair banquets. The Women's Institutes became known for their catering services, which allowed them to raise funds while also fostering rural sociability and community service. Agricultural societies eagerly sought their help to provide affordable meals for the burgeoning crowds that attended fairs. Many fairs and communities now had their own halls, which they used to host dinners, so hotel space was no longer necessary. In 1913, the Shedden Women's Institute organized the Shedden Fair's banquet, and they continued to do so in subsequent years in addition to selling food from booths during fair time.[9] When agricultural societies established "lady director" positions on the fair board, female members often took up organizing lunches and dinners and were the ones who served their male counterparts, who still expected women to play a "helpmate" role.[10]

Fundraising was the main reason women's groups served meals on the fairgrounds. The First World War was a catalyst for women using fairs to fundraise. Groups such as the Women's Institutes of Ontario, the Imperial Order Daughters of the Empire, and church Ladies' Aid Societies spent considerable time preparing and selling food to raise money for the war effort.[11] At the 1914 Wallacetown Fair, the IODE served lunch in the Fair's hall and made $95, which they donated to the Red Cross Fund.[12] Two years later, at the same fair, the Women's Institute "served excellent meals" which earned $90 for the Red Cross Fund.[13]

Women also expanded their mobility and visibility by taking their fundraising activities into the midway, where they organized refreshment booths alongside other concessionaires. At the 1915 Norfolk County Fair, both the IODE and the Red Cross managed refreshment booths selling fruit, soft drinks, wieners, and biscuits, and the Ladies' Aid Society of the local Baptist Church served hot meals in the dining hall.[14] During the Second World War, women's groups again used food to fundraise. At the 1944 Woodbridge Fair, sixteen of the Smithfield Friendship Ladies were "back with their little food counter (All Profits for War Work), and they sold about 100 pies, milk cans of tea and coffee, and yards of sandwiches. They baked the pies themselves."[15] During peacetime, women also sold food at fairs to fundraise. The Shedden Women's Institute continued to manage a food booth after the First World War to support various other causes. They continued to serve food in the interwar period, reporting in 1925 that they made a profit of $48.10 from meals served at the fair, and used the money to buy batten and lining for a quilt to send to a needy family.[16]

Agricultural societies relied on women to host luncheons, and women made extra efforts if these luncheons honoured special guests or dignitaries. In 1938, the Rotary-Kiwanis Luncheon in the Legion Hall was held on the third day of the Aylmer Fair, and Ontario Premier Mitch Hepburn was the distinguished guest. The Ladies Auxiliary of the Legion served more than one hundred people that year, which resulted in some having to eat "in the basement and hallway of the hall, standing up."[17] The next year, the Ladies' Auxiliary to the Canadian Legion again hosted the noon luncheon, and the *Aylmer Express* declared the meal a splendid success.[18]

Providing food for fairgoers was a natural extension of women's duties in the home. By the mid twentieth century, many women's groups hosted food booths at fairgrounds. In 1950, a newspaper article on the Bolton Fair emphasized that "Women's organizations are to the forefront at every well-conducted Fall Fair, and Bolton was no exception. Here members of the Women's Guild of the Presbyterian Church are hard at work satisfying the needs of the "inner-man."[19] A photograph accompanied the article that showed two women serving pie to customers, one of whom was particularly eager-looking. Men expected women to take care of their needs – especially when those needs involved food. Women's responsibility for feeding their families was extended to feeding their communities at fair time.

Sometimes competition between organizations ensued as each group looked to entice visitors. At the 1949 Woodbridge Fair, the Presbyterian Ladies' Aid booth was jam-packed with crowds of fairgoers.[20] Not to be outdone, at the other end of the grandstand was the United Church Women's Association. The newspaper reported that they also "had a busy day in their booth under the grandstand where they sold coffee, sandwiches, pie and other foodstuffs."[21] Diane Tye notes that production of food "brought women together and allowed them to contribute in concrete and very meaningful ways to their communities." She also highlights that it served to legitimate "their position as unpaid domestic workers, who were often pressured to perform important, but invisible, work that church and community members – women included – did not recognize as work."[22] While it is true that a great deal of women's volunteerism has gone unrecognized and was ignored as "real work"[23] at fairs, it was visible for the whole community to see and appreciate. And studies of rural women's volunteerism have shown that they themselves saw their work as critical to their communities' social and economic vitality.[24]

Women's food and beverage tents and booths competed against concession stands run by midway companies or other professional retailers who offered fairgoers everything from hot dogs, candy, ice cream, and soft drinks to tea, sandwiches, and full-service beef dinners. At early fairs, food and drink stands also sold "intoxicating liquors," which some people complained were indulged in "too freely" and caused quarrels. This led to the explicit banning of such sales.[25] Later on, most fairs were dry, but competition between food stands was often fierce. At the 1921 Aylmer Fair, "dozens of booths serving hot dogs, ice cream, soft drinks, etc.,"[26] were competing for fairgoers' patronage. The food provided by commercial concessionaires was very well liked among youth because they tended to serve candy and other novel treats. Children and teenagers, and even parents, were drawn to the midway, the space on the fairgrounds where visitors could enjoy sideshows, games of chance or skill, carnival rides, and other amusements,[27] so it was no surprise that they also indulged in ice cream, cotton candy, soft drinks, and other fast food and treats there.[28] Despite the competition women's booths received from the professional vendors, their stands were still popular because of their well-priced, wholesome, and delicious meals. The *Aylmer Express* notified Elgin County Fair fairgoers in 1932 that the Gleaners Class of St Paul's Church and the Aylmer Women's Institute would be serving food at reasonable prices.[29] Local newspaper fair reports thanked the "ladies" who "worked day and night serving many a hungry and thirsty fair-goer."[30]

Women also used the fairgrounds for fundraising in other ways. The women's auxiliary of the hospital in Aylmer conducted a tag day at the 1921 fair where "a goodly sum was raised by the ladies."[31] Charitable organizations often used tag days to solicit money in return for a tag inscribed with the group's slogan.[32] Individuals who purchased a tag for a small charitable donation made their support for a cause or association known. In 1939, the Aylmer Branch of the Canadian Red Cross ran a booth outside the Aylmer Fair's Crystal Palace and "had taggers mingling with the crowds throughout the duration of the Fair" to raise funds to "carry on efficiently" with their war work.[33] The newspaper also reported proudly that the Red Cross booth was attractively decorated with patriotic symbols such as the Union Jack and Red Cross flags.[34] Despite cold winds and rain, the Red Cross Taggers collected $277 in just two days. Fairgoers had "made a wonderful response to the Red Cross appeal for funds and registration of workers," and one woman, Mrs H.J. Hart, the winner of a car raffled off at the fair, shared her good fortune by donating $100 to the Red Cross.[35]

At fairs, women raised cash, advertised their charities, and encouraged others to get involved in their campaigns. Women's Work Committees of the Canadian Red Cross displayed items of war relief that had been made specifically to ship to Britain. The Aylmer and Malahide Red Cross showcased "a number of beautiful quilts, knitted goods, hospital supplies, articles of clothing, jam, etc." at the 1941 Aylmer Fair and posted a display card that announced that the Aylmer, Lyons, Luton, Lakeview, Kingsmill, Mapleton, and Bayham Women's Institutes had "made 5,800 lbs of jam for overseas and that 3360 lbs. [had] already been shipped through the local branch of the Red Cross" that year.[36]

Dances and concerts were also popular fundraisers. These events raised money for charity groups as well as encouraging sociability and community spirit. At the 1881 East Grey County Fair in Flesherton, the Ladies Aid Society held a concert in the Town Hall on the evening of the fair that was "a grand success" in raising money for the group.[37] At the 1907 Mount Forest Fair, tickets for the concert cost twenty-five cents. The program included a performance from the Mount Forest Concert Band, solo pianos, a Scottish comic song, humorous readings, and costume character songs.[38] In 1940, the Erin Village Red Cross hosted a successful concert the night of the fair. The show featured a play entitled "The Girl from Out Yonder" performed by the community's youth, and raised a profit of $20.[39] Although not an enormous sum of money, $20 was not insignificant and could buy a lot of yarn to knit soldiers' apparel.[40] Every little bit of money raised was important for charity groups, and their activities were also crucial for bringing the community together.

Rural women were often responsible for fostering community spirit and encouraging residents to do their part in supporting projects of local, national, and even international importance. Women's fundraising efforts could be as modest as raising money to make a quilt or as complex as coordinating aid for wartime efforts. Agricultural societies benefited from women who were willing to devote time and energy to feeding and entertaining fairgoers in return for raising money and awareness for causes they held dear. The women who participated in these efforts also received recognition for their work which helped cement their status in the community.

Special Attractions: Women Who Performed

Fair work was not always about serving others. Carnival shows were widely popular in North America by the late nineteenth century, and women often headlined as dancers, acrobats, and other performers.[41] That women were encouraged or employed to perform publicly signalled an acceptance of women's place in the spotlight. However, their performances differed considerably from those of local women. Professional entertainers were able to transgress traditional boundaries of femininity in a way that local women simply could not. Initially, township and county agricultural societies responded to fairgoers' desire for additional entertainment by adding women's driving and riding classes. However, when the public demanded more spectacle, women's performances grew to include many other types of activities. Whether female performers were seen as rare phenomena or as representative of broader social change, their visibility performing centre-stage advanced their public presence.

Throughout the nineteenth and twentieth centuries, agricultural societies promoted their fairs as opportunities for education, entertainment, and sociability. Still, community members often had difficulty agreeing on what kinds of entertainment should be allowed. The nineteenth century was when "the industries of spectacle" experienced a mass cultural explosion in North America.[42] Agricultural societies, like society generally, grappled with including entertainment sensational enough to draw in crowds which did not support their ideology.[43] In 1859, a reporter for the *York Herald* published an article that enthusiastically supported a local councillor who had "introduced, and carried a by-law to prevent, by a stringent license, the exhibitions of horse-riding, puppet shows, &c.," to "preserve and advance public morality." The reporter argued that "every exhibition that does not blend instruction with amusement" was disgraceful, and local officials had a responsibility to "see to it that our morals are not exposed to unnecessary temptation."[44] Rural reformers believed agricultural societies had to decide which amusements were intellectually and spiritually uplifting. While some agricultural society members thought more novel attractions were needed to increase gate receipts,[45] others lamented the existence of events that diverted crowds from the fairs' educational exhibits.[46]

Female performers were especially problematic for some fair organizers. H.B. Cowan, the provincial superintendent of Agricultural and Horticultural Societies, questioned how "women contortionists" could be an acceptable form of entertainment, and argued that their inclusion was "a disgrace, rather than a benefit, to the farmers of Ontario."[47] Some journals, such as *The Farmer's Advocate*, agreed with Cowan, insisting that only innocent attractions and educational features were in good taste, and congratulating fairs that forbade "serpent eaters" and "skirt dancers."[48] Female dancers, like other vaudeville performers, were thought to represent the moral decay and "vile culture of cities which hosted the larger exhibitions."[49]

However, most county and township fairs in Ontario maintained modest forms of amusement, and female performers were usually local women – amateurs, not professionals – whose presence was considered wholesome and uplifting. At the 1880 Burwick Fair, women were a central part of the dinner performance during closing celebrations. In the evening, a "grand concert" was held in the local Orange Hall that included a Woodbridge Brass Band featuring "lady singers" Mrs Machie and Miss Cunnington. Miss M.E. Fielding read a selection from "King John" and recited, by special request, "The Fate of McGregor." The entertainment portion of the evening was closed by Mr Meek, who thanked the performers and "spoke in high terms of the singing of … the elocutionary powers of the lady reader of the evening who he said possessed reading powers of considerable dramatic worth."[50] The 1891 Eramosa Fair Concert "drew a full house" to hear Miss K. Strong perform several songs, "much to the delight of the audience who appreciated her fine voice," including a solo entitled, "A Bird from over the Sea," as well as two duets with Mr Fax: "Master and Pupil," and "Learning to Read."[51]

Women also took centre stage to give educational talks to agricultural society members and fairgoers typically at a local theatre or a fair dining or exhibition hall. A local newspaper reported that the evening's speeches at the 1870 Minto Fair were good, "especially by a lady advocate of 'woman's rights,' and being strictly on teetotal principles, passed off very pleasantly."[52] Although the woman's speech focused on teetotalism rather than women's rights, the agricultural society's decision to include her illustrated that they were not deterred by her connection to the more controversial movement and acknowledged her "right" to express her opinion in a public venue.

Agricultural societies also sought women who exhibited unique, if not exotic, identities and talents. The *Liberal* reported that the highlight for visitors to the 1881 Markham Fair was witnessing "Mrs. Flanders, otherwise Aoewentaiyouh (Beautiful Land)," who "appeared in the pavilion in full Indian costume. This lady is a sister of Dr. Oronhyteka (Burning Sky)."[53] The *Globe* also reported on "Mrs. Flands, of Hagarsville," who they described as an "Indian lady" who exhibited in the "Ladies' Work" classes and won many prizes. The *Globe* also reported that she attracted a great deal of attention when she appeared in "beautiful Indian costume" before fairgoers.[54] Oronhyatekha, also known as Dr Peter Martin, a famous Mohawk (Kanyen'kehà:ka) physician, was a member of the Six Nations who led a rather extraordinary life from the time he was born in 1841 to his death in 1907.[55] Although little is known about his sister, it is recorded that their family was large and that the female members were skillful at fashioning clothing.[56] White settler agricultural society members applauded First Nations women for accepting "civilized practices" such as excelling at Westernized domestic pursuits, yet still identified Indigenous peoples as fundamentally different because of their "otherness." Thus Aoewentaiyouh appealed to white people who wanted to see First Nations peoples assimilated, yet were simultaneously fascinated with the exoticism they symbolized.[57] People also exhibited collections of international curiosities; for example, Mrs Robert Webster displayed Indian, Japanese, and Chinese curios at the 1891 Centre Wellington Fair.[58] These examples illustrate late-nineteenth-century Ontarians' desire to consume other peoples' knowledge, albeit in often superficial and limited capacities.[59] The entertainment and "colour" that racial performance and display brought to the fairgrounds were appealing to white people when they supported popular attitudes.[60]

By the twentieth century, the number of female performers employed at fairs continued to grow. However, they were very different from the guest speakers and amateur singers and dancers who had performed at earlier fairs. Instead of providing educational speeches and enlightening musical performances, these entertainers appealed to fairgoers' desire for more scandalous forms of entertainment. The *Aylmer Sun* reported that the circus attraction at the 1902 Aylmer Fair, which was "surely to please," included the Eight Cornellas, the Flying Moorer, and Mlle Marjorie's troupe of trained dogs.[61] In 1914, the Norfolk County Fair's main attraction was Miss Dorothy DeVonda, a skydiver whose performance the previous year was so popular that they hired her again. Despite suffering "slight injuries in an accident to her parachute at Dunnville recently,"[62] DeVonda made a "spectacular parachute drop from a height of 8000 feet," according to the *British Canadian* newspaper.[63] The same article described DeVonda as an "accomplished and daring young lady" who capably "performed her dangerous act."[64] Some agricultural societies were willing to employ women who performed dangerous feats if it guaranteed paying visitors. Still, neither the agricultural society member nor other fairgoers would have wanted the women in their family to act in such ways.

Local women continued to perform at fairs, especially in the ever-popular musical and dance recitals and pageants, which appealed to rural people's folk traditions or patriotic sentiments. Many of these performances emphasized residents' history and reinforced their settler colonial heritage. In John M. McKenzie's collection *Imperialism and Popular Culture*, Penny Summerfield notes that music halls helped reinforce imperial ideals, and many of the performances at fairs served similar functions.[65] For example, at the 1923 Norfolk County Fair, an audience of over 2,000 jammed the grandstand to watch the "Historical Pageant of Norfolk County," which featured "Hundreds of Norfolk's Prettiest Girls" in scenes that included "Pioneers in Covered Wagons," and "Governor Simcoe's Visit."[66] The pageant was praised for its "picturesque close with Miss Canada surrounded by her people and Miss Norfolk as one of her maids of honor." Local resident Marguerite Clark portrayed Miss Norfolk and Mary Spencer played Miss Canada.[67] At the 1924 Erin Fair, the Guelph Pipe Band furnished the musical program and accompanied a performance from a "Wee Lass," who "stepped to several good old Scotch and Irish reels and jigs, and received a real applause."[68] Many residents of Guelph were of Scottish and Irish descent, and they enjoyed performances that affirmed their heritage. Similarly, the popular "Old Time Fiddlers' Contest" at the 1938 Erin Fair attracted many contestants. Mrs Moffet of Glencross "got a 'big hand' from the crowd" and was awarded one of the top honours in the competition.[69] The *Erin Advocate* reported that the "step dancing and square dancing competitions also drew the attention of a large portion of the crowd" and that the "old time square dance" was especially popular. The dance, "gracefully performed by the four couples from Camilla, who won fame at Toronto Exhibition" reportedly allowed older residents to recall "the happy days of the farm home dance parties." At the same time, "the young generation got a big kick out of comparing the grace and style of the old time dance with the big apple and other fads of the day."[70]

As the twentieth century progressed, agricultural societies still drew fairgoers with modest forms of entertainment. Nevertheless, societies that wanted to attract visitors from further abroad believed they needed to employ professional entertainment. Some of the female performers who helped to entice large crowds to the 1935 Norfolk County Fair included the "Four Dancing Dolls," "Beth Watson, Canada's Premier Scotch Piper & Dancer," "Myrtle Collins, Acrobatic Dancer," "Joyce Brown, Comedy Singer and Dancer," "Ruth and Joyce, Adagio Dancers," and "Jan and Merle, Adagio and Apache Team – Stars of the National Motor Show."[71] The 1938 Aylmer Fair featured more than "16 big acts with more than 50 people," which included the "high-class" vaudeville acts the Goldettes and Dainty Estrellita. The Goldettes were "beautiful dancing girls in gorgeous costumes," while Dainty Estrellita was "the toe dancing star of her age."[72] These "beautiful dancing girls" and the other musical, acrobatic, and comedic acts drew packed the grandstands to the point that many fairgoers found standing room only.[73] In 1941, the show in front of the grandstand was again one of the main attractions, and among the trained animals, high-wire performers, acrobats, magicians, and clowns, the "dance numbers by gorgeous girls" were touted as "two hours of real good fun and entertainment."[74]

In the postwar period, professional female entertainers continued to be featured. At the 1949 Orono Fair, fair organizers hired Barbara Fairchild and her horse War Paint to wow the crowd with her trick-riding exhibition. The newspaper report emphasized that the "Pretty Miss Barbara Fairchild" gave a skilled performance, but also an "eye pleasing" one.[75] Even though female equestrians were more common by the mid twentieth century, Barbara Fairchild's trick riding transgressed gender roles by demonstrating her competency in a dangerous form of the sport. Like Dorothy DeVonda, whose skydiving act pushed the limits of acceptable female behaviour, Fairchild was able to perform such "masculine" skills because she was an outsider. Local women who indulged in similar acts would have been transgressing acceptable standards of morality, propriety, and refinement, aspects of good taste that were encouraged and rewarded in exhibits such as food, flowers, and fancywork. To ease the anxiety created by women who subverted gender norms, reporters attempted to "ground them in the conventional world" by emphasizing their more traditional feminine attributes.[76] For instance, reporters stressed Fairchild's attractiveness as a way to reassure fairgoers of her femininity.

Other professional entertainers profited from society's appreciation for female beauty and sexuality. Although local women were expected to be modest in dress and behaviour, "outside" women were engaged to display their "gorgeous and shapely" bodies. Such performances were not without controversy, but professional shows such as the "Vive Les Girls" performance at the 1958 Norfolk County Fair profited from the public's desire to see "beauty and form in motion."[77] Not all Ontario fairs hired professional performers. As *Canadian Statesman* columnist Ed Youngman explained, when comparing the Orono Fair with the Canadian National Exhibition in 1963, "[the CNE] quit 'girlie' shows this year; we never started them; always being content to do our ogling in the open air, and using our imagination as to what's under the dresses."[78]

Fairs across Ontario featured female entertainers in front of the grandstand or in the concert or exhibition hall. Some were local women whose performances typically conformed to normative feminine behaviour and accomplishment. Other women – usually outsiders who were professional entertainers – demonstrated that women could do daring, physical acts or profit from society's not-so-hidden obsession with sex and beauty. Some women sought opportunities to show their talent, others simply to make some money; some women supported normative gender ideals and others transgressed them. Although all performances enlarged women's access to the public stage and affected societal views of women to some degree, not all women had the same opportunities to move beyond traditional notions of femininity.

Other Contests for Fair Women

Beyond entertainment and food, agricultural societies encouraged local women to display their skills by competing in contests, most of which were added in the twentieth century. These contests were not about production but about performance. Women won awards for various skills such as musical talent, beauty, athleticism – even milking a cow. These contests were entertaining, but they also displayed valued abilities. Unlike professional female entertainers, who were expected to be amusing, local women showcased talents and traits that supported normative behaviour. These competitions were often age-related and carried value-laden messages. Contests for the oldest person on the fairgrounds rewarded senior women for their loyalty, Fair Queen contests rewarded young women for their beauty and energy, and baby contests rewarded married women for their childbearing. Agricultural societies organized these competitions at little cost, but they still attracted significant participation and attention from fairgoers. When women entered these contests, they were active participants who decided their involvement provided some sort of personal or communal benefit.

Women's participatory longevity was honoured at fairs; even before official competitions were held, local newspaper reporters commented on long-returning visitors or exhibitors. The Dutton *Advance* notified readers that one of the oldest fairgoers at the 1916 Wallacetown Fair was Mrs Helen Gunn of Wallacetown, who had never missed a fair. This year, "owing to her great age, she was unable to go about as usual, but, seated in an auto, was keenly interested in everything that transpired."[79] As an agricultural society matured and the annual fair aged, it became a badge of honour to be recognized as a long-time supporter, and fairs hosted regular competitions in which women were typically the winners. In 1946, eighty-six-year-old Miss Sarah Patterson and eighty-year-old Mrs Sarah Carrevan were awarded prizes for being the oldest persons at the Smithville Fair (see Figure 6.1).[80] At the 1950 Coe Hill Fair, Mrs Mary Whitmore, eighty-two, and Mrs Anne Whitmore, eighty-eight, won the honour.[81] At the 1978 Markham Fair, "several hundred senior citizens "signed in" at the special historical booth," where it was discovered that the oldest visitor to the fair was ninety-four-year-old Norah Macklem of Stouffville. Mary Rose of Markham was runner-up at ninety-two years of age.[82]

Other contests focused on how women performed specific tasks. In 1934, some OAAS delegates expressed concern that the entertainment on fairgrounds was eclipsing their agricultural exhibits. To remedy the situation, one fair director, Malcolm Calder, encouraged agricultural societies to host milking contests for women, arguing that the Beaverton Fair milking contest had proven a big attraction with fairgoers.[83] The next year, OAAS directors suggested that milkmaid contests and chicken-plucking contests were "suitable" attractions because they combined education and entertainment.[84] The appeal of these types of events obvious. They fulfilled agricultural societies' mandate of educating rural women yet allowed for the competition that interested fairgoers.

Other popular contests judged women on their public speaking, singing, dancing, or musical talent. Although these competitions were not agricultural, they showcased valued elements of femininity in rural society. Women's

Figure 6.1 Winners of the Oldest Persons at Smithville Fair
Miss Sarah Patterson, eighty-six, and Mrs Sarah Carrevan, eighty, won the prizes for the "Oldest Persons" at the fair. *Hamilton Spectator*, "Fair," 1946. Local History & Archives, 32022191073309, Hamilton Public Library. Image courtesy of the Hamilton Public Library.

contests showcased hard work and talent, but also aspects of middle-class refinement. The ability to sing a tune, play an instrument, or speak eloquently on a topic demonstrated both a woman's skill and her respectability. At the 1897 Erin Fair, Miss Alma Lamont, who was only twelve years old, took first prize in the Highland fling dance competition, in which she had "looked charming in her Highland dress."[85] In 1924, the Erin Fair held a public-speaking contest that was open to the pupils of the township, and the top three prizes all went to young women.[86] At the 1928 Georgetown Fair, the first- and second-prize winners in the public-speaking contest were young women also, and Margaret McMaster of Glen Williams beat John Creighton of Georgetown to win first prize and a silver cup in the piano competition.[87]

By the late nineteenth and early twentieth centuries, sporting events were acceptable places for the sexes to socialize, and whole families often attended.[88]

Women had more opportunities to engage in sport, including ice skating, croquet, bicycling, baseball, and rowing. In 1886, the *Farmer's Advocate* advertised rowing for girls as "a healthy and invigorating exercise" and approvingly reported that "the fact that 'our girls' are developing skills of this kind is a very satisfactory sign of the times." The article lamented how in "former days, the rules upon which they were brought up were peculiarly restrictive, and few outdoor amusements were open to them; but now, the desirability of their having more invigorating recreation ... is becoming generally admitted."[89] By the twentieth century, changes to women's fashion and the promotion of their physical education encouraged young women to participate more in sports. Competitions became more "vigorous, competitive, and organized."[90]

Fair organizers incorporated sporting contests in their program as another way to draw crowds. They encouraged young women to compete as a way to embrace the expanding North American sporting culture and to keep young people interested and actively participating in the fair. At the 1910 Weston Fair, boys and girls competed separately in foot races, wheelbarrow races, sack races, potato races, bicycle races, and relay races. Long-jump competitions and three-legged races were offered for boys only.[91] Although girls and boys did not compete against one another, competitions for girls still encouraged their athleticism. Some fairs limited races and other sporting events to boys and men,[92] but most fairs included events for women.

Young women also played team sports at fairs. At the 1911 Newmarket Fair, an "interesting feature" was the basketball match between the Newmarket and Richmond Hill girls' high-school teams. The Richmond Hill team won the game by a score of 26 to 19.[93] The growth of softball in Canada paralleled that of the United States, and by the 1940s, softball was lauded as the number one sport for women in North America.[94] In Canada, women's softball leagues flourished in the 1930s and by the 1940s, when the All-American Girls' Softball League was established, scouts were recruiting players from across Canada.[95] Generally, women's league players retired from the sport once married, but in some professional leagues women continued to play after marriage, and even after having children.[96] Fairs held women's softball games and tournaments across Ontario. Their pervasiveness reflected softball's acceptance as a respectable leisure activity. At the

1928 Georgetown Fair, the final game in the women's softball tournament attracted a crowd.[97] One of the main highlights at the 1931 Elgin County Fair was the softball tournament in which men's and women's teams competed and which attracted a large crowd of spectators.[98] In 1940, the Puslinch Agricultural Society offered a winning prize of $1.25 to the girl who won a softball-throwing competition.[99] Noncontact sports such as basketball and baseball maintained a level of decorum for female players yet showed great skill and athleticism.

Fairs also offered competitions that judged young women on their physical appearance, not just their physical abilities. Some competitions were fairly innocent, such as the "Freckle Contest" at Lion's Head Fair, which was deemed "a big hit."[100] A competition for "the best head of unbobbed hair on girls 8 to 18 years" was held at the 1925 Thorold Fair. The *Thorold Post*'s editor, John H. Thompson, sponsored the prize money to protest the increasing number of women with short hair. He asserted that if he "had a thousand girls, not one would have bobbed hair," arguing that the "Creator made female hair long, and I am in favor of leaving it that way." Despite the problem with Thompson's logic, the competition itself was a success. Over nineteen girls entered, and eleven-year-old Gertrude Wallace, "whose lovely head of light blonde wavy tresses was considered best," won the competition. The judges considered the length, texture, and care of the hair when making their final decision, making it clear that the young women's "faces did not count."[101]

The most popular contests judged entrants' beauty and manners. Beauty contests and later "Fair Queen" contests developed over the late nineteenth and twentieth centuries, and often only young, single women were eligible. Although most Fair Queen competitions emphasized that they were not just about beauty, it was clear that physical attractiveness was central to the judging. At the 1911 Cooksville Fair, one of the "most amusing competitions" was a contest in which a donor awarded a fruit cake to "the best looking girl." A newspaper report noted that "evidently it is a cake of great renown, for as pretty a bevy of damsels congregated as could be wished."[102] This event was likely influenced by similar contests that began in the United States in the late nineteenth century. American beauty contests started as a way to attract publicity and tourism to resort towns and community festivals.[103] The

Toronto Industrial Exhibition had tried to host a beauty contest in 1884, convinced that Canadian beauty could not be surpassed, but despite the organizers' promise to keep contestants' identities secret except for the two winning entries (judged by photographic portraits), the competition failed because women "were reluctant to court public view deliberately."[104] In the twentieth century, that reluctance began to dissipate as women moved more into the public eye.

However, it was not until the postwar period that beauty and Fair Queen contests proliferated. Beauty contests had shifted away from their side-show roots to become more professional and career- and scholarship-oriented competitions. In doing so, they also became part of the culture that built on the tradition of middle-class civic boosterism.[105] The Miss Canada Pageant began in Hamilton in 1940 but moved to Toronto in 1949, where it was renamed the Miss Canada Health, Talent and Beauty Pageant. A reporter for *The Globe and Mail* wrote in 1949 that the contest was no longer "just a pulchritude parade." Although he explained women still wore bathing suits, which revealed "legs galore," he also noted that "the whole tone of the thing is being uplifted. Beauty alone doesn't win the prize (it says here) without health, talent, poise, bearing, grace and ... er ... looks."[106] The reporter's tongue-in-cheek comments suggested that he was not completely convinced that the pageant was transformed, but he did note that the prizes were worth significantly more than previously. Before the winner won $200; now Miss Canada received $3,000 in scholarships, a diamond ring worth $1,000, and more than $1,000 worth of other gifts, in addition to a trip across Canada.[107]

Local fairs recognized the appeal of these events and the contestants they attracted. Miss Canada of 1951, Marjorie Kelly, visited the 1951 Norfolk County Fair the fairgrounds. The *Reformer* reported that the North Walsingham native was "the living manifestation of the *beauty, culture, co-operativeness* and *enthusiasm* that we of Norfolk are so proud, particularly among our youth" [italics added]. It further noted that the community "appreciated her presence at the fair, and her readiness to share her talents with others." Indeed, the newspaper reported that Kelly's "words of acceptance of the scroll of honor, presented to her by the Warden on behalf of the people of Norfolk showed her to be unspoiled by her successes, proud of her community, her county, and intensely Canadian. As she sang 'Land of Hope and Glory' we knew why she was chosen as Miss Canada and no more appropriate expression of a person's feelings towards their country and way of life was ever made."[108] Kelly had grown up on a tobacco farm in the far west end of Norfolk County and credited the hard work she did on her family farm with giving her the discipline to succeed and "never give up on anything I started." She went on to compete in the Miss America pageant and moved to the United States after meeting her husband,[109] but in 1951 she was the perfect example of a young, local woman whose physical appearance, comportment, and speech epitomized the proper womanhood that community leaders hoped all female residents would emulate.

Fair beauty contests existed within the larger framework of global beauty pageants used to encourage ideals of womanhood and citizenship. Pageants portrayed, and ultimately judged, racial, ethnic, and cultural difference.[110] When agricultural societies adopted "Fair Queen" competitions, the same feminine characteristics were put on display. The young women who won these titles never won the prestige that national beauty titles bestowed, but they achieved a certain degree of celebrity in their own communities. By the 1960s, the OAAS sponsored a provincial competition at the Canadian National Exhibition where local winners competed for a provincial title.

By the 1960s and 1970s, most fairs in Ontario had some form of Fair Queen competition. Some of those contests specified beauty as a fundamental component of evaluation, while others emphasized a woman's personality. The 1965 Orono Fair Beauty Queen Contest was sponsored by an Oshawa company, Smith Beverages, and was open to contestants who were sixteen years of age and older and resided in Durham County. Short formal dresses were worn by all contestants and the "most beautiful" was crowned "Miss Durham Central."[111] The "Miss Durham Central" competition was labeled a beauty contest, but most fairs refused the Beauty Queen title, instead opting to crown their winner Fair Queen. Beauty was just one – albeit important – element of the judging at these competitions. For example, the 1965 Miss Acton Fall Fair competition advertised that poise and personality were just as important as appearance.[112]

The eligibility of contestants often involved age and residence. When it began in 1963, the Miss Acton Fair competition was the first of its kind in Southern Ontario, and the Canadian National Exhibition (CNE) reportedly modeled its Fair Queen contest after it.[113] In 1965, the Miss Acton Fair contest was open to "single girls, 15 years and over (no age limit), from the town of

Acton, and four townships, Erin, Eramosa, Esquesing, and Nassagaweya."[114] The eligibility rules ensured that women were young, unmarried, and lived in the community. Racial, ethnic, and cultural differences among contestants were not publicized, either because they did not exist in communities where most contestants were likely of European-Canadian extraction, or because organizers and the media chose to ignore whatever diversity existed. Pageant organizers often argued that they restricted entry to single women because married women did not have the time to devote to their duties as Queens,[115] but organizers probably wanted contestants to represent purity and wholesomeness – virtues that only young, chaste women could claim. Cooperativeness and youthful enthusiasm may have also been easier to find in younger contestants. Organizers often selected important figures from their communities as judges, though sometimes they were able to secure more broadly recognized public figures or celebrities. The first Acton Fair Queen contest in 1963 acquired three well-known personalities: Norm Marshall, a popular Hamilton radio and television sportscaster; Clifford Muir, the manager of a Guelph radio station; and Big Al Jones, a Kitchener television station entertainer.[116] They were all men, and although they may not have all privileged the same characteristics, it is likely that conventional notions of femininity and beauty influenced how they selected winners.

Contestants for the Miss Acton Fair Contest participated in an advance competition parade where they wore "street togs, dresses and high heels."[117] Following the parade, they were interviewed by the judges. To build drama and draw in crowds, the results of the contest were kept secret until the crowning ceremony on the first night of the fair. The winner won most of the prizes, but the runners-up, her "ladies-in-waiting," also received "part of the prize loot," which included cash, vouchers, and merchandise from local merchants and businesses.[118]

Young women participated in these contests for more than prizes because the title of "Fair Queen" bestowed approval from their peers and the broader community. The *Georgetown Herald* reported emphasized the attention the contestants at the 1976 Miss Acton Fair received: "Though the audience jammed into the Acton Arena Friday expressed its enjoyment of the variety show by clapping and cheering, an atmosphere of tension lingered. The crowd was waiting. They wanted to see which of the 13 girls entered in the Miss Acton Fall Fair competition would receive the crown."[119] The winning contestant that year, Charmaine Bigelow, was described as an aspiring dental hygienist who was "a tall, slim girl with long, straight light brown hair and large innocent eyes which sparkled with tears" when the master of ceremonies announced her as the winner.[120] As part of her duties, she attended local functions as the Acton Fair's representative, attended the OAAS convention in Toronto, and competed against Fair Queens from all over Ontario for the provincial crown at the CNE. Bigelow's prizes included roses, trophies, a silver tiara, and a red velvet red cape, in addition to gifts and vouchers from local merchants. They did not include the scholarships or endorsements associated with larger beauty pageants but still, when asked why she had wanted to enter the competition, she explained that she had wanted "to make her family proud."[121] This was also the reason Marilyn Ilott, the 1979 Markham Fair Queen, decided to compete. She was a student in fashion design at Seneca College and a part-time model, and she was described as a "pretty brunette with a sparkling smile and pleasing personality."[122] Her grandmother and sister had encouraged her to compete, and when she won, she expressed her joy that her grandfather, who was turning ninety-one years old, was there. She told reporters that she entered the competition for her grandparents, Mr and Mrs Reuben Meyer, who were reportedly "two of the happiest people attending the event."[123] The young women who participated, the crowds that amassed to watch, the local merchants who donated prizes, the directors who organized, and the agricultural societies who supported the contests all championed the idea that winning a Fair Queen title was a significant accomplishment. Perhaps the clearest indication that women enjoyed the honour bestowed by these titles were the jubilant smiles they wore when being crowned (see Figure 6.2).[124]

Women also competed for the title of Dairy Princess, Grape Queen, Tobacco Queen, and Queen of the Furrow, to name just a few such contests. Many took place at fairs, or the winners attended fairs to represent their industries or associations. Typically, the women who entered these contests had to have knowledge and/or skills related to the groups they hoped to represent. For instance, the Dairy Princess competition evaluated how contestants milked a cow, while the Queen of the Furrow was expected to exhibit skill with a tractor and plow. Dairy Princess competitions were especially popular. Each county selected a winner to compete in the provincial competition at the CNE, and the provincial winner went on to compete at the

Figure 6.2 The Shedden Fair Queen
Fair Queen competitions were popular, and women were honoured to be awarded their titles. "Shedden Fair – Fair Queen," *St. Thomas Times-Journal*, 30 June 1973. *St. Thomas Times-Journal* fonds, C8 Sh3 B1 F19 19, Elgin County Archives. Image courtesy of Elgin County Archives.

National Dairy Queen contest.[125] These contests differed from Fair Queen contests because they were open to married women. Winners attended fairs to distribute prizes at dairy cattle shows, visited schools to teach students about dairy farming and products, and travelled on international dairy farm and show tours. Dairy Queens were given significant responsibilities to represent the industry at local, provincial, national, and international functions.[126] The Dairy Queen competition was acknowledged as a different type of contest because "Instead of bathing-suits, considered standard equipment for most potential beauty queens, wardrobes included all-white uniforms of slims, shirts, caps and shoes with ankle socks," and prizes included "such gifts as sanitizers for milk tanks and filters for milk strainers."[127]

It would be unfair to say Fair Queen contests were just about beauty, especially since many of the later incarnations of this competition required contestants to be knowledgeable about farming in the region and capable spokespeople for the fairs, but it is clear that the amount of skill required to be Fair Queen did not compare to that involved in becoming a Dairy Princess. One might question whether or not competitions that emphasized beauty and traditional notions of femininity improved women's status on the fairgrounds. The publicity women received as the public face of an agricultural or community organization, however, suggests that their participation represented a further acceptance of women as essential contributors to fairs, albeit in a thoroughly gendered way.

Showcasing Motherhood: Women and Baby Shows

Another significant contest at fairs was the baby show, which was primarily about celebrating motherhood. By the early half of the twentieth century, many fairs encouraged mothers to put their children on display, often on the grandstand in full view of fairgoers.[128] Baby shows reflected how mothers' skills were increasingly scrutinized in the twentieth century. In the late nineteenth century, state fairs in the United States adopted baby shows as a means of promoting the nation's success in childrearing, and it was not long before Canadian fairs followed suit. It was believed that a baby's measurements indicated their normalcy, and thus health, and so a baby's height and weight, as well as circumference of the head, chest, and abdomen, and the length of arms and legs were all assessed.[129] Fair organizers in Ontario who promoted baby shows were likely also influenced by the "Better Babies' Movement" that sought to teach women how to be expert mothers.[130] By the time of World War One, such reformers made a concerted effort to improve Canadian health by modernizing childrearing methods.[131] By the 1920s,

Canada's emerging welfare state was beginning to formalize definitions of proper parenting.[132] Medical professionals advocated for "scientific motherhood" in the form of "expert tutoring and supervisions of child-rearing duties."[133] New hospitals, childcare clinics, visiting nurses, and school programs all worked to advance ideas of proper childcare. Baby shows were adopted at agricultural fairs as part of this movement. In the same way that fair competition was used to educate men about raising livestock, baby shows encouraged mothers to raise healthy children. Fairs also promoted good childcare practices through health exhibits and displays about infant and preschool health.[134]

Baby contests also promoted the idea that the maintenance of health standards was a woman's individual responsibility.[135] The prize a baby received was believed to be related to the skill and care of its mother. The belief that women were responsible for raising healthy citizens was not new to the countryside. In 1846, the *British American Cultivator* explained that it was a woman's job to ensure the physical, spiritual, and moral growth of her children:

> She is responsible for the nursing and rearing of her progeny; for their physical constitution and growth; their exercise and proper sustenance in early life ... She is responsible for the child's *habits*; including cleanliness, order, conversation, eating, sleeping, manners, and general propriety of behavior ... She is responsible for their deportment. She can make them fearful and cringing; she can make them modest or impertinent; ingenious or deceitful; mean or manly; clownish or polite ... She is to a very considerable extent responsible for the temper and disposition of her children. Constitutionally they may be violent, irritable, or revengeful; but for the regulation or correction of these passions a mother is responsible.[136]

The article laid the blame on mothers for any of their children's failings, which were "a living monument of parental disregard; because generally speaking, a mother can, if she will, greatly control children in these matters."[137]

But although baby shows eventually gained considerable popularity, they were not always received with such fervour. An article published in the *Farmer's Advocate*, reprinted from *Moore's Rural New Yorker*, satirized American agricultural societies for believing that "exhibitions of the fairest girls, the prettiest and fattiest babies, the youngest mothers of the largest families," could serve to "awaken a profound interest in Agriculture and a love of Rural life."[138] "Susan, from Streetsville" wrote a letter to the editor of the *Brampton Conservator* to protest the 1878 Cooksville Fair baby show. She argued that "to exhibit human creatures like so many cattle or pigs, I consider degrading, and feel sorry that such low vulgarism should be borrowed from Yankeedom. No mother, I think, will be found to compete in this department."[139] Despite her protest, the show went on and was deemed a success by the agricultural society. The prize-winning "best and handsomest baby under eighteen months old" won a silver medal valued at $10, donated by seedsman William Rennie of Toronto.[140] The newspaper reporter stated matter-of-factly that the Cooksville Fair was "just the place to offer a silver medal for the handsomest baby," because "the young ladies who attend Cooksville Fairs, are the prettiest, sweetest-looking and most stylishly dressed."[141] By exhibiting their children, women were also exhibiting themselves. Whether they chose to compete in such competitions because they believed they had superior parenting skills or the most handsome children, the opportunity to be acknowledged by their peers and win a prize was too hard to resist.

By the First World War, the crowds who came to baby shows were meant to compare these model children to their own, recognize any deficiencies in their offspring, and seek expert advice on how to remediate any shortcomings, but spectators were not always satisfied with judges' evaluations. They were more interested in the spectacle, the judgement of their peers, and the sight of the happy little tots who decorated the stage. Twenty-two mothers entered their children in the baby show at the 1914 Norfolk County Fair, which was described as the most popular attraction at the fair. The reporter covering the event explained that "it was taken for granted that each and every one honestly believed that their own baby was "the best thing that ever happened," and deserved first prize" but unfortunately he had to relate that "only three passed the judges eagle eyes. Spectators of the show agreed that the most winsome and pretty baby of the whole coterie was the cherub who drew third prize."[142]

The judges' identities and why their evaluation of the babies conflicted with that of the audience are both unknown. If the judges were physicians and nurses, as was common during this time, perhaps their training led them to believe that the healthiest child was not necessarily the prettiest. Or perhaps the evaluation of babies was much more subjective than the organizers or judges liked to admit. Baby shows, like the women's exhibits discussed in previous chapters, illustrated the limitations of visual learning. Agricultural societies relied on what could be seen to teach fairgoers, but often seeing was not sufficient.

At the 1916 Wallacetown Fair, when the judge, Magistrate Hunt, went about selecting a winner from the twelve "handsomest and cutest little Canadians" he had ever seen, he confessed he was "up a tree." He noted that although he had adjudicated on the most intricate legal disputes, he could not make a decision and instead asked the two ladies assisting him to award the first prize.[143] If the judge was a local resident, the prospect of facing a bevy of disappointed mothers after the show was likely frightening. Magistrate Hunt may have realized that flattering the children and expressing difficulty with the task would not save him from this possibility. Whether his female assistants were nurses, better equipped to handle the decision, is not clear.

Women were expected to be knowledgeable child carers, but baby shows did not challenge the authority men had to evaluate their skills – maternal or otherwise. By the 1930s, most baby shows employed male physicians to judge the competition with the assistance of local nurses (see Figure 6.3). Fair organizers likely felt that employing medical professionals helped support their claim that these contests were scientific assessments of infants' health. The director of the show at the 1938 Aylmer Fair was Dr H.G. McLay, and his committee was composed of Fred R. Barnum and Ralph O. Standish. Mrs E.S. Livermore was the "hostess" at the Crystal Palace, where the show took place, while Mrs Byde Parker was the official weigher and Miss Bancroft was the official stenographer. Miss Frey, Mrs L.D. Stocks and Mrs Elvin Wisson were the nurses who aided Dr McLay in his evaluation of the contestants.[144] At the 1938 Mount Forest Fair, two doctors from Toronto, Dr Geo. Gardiner and Dr W.T. Noonan, judged the event with the help of public-school nurses who weighed and measured more than fifty contestants.[145] The women assisting and the female nurses employed often did most of the work, but the male doctors and directors were officially in charge. Unless men such as Magistrate Hunt relinquished their power, they were the ones supervising these events, and they were usually the ones passing judgement.

From Church Booths to Baby Shows 237

Figure 6.3 Baby Show Judges at Murillo Fall Fair
The judges of the baby show at the Murillo Fall Fair included Dr Caldwell (background), Mrs Jack Dawson (first on the left), and Anne Sinclair (nurse). "Group Photograph," 1939. P1783, Thunder Bay Public Library. Image courtesy of Thunder Bay Public Library.

The beliefs that medical practitioners knew best, and that a child's health could be determined by their measurements, build, attractiveness, and behaviour was used to promote these shows as educational, rather than simply entertaining. But the popularity of these "attractions," the thronging crowds, the disagreement about the winners, and the promotion of the "beautiful types of babyhood,"[146] suggest that entertainment was a significant part of their appeal. Promoting beautiful infants was the reason men such as John S. Martin, the Ontario Minister of Agriculture, arranged for the Ontario Picture Bureau to attend the 1925 Norfolk County Fair so that the baby show would be "captured on celluloid for all the world to see."[147] In 1931, the Springfield Fair baby show drew "the crowd from every other part of the grounds."[148] In addition to the classes for babies, the local Women's Institute sponsored a prize for "the best looking mother with babe under one year," which was won by Mrs L.C. Murphy and her daughter, Viola Betty. The show also featured prizes for the most recently married couple and the oldest couple attending.[149] Earlier competitions, such as the 1914 Wallacetown Fair competition for the largest family, also encouraged large rural families.[150] At the 1919 Aylmer Fair, the "best and healthiest family under 10 years of age" was awarded a prize.[151] The focus on strong-bodied soldiers during wartime resulted in efforts to improve health in Ontario.[152] These competitions allowed fair organizers to promote their ideal family by providing a strong visual representation to the community and "all the world" that rural people were central to the strength of the nation.

Baby shows persisted in the postwar period and reports continued to emphasize the excellent job rural families did in raising their young. A series of photographs titled "Fall Time in Ontario" published in newspapers across Ontario in the 1940s and 1950s featured images taken at fairs. In a 1949 edition, photographs of children at fairs were featured to illustrate that "nowhere in the world is there a sturdier or more photogenic younger generation than right here in Ontario."[153] A province-wide report of the 1950 Fergus Fair included numerous photos of the large baby contest that year, and noted that "They raise fine crops in Wellington County, but none finer, or in which they take more pride, than babies such as these."[154] The display of happy, healthy children at fairs was used as evidence of the success of rural families in raising the next generation of Canadian citizenry. Raising a strong family was considered just as important as breeding an improved herd of livestock or growing a superior field of crops.

Even in the 1970s, the popularity of baby shows did not wane, despite criticisms that they were out of step with modern society. Isabella Bardoel, a reporter from *The Globe and Mail*, attended the popular Rockton Fair to report on the baby show that year. Bardoel described how "Hundreds of anxious parents and grandparents jammed into one of the fair's largest tents ... to view and stew over the largest crop of infants from the area ranging from 3 months to 18 months." She noted that every passerby was drawn to watch the competition: "Log sawing, nail driving, horse and wagon demonstrations and tractor pulls didn't have the drawing power of about 80 dribbling babies."[155] Jeanette Jamieson, the first female president of the Rockton Agricultural Society, was interviewed, and she explained the society had thought about dropping the event because "Some people criticized the idea

Figure 6.4 Baby Show at the Aylmer Fair
The newspaper caption reads, "The Heat Was On – At the Aylmer Fair Baby Show on Saturday, more than 100 babies, along with their moms and other anxious relatives, packed the upstairs of the arena to wait for the judges' decisions. Due to the lack of air-conditioning, the contest was hot, in more ways than one." "Aylmer Fair – Exhibits, Winners, and Baby Show," 1978. *St Thomas Times-Journal* fonds, C8 Sh2 B2 F3 74k, Elgin County Archives. Image courtesy of Elgin County Archives.

of judging babies," and "to every mother her baby is the best – and that's the way it should be," but because the event continued to be as popular as ever, the society decided to carry on.[156] The prizes were not what drove participation – that year the first prize was $5, second prize a small box of baby food, and the top three babies won ribbons – but rather it was mothers' pride in their children that drove them to compete. Mary Lee Rainy of Cambridge, nineteen, entered her daughter Sarah Jane, who won in the girls' twelve- to eighteen-month category at the Rockton Show. Her excitement could not be contained when she declared, "My mother will just die, she'll be so happy."[157] Bardoel noted that the expressions of the mothers whose babies did not win the prize "clearly showed the opposite reaction," but nevertheless the event continued because of parents' desire to showcase their children to their community (see Figure 6.4).

By this time, most shows no longer had doctors presiding as judges, although nurses continued to be involved, perhaps in order to maintain the appearance of some scientific or medical basis for evaluation. Overall, the baby shows of the 1970s had a lot in common with those that came a century earlier. They gave proud parents the opportunity to display their childrearing and agricultural societies a way to illustrate that rural families were succeeding in raising the next generation of Canadians. The concept of success expressed, however, was based primarily on an infant's appearance – their measurements, build, attractiveness, and mannerisms. The baby show was another event supporting the idea that appearance was a legitimate means for determining merit.

Female Fairgoers: Seeing and Being Seen

Women showed themselves simply by attending fairs. Many women who went to Ontario fairs never organized an event, entered a contest, worked a booth, or performed on a stage, but they used the fair to meet with friends and family, to make new acquaintances, learn new skills, discover new products, or simply to be entertained. Whatever they were doing on the fairgrounds, they were seeing and being seen.

Fairs provided opportunities to meet with friends and neighbours who were interested and engaged in similar pursuits. In his investigation of twentieth-century agricultural fairs and ploughing matches, David Mizener argues that the "exchange of greeting and the conversations between family, friends, and acquaintances at these events played a crucial role in renewing ties that formed the basis of kinship networks and communities."[158] Indeed, the "happy delighted groups," reported the *Brampton Times* in 1869, would find the Peel County Fair to be "a great County re-union, and not a few from a distance will embrace the opportunity of meeting their old friends and connexions."[159] The newspaper also reported the next year that the "yeoman of Peel will turn out en masse [to the fair] ... for the very good and sufficient reason that their 'roof-trees' would quiver with family indignation if they were even to hint at remaining at home."[160] In 1901, the social element of the West Elgin Fair was considered one of its best features. It was reported that its "long years of usefulness and prosperity have drawn around it many friends to all parts of the adjoining townships, who make it an unfailing point to meet annually to exchange greetings."[161]

Young people also attended fairs to interact with the opposite sex. For a single woman or man who had exhausted their immediate circle of prospects, fairs were important places to seek out a future husband or wife. Township and County fairs brought families together who may have had limited contact otherwise, and many young adults saw fairs as a chance to encounter the opposite sex in a public setting. When nineteen-year-old John Ferguson attended the 1869 Edmonton Fair in Peel County, he noted in his diary that he was disappointed that the cold and wet day had resulted in the "girls [being] rather scarce." The fair was not a complete disappointment, however, because he considered those young women who were in attendance to be a "scarce but select" group.[162] A study of the 1871 Peel County Fair reveals that the event attracted a disproportionately high number of young women of marriageable age in fair competitions. This likely reflected their desire to showcase their competency in domestic pursuits, which would have been attractive to potential suitors.[163]

Women did not have to compete to be noticed. Simply being there was enough to draw attention. Women who competed in beauty contests or performed specific talents and skills obviously increased their visibility, but women's beauty was often noted when they walked through the fairgrounds. As discussed in earlier chapters, women's presence often elicited reporters' praise for their attractiveness, and whether because of their "brilliant" and "gay" dresses,"[164] or their "blooming" "rosy-cheek[s],"[165] women enhanced the fairgrounds, and men took notice. When women exhibited items such as bread and fancywork, they received respect for their skill and taste. When women walked through the fairgrounds, however, their behaviour and appearance was evaluated, and they knew it. For example, before Whitby Township resident Frances Tweedie attended the 1866 Scarboro Fair, she and her friend Jennie spent time working on their outfits, and she spent the week of the fair socializing and visiting with friends, especially those of the opposite sex, paying little regard to the exhibits on display.[166] In 1909, an anonymous writer published a poem in the *Erin Advocate* describing how he went to the Erin Fair with the explicit purpose of seeing the girl he loved.

When I go up to Erin Fair
I hope the girl I love is there;
She is a lovely sight to see
She comes from pretty Comingsby
She is a peach, she is a pear
A sweet muckmelon, rich and rare;

When I go up to Erin Fair
I hope the girl I love is there;
They'll race their horses all in vain
I will not see the roots and grain,
I will not see the squash – a beaut –
Nor glimpse the honey or the fruit;
Fruit, pumpkins, crazy quilts, avaunt,
My girl is all the show I want.

She's an inducement and a lure;
Blindfold me and I'll find her sure;
And if ten thousand filled the park,
I'd find my sweetheart in the dark,
I hope the girl I love is there,
When I go up to Erin Fair.[167]

While some connections at fairs were instant, others developed over time. Agricultural society members who served during fairs often created lasting relationships, many platonic but others romantic. In a 1979 newspaper article, newly married couple Charles and Violet Barrett, both in their eighties, related that their relationship began with their local fair involvement. Charles Barrett was the secretary-treasurer of the Caledon Agricultural Society from 1936 until 1957, when he was married to his first wife. Violet's first husband, Ben Bull, was a previous fair president. Violet started exhibiting in 1927 and was the director and committee member of the Children's School Department for many years. The couple had known each long before their marriage, but it was not until Violet's first husband passed away, followed soon after by Charles' wife's death, that the two reconnected and eventually married. Charles had taken some time away from fair duties, but because

Figure 6.5 Big Crowd at Norfolk County Fair
"A Snapshot of the Big Crowd," Norfolk County Fair, Simcoe, 16 October 1913.
Norfolk County Fair Collection, C-S-35, Eva Brook Donly Museum and Archives.
Image courtesy of the Eva Brook Donly Museum and Archives.

Violet was the convener of the children's section that year, he was "back at the fair helping out his new wife."[168]

As Figure 6.5 illustrates, fairs were places where "multitudes from all points of the compass"[169] flocked together, and often friendships and courtships developed. Women young and old met with family, friends, and peers to enjoy the festivities of the day, but also to create, sustain, and build relationships for the future.

Conclusion

Fairs offered women the chance to take in the exhibits, be entertained, earn money and awards, win accolades and fame, see old friends, and establish new relationships. But whether they were fundraising, competing, courting, or earning a living, they were making themselves and their passions visible. Even women who never organized an event, entered a contest, worked a booth, or performed on a stage had an important role to play. When they purchased a pass and made their way to the hall to see the women's exhibits, bought a piece of pie or a concert ticket, or simply found a seat in the grandstand to watch an entertainer perform her act, they helped support other women's work.

Women increased their mobility and presence on the fairgrounds by moving beyond the dining, concert, and theatre hall to find space to situate themselves in front of the grandstand, the midway, and everywhere else in between. As fairgoers, women had long traversed most areas of the fairgrounds – whether they were welcome or not – but as performers, contestants, and volunteers, more space became available to them. Women's increased visibility also reflected their enlarged presence in society; fairs offered opportunities for women to push the boundaries of some social mores at the same time that they reinforced others. Woman's ability to challenge or confirm gendered notions of womanhood at fairs illustrates how conflicted these concepts were in the first place. But women were not just visible and mobile, they were also increasingly vocal, and in the 1980s and 1990s they continued to push the limits of their fair experiences.

Conclusion

The story of women's involvement in agricultural societies and fairs involves more than lists of fancywork, accounts of judging mishaps, or tales of female balloonists. The things women made, the activities they took part in, and their service on fairgrounds allow for a better understanding of rural women's values, beliefs, and day-to-day lives. While some people might argue that fairs have become nostalgic remnants of a rural past, this is not so, in no small part because of women's participation. Rural populations have declined and the members of farming communities are shrinking.[1] This has affected many voluntary organizations, including agricultural societies. For this reason, some fairs may be more celebratory of traditional pursuits than their predecessors, who advertised themselves as progressive. However, they continue to be microcosms of the rural world, and celebrations of yesteryear live alongside efforts to improve the present.

As in the past, women are crucial to agricultural societies' and fairs' successes today. Traditional notions of womanhood still exist, but so do new conceptions of women's roles. This book demonstrates how the expectations placed on women have been challenged and changed, often because of women's own agency and initiative. This lesson is valuable for all those – not just rural women – who continue to face barriers in their private and public lives. This story can inspire them to work on behalf of their interests and those of their communities.

The challenge of any book attempting to make useful generalizations from a specific group is that the complex nature of people's lives is not always evident.[2] For example, this work does not represent all the activities and traditions found in the countryside. The geographic parameters and source material used here have contributed to an analysis that focuses on women who were typically white, Anglo, English-speaking Ontarians. Agricultural societies maintained a commitment to serving what they perceived as agrarian interests over other rural interests. So this book details why women actively participated in agricultural societies and fairs, rather than why some women did not. As a result of this focus, this analysis of women's experiences is less complicated than it could be by the intersection of other important axes of social relations, including class, occupation, race, and ethnicity.[3]

Comparing rural women's experiences with what we know of urban women is difficult. First of all, the term "rural" itself is contested. In addition, the scholarly literature on urban women during this period is more developed than that on women who lived in the countryside. And finally, historical feminist scholarship in Canada has emphasized the diversity of women's experiences across the country. So trying to identify shared experiences among rural women and compare them to those of urban women is daunting.[4]

Still, this study suggests some comparisons. Rural women's service to their communities, even in organizations that were often hesitant to accept them, highlights both their commitment to maintaining the fabric of kinship and community and their desire to play a public role at perhaps the most important social gathering of the year. Rural women understood interdependence from their household experiences, and they chose to emphasize cooperation over conflict when participating in agricultural societies and at fairs. Without romanticizing this cooperation, one can still acknowledge that rural women, and rural society generally, valued mutuality and shared work. Unlike many urban women during the period of this study, rural women often worked side by side with their husbands. They had a level of intimacy with their spouses' labour that urban women did not.

Furthermore, in the postwar period, urban women entered the paid workforce in more significant numbers and were attracted to the Women's Liberation Movement and its calls for equal wages for equal work. Meanwhile, most rural women still worked alongside their husbands on family farms, and their ambitions were tempered by the need to place family and community before feminist ideology, especially since these social networks were fundamental for emotional, social, and financial support.[5] Female members recognized unfair practices in agricultural societies, however, and they negotiated those unequal distributions of power by creating separate women's divisions where they could enjoy the camaraderie of other women, discuss ideas and activities that interested them, and build confidence in their ability to provide leadership on fair boards. This book supports scholars who argue that, although the patriarchal system constrained such women, they employed feminist strategies in response, including "female cooperation, collaboration, and female empowerment."[6] They sought out shared interests at fairs, developed ways to serve those interests in agricultural societies, and inspired and enabled other women to pursue their goals on the fairgrounds. They cultivated community throughout this process.

Some people might argue that these women's decision to emphasize separation over integration impeded their progress in achieving more authority within agricultural societies. Still, although some women's divisions remained beholden to the men for operating funds, by electing their own ruling executive and conducting their own meetings these women displayed a gender consciousness that allowed them to advance their interests with limited male interference. The confidence women developed in their own departments also gave them the courage to put their names forward for board executive positions, including that of president, later on. Women who were elected to such positions achieved a level of respect and appreciation from their peers despite their sex. Gendered assumptions were never wholly erased in agricultural societies, but men could act as allies for women who wished to expand their participation.

Women's volunteer work also found recognition at fairs. Researchers of volunteerism have noted that women's efforts have often not been considered "real work" by society,[7] but the service women gave at fairs was visible for the whole community to appreciate. Rural women were not afraid to recognize the value of their work and to press for others to acknowledge it as well. The "culture of the eye" that Keith Walden describes in nineteenth-century fairgoers' understanding of objects as illustrative of social status applies to women's exhibits at local fairs. It can also describe visual confirmations of women's volunteerism.[8] At fairs, people witnessed women's voluntary work, such as their fundraising for community, national, and international causes, and this strengthened their authority in the community and led to increased responsibilities as engaged citizens. This study agrees with the literature that highlights women's use of fairs as spaces for public recognition.[9]

Furthermore, women's involvement in fairs reveals a great deal about the boundaries of acceptable womanhood in Ontario during the period in question. The competitions women entered and the things they exhibited illustrate rural people's ideas about feminine identity. From dairy produce to handicrafts, women's exhibits showcased their commitment to supplementing household income and nurturing their families, while emphasizing both their respectable status and their understanding of beauty and refinement. Rural women exhibited items that demonstrated their identities as both industrious housewives and sophisticated people. The things they made for fair competitions or from the awards they received served to memorialize their past achievements, create bonds with their families and communities, and provide visual representations of their beliefs and interests. By competing at fairs, women publicly demonstrated industry, thrift, taste, and respectability, which helped to entrench these ideals in the countryside. In other words, fairs were sites that validated women's domestic activities by giving them a public venue for expression.

Related to women's public role was the importance of appearance. Visual emphasis was widespread at fairs, but their focus on women's appearance in the nineteenth and twentieth centuries was especially gendered. Female beauty was considered desirable by male organizers. The work women displayed at fairs was often valued first and foremost for its appearance, and women's presentation of attractive work was related to their own morality and taste. When women took up traditionally masculine pursuits and moved into masculinized spaces, appearing feminine and respectable was often necessary. Women challenged the privileging of their looks by showcasing skills such as horsemanship and livestock showmanship, but even in these areas their abilities were judged in conjunction with their appearance.

Despite such limitations, however, fair exhibitors continued to enlarge the female sphere and challenge the status quo by increasingly showcasing talents that challenged traditional notions of femininity. Although agricultural societies' perception of womanhood was primarily influenced by what was displayed in the exhibition hall rather than in the show ring, women increasingly participated in competitions traditionally reserved for men. Horse and livestock classes allowed them to demonstrate skills not typically feminine, which led to a reshaping of gender boundaries. Women claimed traits that had previously been perceived as masculine, such as power, courage, knowledge, and leadership. Although initially uncomfortable with women's entry into this world, men came to value what women contributed to the show ring through their hard work and dedication.

Women also participated in contests and voluntary work that increased their visibility in spaces previously limited to men. They moved beyond dining and exhibition halls into the midway and the grandstand not merely as fairgoers but as active participants, taking advantage of opportunities to enlarge their sphere on the fairgrounds. Although many of their activities reinforced traditional ideas of female behaviour, fairs also presented visions of women that dismantled such socially constructed notions.

The challenges women faced and continue to face in agricultural societies and fairs, and indeed in society more broadly, should not be ignored. Still, it is useful to listen to women's own remembrances, which emphasize the positive experiences they had, including the friendships and skills they acquired over the years and the joy and personal meaning they found in fair activities. For the most part, women who participated felt deep affection for these groups and events and for the people connected to them.

In 1980 and the decades that followed, women continued to struggle against sexist ideas about their proper role in agricultural societies.[10] Only a small minority of agricultural societies had allowed a woman to serve as president, and many members, including some women who actively sought to increase female involvement on fair boards, still believed that their leadership was better served by men. But other women fought for access to all departments and spaces on the fairgrounds. Still, it would not be until the late 1990s when most agricultural societies had had at least one female president or had women participating in management committees. For the societies that later achieved gender balance or saw a majority of female members enlist, criticism that "too many women" now served on fair boards was sometimes heard. Conversely, some people found it strange that men started to enter handicraft and culinary arts competitions in the homecrafts department and began winning awards for traditionally feminine work.[11]

Ultimately, this book agrees with other studies that endorse the term "feminisms" and understand the definition of feminism as capacious.[12] We need to allow women in the past to claim a feminist history based on their deconstruction of notions of femininity and their work to empower and improve the lives of other women. The hundreds of thousands of women who participated at fairs across Ontario in the nineteenth and twentieth centuries believed that their work and their abilities were worthy of attention, and their involvement heightened women's status in both the private and public spheres. They also took advantage of opportunities caused by war, economic depression, or social change to achieve more authority. They used that increased power to expand and support their interests and activities.

Women took advantage of fairs to create friendships, serve and promote community causes, showcase and develop skills and talents, make family heirlooms, assert social authority, earn some cash, or simply enjoy a good spectacle. They benefited from the multifarious nature of fairs as places where women could either support or dismantle notions of feminine behaviour and proper womanhood. The complexity of their participation in fairs illustrates the complexity of rural women's lives in general, and the reality that they often straddled multiple lines of identity.

Notes

Introduction

1 A "democrat" wagon is a light farm wagon with two or more seats, drawn by two horses.
2 M.O.H., "At a Country Fair," *Globe* [Toronto], 5 November 1898.
3 Ibid.
4 For a useful consideration of the term "rural," and what experiences defined rural life and created a coherent rural identity in the nineteenth and twentieth centuries, see Sandwell, *Canada's Rural Majority*, 9–25.
5 Walden, *Becoming Modern in Toronto*, 59.
6 Ibid., and Heaman *The Inglorious Arts of Peace*.
7 Ibid., 269–70.
8 Ibid., 259, 269–70.
9 American scholars who provide a strong analysis of women's fair experiences include Kelly, "The Consummation of Rural Prosperity and Happiness," 574–602 and Borish, "Women at the New England Agricultural Fair in the Mid-Nineteenth Century," 155–76.
10 David Mizener's PhD dissertation on Ontario fairs in the twentieth century argues that agricultural fairs and ploughing matches were platforms used to articulate and confirm ideas about rural identity. See Mizener, "Furrows and Fairgrounds."
11 Sandwell, *Canada's Rural Majority*, 4, 73. In 1976, the rural population in Canada fell for the first time. In Ontario, the rural population grew over

the course of the early twentieth century, then remained relatively stable between 1921 and 1951.

12 It is understandable that Canadian historians such as Graeme Wynn, Ross D. Fair, and Daniel Samson have produced excellent studies of agricultural societies in the eighteenth and nineteenth centuries but found few records that promote a sustained investigation of women's experiences in these organizations. They have convincingly argued that these groups were dominated by men more intent on creating middle-class masculine identity and authority than reaching a broad base of rural community involvement. See Wynn, "Exciting a Spirit of Emulation," 3–51; Fair, "Gentlemen, Farmers, and Gentlemen Half-farmers"; and Samson, *Spirit of Industry and Improvement*. Works on agricultural societies that do not provide an examination of women's participation include Jones, *History of Agriculture in Ontario*, and Cochrane, "Agricultural Societies, Cattle Fairs, Agricultural Shows and Exhibition of Upper Canada Prior to 1867." The Ontario Association of Agricultural Societies has published works that discuss women's involvement, but without a great deal of analysis: Ontario Association of Agricultural Societies, *The Story of Agricultural Fairs and Exhibitions*, and Scott, *A Fair Share*.

13 See Parr, "Gender History and Historical Practice," 354–76, and Scott, *Gender and the Politics of History*.

14 Sandwell, *Canada's Rural Majority*, 9; see also Parkins and Reed, "Toward a Transformative Understanding of Rural Social Change," 11–13.

15 Sandwell, *Canada's Rural Majority*, 9–10.

16 Ibid.

17 Parkins and Reed, "Toward a Transformative Understanding of Rural Social Change," 13.

18 Scholars studying Women's Institutes in Ontario provide useful analyses of women's roles in rural reform. See Kechnie, *Organizing Rural Women*; Halpern, *And on That Farm He Had a Wife*; Ambrose, *For Home and Country*; Ambrose and Kechnie, "Social Control or Social Feminism?" 222–37; Crowley, "The Origins of Continuing Education for Women," 78–81. For other important scholarly research on Women's Institutes, see Ambrose, *A Great Rural Sisterhood*, and Andrews, *The Acceptable Face of Feminism*. For a broader consideration of rural reform movements in North America, see Taylor, *Fashioning Farmers*; Kline, *Consumers in the Countryside*; Holt,

Linoleum, Better Babies and the Modern Farm Woman; Jellison, *Entitled to Power*; Danbom, *The Resisted Revolution*; and Bowers, *The Country Life Movement*.

19 Gal, "Grassroots Consumption," 149–51. Rural women devoted a significant amount of time to the canning of fruits and vegetables and the manufacture of a variety of preserves in the first part of the twentieth century. But whether canning was already popular and fair classes simply reflected this, or agricultural societies' promotion of the work encouraged more women to can, one cannot know.

20 Ferry, *Uniting in Measure of Common Good*, 199; 224.

21 See Kechnie, *Organizing Rural Women*; Halpern, *And on That Farm He Had a Wife*; Ambrose, *For Home and County*; Ambrose and Kechnie, "Social Control or Social Feminism?"; Crowley, "The Origins of Continuing Education for Women"; Ferry, *Uniting in Measures of Common Good*; Badgely, *Ringing in the Common Love of Good*; Hann, *Farmers Confront Industrialism*; Irwin, "Government Funding of Agricultural Associations"; Sanmiya, "A Spirit of Enterprise." For another useful Canadian source on the United Farmers' Movement, see Rennie, *The Rise of Agrarian Democracy*.

22 This book was also influenced by the extensive literature on women's religious volunteerism in Canada, including such works as Christie and Gauvreau, *A Full-Orbed Christianity*; Kinnear, "Religion and the Shaping of 'Public Woman'"; Van Die, "Revisiting 'Separate Spheres'"; Marks, *Revivals and Roller Rinks*, 22–51.

23 See Grey Osterud, *Bonds of Community*, and Neth, *Preserving the Family Farm*. Grey Osterud and Neth argue that American farm women and men in the nineteenth and early twentieth centuries forged bonds of interdependence in the household and the community. This contrasts with urban women's historians who argue that middle-class women developed a distinctly female sphere. See Cott, *The Bonds of Womanhood*, and Smith-Rosenberg, *Disorderly Conduct*, for examples of urban studies.

24 Grey Osterud, *Bonds of Community*, 9, 275.

25 See Neth, *Preserving the Family Farm*.

26 Fink, *Agrarian Women*, 28.

27 A great deal of literature has been written about this topic in Western Canada. See Bye, "'I Like to Hoe My Own Row,'" 135–67; Bye, "'I Think So Much of Edward,'" 205–37; McManus, "Gender(ed) Tensions in the Work

and Politics of Alberta Farm Women," 123–46; Rollings-Magnusson, "Canada's Most Wanted," 223–8; Rollings-Magnusson, "Hidden Homesteaders," 171–83; McCallum, "Prairie Women and the Struggle for a Dower Law," 19–33; Jahn, "Class, Gender, and Agrarian Socialism," 189–206; and Cavanaugh, "The Limitation of the Pioneering Partnership," 198–225. For Ontario, see Crowley, *Agnes MacPhail and the Politics of Equality* and Carbert, *Agrarian Feminism*. In Atlantic Canada, Carbert argues that the tendency of rural communities to "maintain the status quo" and for rural women to be reluctant to stand for elected office, are results of decades-long economic and political factors; Carbert, *Rural Women's Leadership in Atlantic Canada*.

28 Grey Osterud, *Bonds of Community*, 276.
29 Halpern, *And on That Farm He Had a Wife*, 3.
30 Kechnie, *Organizing Rural Women*, 3, 60, 81.
31 See Ambrose, *A Great Rural Sisterhood* and *For Home and Country*; and Ambrose and Kechnie, "Social Control or Social Feminism?" 234–7.
32 Carbert, *Agrarian Feminism*, 10.
33 Andrews, *The Acceptable Face of Feminism*, xiii, 166–9.
34 Prown, *Art as Evidence*. Prown emphasizes object-centred approaches to material culture. His method of analysis requires researchers to engage in the precise description of objects, which requires a certain level of detailed knowledge about the material features of the object.
35 Kingery, "Introduction," 1.
36 For an excellent discussion of the range of approaches employed in material culture see Harvey, "Introduction: Practical Matters," 1–23.
37 Ulrich, *The Age of Homespun*.
38 Turkle, *Evocative Objects*, 5.
39 Ibid.
40 Tobin and Goggin, *Women and Things*, 1.
41 The oral interviews were originally conducted for my PhD research at the University of Guelph.
42 Thompson, *The Voice of the Past*, 8.
43 Ibid., 21.
44 Tumblety, *Memory and History*, 7.
45 Ibid.

Notes to pages 16–20

46 High, *Oral History at the Crossroads*, 20.
47 Thompson, *The Voice of the Past*, 22.

Chapter One

1 "Presentation to Mrs. Velda Dickenson," in *One Hundred and Twelfth Annual Report of Ontario Agricultural Societies*, 97–8.
2 Historians who discuss this include Halpern, *And on That Farm He Had a Wife*; Ambrose, *For Home and Country*; Ambrose and Kechnie, "Social Control or Social Feminism?"; Crowley, "The Origins of Continuing Education for Women"; and Andrews, *The Acceptable Face of Feminism*. Important sociological works that discusses rural women's volunteerism include Petrzelka and Mannon, "Keepin' This Little Town Going," 236–58; Little, "Constructions of Rural Women's Voluntary Work," 197–210; Abrahams, "Negotiating Power, Identity, Family, and Community: Women's Community Participation," 768–96.
3 Halpern, *And on That Farm He Had a Wife*, 8–9.
4 Black, *Social Feminism*, 53. See also Halpern, *And on That Farm He Had a Wife*; Carbert, *Agrarian Feminism*; and Ambrose and Kechnie, "Social Control or Social Feminism?"
5 Fair, "Gentlemen, Farmers, and Gentlemen Half-farmers," 49. Fair notes that the name of the first society is unclear. In an official letter sent by Simcoe's secretary it was called the "Agricultural Society of Upper Canada" and the same name was used in some of Ontario's fair history, but it was also addressed as the "Agricultural Society of Niagara" and the name "Niagara Agricultural Society" was used often in colony newspapers and is inscribed on the 1966 historical plaque that commemorated its establishment. Therefore, I have selected to use this name in my study.
6 Johnson, *Dictionary of the English Language*.
7 Samson, *The Spirit of Industry and Improvement*, 59–60.
8 "What Is Agriculture," *Farmer's Advocate* 3 (November 1868): 164.
9 When the Ontario School of Agriculture and Experimental Fair first opened its doors in 1875, it defined its purpose in gendered terms as being to give "a thorough mastery of the practice and theory of husbandry to young men of the Province engaged in Agricultural and Horticultural

pursuits." *Annual Report of the Ontario School of Agriculture and Experimental Farm*, 24.
10 "Trades for Women," *Farmer's Advocate* 12 (June 1877): 140.
11 "Go to the Fairs – Why?" *Brampton Conservator*, 7 September 1877.
12 Scott, *A Fair Share*, 17–18.
13 Samson, *Spirit of Industry and Improvement*, 12.
14 Ibid., 55.
15 Ibid., 75–6.
16 Ibid., 251. Fair, "Gentlemen, Farmers, and Gentlemen Half-Farmers," also identifies women, but determines they were largely excluded. Older works that exclude any discussion of women include Jones, *History of Agriculture in Ontario*, and Gates, *The Farmer's Age*.
17 Samson, *Spirit of Industry and Improvement*, 272–3.
18 For an analysis of the religious and political discourses on gender, and the complex relationship between public and private realms in early nineteenth-century Canada, see Morgan, *Public Men and Virtuous Women*. There is an extensive literature on women's public presence in nineteenth-century America. See Lasser and Robertson, *Antebellum Women*; Jeffrey, *The Great Silent Army of Abolitionism*; Hoffert, *When Hens Crow*; Yellin and Van Horne, *The Abolitionist Sisterhood*; Ryan, *Women in Public*; Ryan, *Civic Wars*, and Ryan, *Cradle of the Middle Class*; Hewitt, *Women's Activism and Social Change*.
19 Mrs Signourney, cited in "The Two Sexes," *Ontario Farmer* 3 (June 1871): 191.
20 "Minute Books Collection, Agricultural Society of Addington County, 1853–1873," B277479, Agricultural Society of the County of Addington fonds, F 1259, Archives of Ontario; "Smith Township Agricultural Society, Minute Book," [1855–], B-76-013, Trent University Archives; "County of Peterboro Agricultural Society, Minute Book," [1855–], accession 1959-004-W0535, "Peterborough Agricultural Society 1868–1875, Minute Book," accession 99-004-W0323, Peterborough Agricultural Society fonds, Centennial Museum and Archives (Peterborough); "Oro Agricultural Society Minutes, Entries," [1859–1886] accession 970-77, R2A S3 Sh4, Simcoe County Archives; "Erin Agricultural Society Minutes, 1862–1941," [Microfilm], Roll 1 A989.97, Wellington County Museum and Archives; "Brant Agricultural Society Minute Book, 1862–1875," A960.066.001, "Brant Agricultural Society Membership Book, 1868–1896," A960.066.002, Brant

Agricultural Society fonds, Bruce County Archives; "Huron Agricultural Society Minutes, 1868–1911," Agricultural Societies Collection, XA1 RHC A0386023, Box 1, File 2, University of Guelph Archival and Special Collections; "Electoral Division of Cardwell Agricultural Society, Minute Book," [1868–1949] accession 988-3; R2B S5 Sh4, Simcoe County Archives; "Minute Book, East-Middlesex Agricultural Society," [1873–1913], B5289-1, Western Archives; "South Wellington Agricultural Society, Minute Book," [1878–], A198338 – MU 102, Wellington County Museum and Archives; "North Walsingham Agricultural Society Minute Book, 1889–1920," accession 10582, North Walsingham Agricultural Society fonds, F 1261, Archives of Ontario.

21 Ontario Association of Agricultural Societies, *The Story of Ontario Agricultural Fairs and Exhibitions*, 10–11.

22 "Premiums Awarded at the Frontenac Cattle Show," *British Whig* [Kingston], 14 October 1835. One class for cheese and another for butter were offered, as well as three "Domestic Industry" classes for flannel, cloth, and socks. The vast majority of classes were for livestock or field crops. Interestingly, the Frontenac Fair premium list is also one of the few during this period that identified married women, not just widowed or single women, as competitors.

23 Ontario Association of Agricultural Societies, *The Story of Ontario Agricultural Fairs and Exhibitions*, 24.

24 Scott, *A Fair Share*, 68.

25 "County and Township Agricultural Societies," in *Journal and Transactions of the Board of Agriculture of Upper Canada*, vol. 1 (Toronto: Board of Agriculture, 1856), 5. The terms of these grants changed throughout the period of this study.

26 "The Agricultural Association," in *Journal and Transactions of the Board of Agriculture of Upper Canada*, Vol. 1 (Toronto: Board of Agriculture, 1856), 26–8.

27 Ibid., 24. The Provincial Agricultural Association's constitution that was read and approved on 17 August 1846, declared that "the objects of the Association shall be improvement of Farm Stock and Produce; the improvement of Tillage, Agricultural Implements, &c.; and the encouragement of Domestic Manufactures, of Useful Inventions, and, generally, of every branch of Rural and Domestic Economy."

28 Marti, *To Improve the Soil and the Mind*, 211–12.

29 Wynn, "Exciting a Spirit of Emulation," 32.
30 Greer, *The Patriots and the People*, 201–2, cited in Heaman, *The Inglorious Arts of Peace*, 262.
31 Heaman, *The Inglorious Arts of Peace*, 262.
32 Ibid., 261–6. Heaman provides an excellent survey of women and exhibitions in central Canada, and the discourse that surrounded their early involvement at fairs, in her chapter "Women and the Political Economy of Exhibitions."
33 William F. Phelps, "The Co-Education of the Sexes," *Farmer's Advocate* 3 (May 1868): 73.
34 Amanda, "Domestic Education," *Canadian Agriculturalist*, 1 August 1849.
35 Valverde, *The Age of Light, Soap, and Water*, 17–18.
36 See Ferry, *Uniting in Measures of Common Good*; Taylor, *Fashioning Farmers*; Kechnie, *Organizing Rural Women*; Ambrose, *For Home and Country*; and Halpern, *And on That Farm He Had a Wife*.
37 Ferry, *Uniting in Measures of Common Good*, 223–4, 233–4.
38 S.A. Laidman, "Prize Essay: Agricultural Exhibitions as Educational Institutions for the Farmer and His Family," *Farmer's Advocate and Home Magazine* 21 (August 1886): 229.
39 "Peel Fall Fair! Another Grand Success!" *Brampton Times*, 6 October 1871.
40 Larger district or country societies existed alongside smaller township societies for much of this period. In the 1868 Report of the Commissioner of Agriculture and Arts of the Province of Ontario, county agricultural society memberships varied; as many as 408 registered members belonged to the South Ontario Agricultural Society (246 members reported for the North Ontario Agricultural Society), while as few as 37 paid members were reported for the West York Agricultural Society (although the East York Agricultural Society had 147 members and the North York Agricultural Society reported 288 members). Larger counties such as Ontario and York often split the county into geographical districts. Township societies were popular in every county. Although these societies typically had smaller memberships, this was not always the case. The same commissioner's report revealed that the Darlington Township Agricultural Society in Durham West had 331 members, more than many county organizations, while the Lochiel and Kenyon Township Agricultural Society in Glengarry County reported only 22 members. The difference in numbers reflects the

fact that some counties had more township agricultural societies than others, while certain areas of the province also had larger agricultural communities and a larger population in general. "Analysis of Reports of Agricultural Societies for the Year 1867," in *Report of the Commissioner of Agriculture and Arts of the Province of Ontario for the Year 1868*, 48–129. Although there had been longstanding arguments against having both county and township societies – some reformers believed that fewer township societies should exist to allow for more robust county societies, while others felt that township societies were more beneficial to local farmers and fostered more equality among members – generally township societies were popular, a trend that continued in the twentieth century. "Agricultural Shows," *Farmer's Advocate* 2 (December 1867): 100; "Agricultural Exhibitions," *Farmer's Advocate* 3 (January 1868): 1; "Township Agricultural Societies," *Brampton Times*, 24 January 1868.

41 "Brant Agricultural Society Membership Book, 1868–1896," A960.066.002, Brant Agricultural Society fonds, Bruce County Archives.

42 "Township of Erin Annual Exhibition, Prize List, 1893," "Prize List, 1896, of the Township of Erin Fall Exhibition," [Microfilm] Roll 1 A989.97, Erin Agricultural Society Records, Wellington County Museum and Archives.

43 "Prize List of the Erin Fall Exhibition, 1916," "Prize List of the Erin Fall Exhibition, 1917," [Microfilm] Roll 2 A19889.97, Erin Agricultural Society Records, Wellington County Museum and Archives.

44 Nineteenth-century newspapers and farm journals regularly published agricultural society annual meeting minutes, which illustrate the focus of their discussions. Useful minute books include: "Minute Books Collection, Agricultural Society of Addington County, 1853–1873," B277479, Agricultural Society of the County of Addington fonds, F 1259, Archives of Ontario; "Smith Township Agricultural Society, Minute Book," [1855–], B-76-013, Trent University Archives; "County of Peterboro Agricultural Society, Minute Book," [1855–], accession 1959-004-W0535, "Peterborough Agricultural Society 1868–1875, Minute Book," accession 99-004-W0323, Peterborough Agricultural Society fonds, Centennial Museum and Archives (Peterborough); "Oro Agricultural Society Minutes, Entries," [1859–1886] accession 970-77, R2A S3 Sh4, Simcoe County Archives; "Erin Agricultural Society Minutes, 1862–1941," [Microfilm], Roll 1 A989.97, Wellington County Museum and Archives; "Brant Agricultural Society

Minute Book, 1862–1875," A960.066.001, Brant Agricultural Society fonds, Bruce County Archives; "Huron Agricultural Society Minutes, 1868–1911," Agricultural Societies Collection, XA1 RHC A0386023, Box 1, File 2, University of Guelph Archival and Special Collections; "Electoral Division of Cardwell Agricultural Society, Minute Book," [1868–1949] accession 988-3; R2B S5 Sh4, Simcoe County Archives; "Minute Book, East-Middlesex Agricultural Society," [1873–1913], B5289-1, Western Archives; "South Wellington Agricultural Society, Minute Book," [1878–], A198338 – MU 102, Wellington County Museum and Archives; "North Walsingham Agricultural Society Minute Book, 1889–1920," accession 10582, North Walsingham Agricultural Society fonds, F 1261, Archives of Ontario.

45 "Agricultural Exhibitions," *Farmer's Advocate* 3 (January 1868): 2. The *Farmer's Advocate* promoted the idea of female exhibitors as well, arguing that agricultural societies should encourage "lady exhibitors, whether it should be in art, fancy or useful department."

46 "Fairs and Expositions: Second Day's Meeting of the Fifth International Convention," *Globe* [Toronto], 29 July 1887.

47 Ibid.

48 "Fall Fairs' Revival: New and Important Changes Planned," *Globe* [Toronto], 22 April 1902.

49 Ibid.

50 "New Style Fall Fair: Many Promoters Advocate Educational Features," *Globe* [Toronto], 18 February 1904.

51 "Men and Women from the Farm: They Were in Evidence at the Exhibition," *Globe* [Toronto], 10 September 1908.

52 Kechnie, *Organizing Rural Women*, 37–60.

53 Danbom, *The Resisted Revolution*, 62; and Fink, *Agrarian Women*, 26.

54 See Taylor, *Fashioning Farmers*, and Halpern, *And on That Farm He Had a Wife*.

55 Taylor, *Fashioning Farmers*, 56.

56 Holt, *Linoleum, Better Babies, and the Modern Farm Woman*, 5.

57 "The Great Fair, Another Success Scored by the People's Favorite," *Advance* [Dutton], 7 October 1909.

58 "The Great Fair, Outlook Bright for a Bigger Fair than Ever," *Advance* [Dutton], 23 September 1909.

59 See Chapters 2, 3, and 4 for a discussion of women's exhibits and display.

60 Sangster, *One Hundred Years of Struggle*.
61 For Canadian works on rural reform, see Kechnie, *Organizing Rural Women*; Halpern, *And on That Farm He Had a Wife*; Carbert, *Agrarian Feminism*; and Taylor, *Fashioning Farmers*. American scholarship on this topic includes Kline, *Consumers in the Country*; Holt, *Linoleum, Better Babies, and the Modern Farm Woman*; Neth, *Preserving the Family Farm*; Jellison, *Entitled to Power*; Fink, *Agrarian Women*; Danbom, *The Resisted Revolution*; and Bowers, *The Country Life Movement in America*.
62 See Young, "Conscription, Rural Depopulation, and the Farmers of Ontario," 289–318.
63 "The New Electorate: The Women of Canada," [Globe Advertisement] *Farmer's Advocate and Home Magazine* 54 (16 October 1919): 1887.
64 "Women in Country Homes Sharing in Improvements: No Longer Does the House Take Second Place to the Barn in the Matter of Decent Equipment – Many Labor-savers Needed," *Globe* [Toronto], 14 May 1919.
65 Kechnie, *Organizing Rural Women*, 3.
66 Ibid., 55.
67 Halpern, *And on That Farm He Had a Wife*, 79.
68 Rodney W.I. Minute Book, June 1906–May 1911, Rodney Women's Institute fonds, R7 S5 Sh1 B2, Elgin County Archives.
69 "May 31, 1912," Rodney W.I. Minute Book, June 1911–May 1917, Rodney Women's Institute fonds, R7 S5 Sh1 B2, Elgin County Archives.
70 "August 27, 2013," Shedden Women's Institute – Minute Book, 1913–1919, C9 Sh4 B1 F1, Eglin County Archives.
71 "September 17, 1915," Shedden Women's Institute – Minute Book, 1913–1919, C9 Sh4 B1 F1, Eglin County Archives.
72 For more examples, see Chapter 6.
73 Christie and Gauvreau, *A Full-Orbed Christianity*, 119; Kinnear, "Religion and the Shaping of 'Public Woman,'" 197; Van Die, "Revisiting 'Separate Spheres,'" 254; and Marks, *Revivals and Roller Rinks*, 22–51.
74 For specific examples, see Chapter 6.
75 "The Rural Fall Fair," *Farmer's Advocate and Home Magazine* 51 (27 July 1916): 1252.
76 "Put New Life into Fall Fairs," *Farmer's Advocate and Home Magazine* 51 (5 October 1916): 1645.
77 "The Rural Fall Fair."

78 Mrs Wm. Todd, Orillia, cited in Jean Fidlar, "Central Ontario Women's Institute Convention," *Farmer's Advocate and Home Magazine* 54 (27 November 1919): 2137.
79 Samson, *The Spirit of Industry and Improvement*, 12.
80 See Comacchio, *The Dominion of Youth*, and Commacchio, "Nations Are Built of Babies"; Christie, *Engendering the State*; and Campbell, *Respectable Citizens*.
81 "September 17, 1915," Shedden Women's Institute – Minute Book, 1913–1919, C9 Sh4 B1 F1, Eglin County Archives.
82 "August 23, 1922," Shedden Women's Institute – Minute Book, 1913–1919, C9 Sh4 B1 F1, Eglin County Archives.
83 "August 29, 1924," Shedden Women's Institute – Minute Book, 1913–1919, C9 Sh4 B1 F1, Eglin County Archives; and "September 24, 1924," Shedden Women's Institute – Minute Book, 1913–1919, C9 Sh4 B1 F1, Eglin County Archives.
84 "July 22, 1924," Shedden Women's Institute – Minute Book, 1913–1919, C9 Sh4 B1 F1, Eglin County Archives.
85 "August 26, 1925," Shedden Women's Institute – Minute Book, 1913–1919, C9 Sh4 B1 F1, Eglin County Archives.
86 "September 30, 1925," Shedden Women's Institute – Minute Book, 1913–1919, C9 Sh4 B1 F1, Eglin County Archives.
87 Mrs Laura Rose Stephen, "Lady Directors on Fall Fair Boards," in *Ontario Department of Agriculture Twentieth Annual Report of the Agricultural Societies and of the Convention of the Association of Fairs and Exhibitions*, 29.
88 Ibid., 30.
89 Ibid., 31–2.
90 Scott, *A Fair Share*, 95.
91 Scott, *A Fair Share*, 95; 105.
92 Reaman, *A History of Agriculture in Ontario*, vol. 2 (Toronto: Saunders, 1970), 19–20.
93 Miss M.V. Powell, "Report of Revision of Prize Lists," in *Ontario Department of Agriculture Twentieth Annual Report of The Agricultural Societies and of the Convention of the Association of Fairs and Exhibitions*, 60.
94 "January 21, 1925," and "March 21, 1925," Beeton Agricultural Society Minute Book, 1868–1949, Electoral Division of Cardwell Agricultural Society…Minute Book, accession 988-3, R2B S5 Sh4, Simcoe County Archives.

Notes to pages 34–7

95 Westover, *Fair Days and Fair People*, 12.
96 "First Hundred Years for Neustadt's Fall Fair: History and Prize List for 1970," A971.020.001, Bruce County Archives.
97 "Report Serious Weed Spread at Fairs' Association Meeting," *Globe* [Toronto], 3 February 1928.
98 Ibid.
99 Scott, *A Fair Share*, 105.
100 Ibid.
101 "Fall Fairs Prospering," *Globe* [Toronto], 5 October 1933.
102 "Prize List, 1931, Collingwood Township Agricultural Society Fall Fair," Ontario Fall Fair Catalogues, XA1 MS A073, University of Guelph Archival & Special Collections.
103 "Splendid Exhibits at Elgin County Fair and Dairy Cattle Show," *Aylmer Express*, 10 September 1931.
104 K. Goodfellow, "Ladies' Exhibits at Fairs & Exhibitions," in *Ontario Department of Agriculture, Thirtieth Annual Report of the Agricultural Societies and of the Convention of the Association of Fairs and Exhibitions*, 47–9.
105 Ibid., 49.
106 "Eighty-Fifth Anniversary, Erin Fall Fair, Prize List, 1935," [Microfilm], Roll 2, A19889.97, Wellington County Museum and Archives.
107 "Prize List: The Lucknow Agricultural Society's Seventieth Annual Fall Exhibition, 1935," Ontario Fall Fair Catalogues, XA1 MS A073, University of Guelph Archival & Special Collections.
108 J.A. Carroll, "Organization to Promote Women's Interests," in *Ontario Department of Agriculture, Thirty-Eighth Annual Report of the Agricultural Societies and the Convention of the Association of Agricultural Societies*, 85.
109 "Ethel Monture Admired at C.N.E. Descendant of Brant Strikingly Costumed, Speaks for W.I.," *Globe and Mail*, 30 August 1939; "Opportunity Lack Blamed for Loss of Indian Talent," *Globe and Mail*, 22 October 1948; "Deplores Stress on Savagery in Indian Tales," *Globe and Mail*, 24 February 1958; "Mrs. Monture Gets New Post," *Globe and Mail*, 13 November 1959; Dorothy Dumbrille, "Famous Canadian Indians," *Globe and Mail*, 2 July 1960; "Indians Need New Image, Writer Says," *Globe and Mail*, 17 July 1964. Ethel Brant Monture was the great-great-granddaughter of the Six Nations Chief Joseph Brant. She had a long history of community service, including roles with Christian organizations, the Women's Institutes, and advocacy for Native peoples.

110 Mrs Brant Monture, quoted in "Fall Fair Group Is Celebrating Silver Jubilee," *Globe and Mail*, 22 February 1962.
111 Mrs Ethel Brant Monture, quoted in *The Story of Ontario Agricultural Fairs and Exhibitions*, 206.
112 Ibid.
113 Mrs W.H. Monture, "Women's Meeting – February 3rd, 1938," in *Ontario Department of Agriculture, Thirty-Eighth Annual Report of the Agricultural Societies and the Convention of the Association of Agricultural Societies*, 84.
114 J.A. Carroll, "Organization to Promote Women's Interests," in *Ontario Department of Agriculture, Thirty-Eighth Annual Report of the Agricultural Societies and the Convention of the Association of Agricultural Societies*, 84–5.
115 Ibid.
116 Ibid., 85–6.
117 Mrs J.K. Kelly, "What Women Workers Are Thinking about Fall Fairs," in *Ontario Department of Agriculture, Thirty-Eighth Annual Report of the Agricultural Societies and the Convention of the Association of Agricultural Societies*, 44–5.
118 Ibid., 45.
119 Andrews, *The Acceptable Face of Feminism*, 15.
120 Ibid., 66.
121 Mrs J.K. Kelly, "The Duties of Women Directors," in *Ontario Department of Agriculture, Thirty-Eighth Annual Report of the Agricultural Societies and the Convention of the Association of Agricultural Societies*, 91–2.
122 Ibid., 92–4.
123 Ibid.
124 Miss B. McDermand, "Co-operation – Women's Institutes Branch," in *Ontario Department of Agriculture, Thirty-Eighth Annual Report of the Agricultural Societies and the Convention of the Association of Agricultural Societies*, 88.
125 Federated Women's Institutes of Ontario, *Ontario Women's Institute Story*, 74. McDermand graduated from Moulton Ladies' College and the MacDonald Institute before being hired to do extension work in Ontario and Alberta. She served as the assistant superintendent of the Alberta Institutes before attending Columbia University to complete a degree in Household Science. She then worked at Cornell University before she was appointed to

the position of superintendent of the Ontario Women's Institutes Branch. She resigned in 1938 before marrying Mr Guy Skinner.
126 McDermand, "Co-operation – Women's Institutes Branch," 89.
127 "Rural Ontario Women Make Entry into Field Hitherto Held by Men," *Globe* [Toronto], 4 February 1938.
128 W.J. Hill, quoted in "Rural Ontario Women Make Entry into Field Hitherto Held by Men."
129 Mrs Monture, quoted in "Rural Ontario Women Make Entry into Field Hitherto Held by Men.
130 "Rural Women Invade Man's Realm in Fair Work," *Farmer's Advocate and Home Magazine* 73 (10 February 1938): 75.
131 Ibid.
132 "Women and the Local Fair," *Farmer's Advocate and Home Magazine* 73 (24 February 1938): 100.
133 Petrzelka and Mannon, "Keepin' This Little Town Going," 236–58. In their study on rural American women's volunteerism, Petrzelka and Mannon found that, unlike suburban or urban women, rural women did not minimize the importance of their volunteer work.
134 "Women and the Local Fair," 100.
135 Mrs O'Leary, "Duties of Judges," in *Ontario Department of Agriculture, Thirty-Eighth Annual Report of the Agricultural Societies and the Convention of the Association of Agricultural Societies*, 106.
136 "Eighty-First Annual Exhibition, West Elgin Fair, Beef and Horse Show, Bacon Hog Fair, Prize List, 1941," ECVF Box 12, File 13, Elgin County Archives.
137 "Rural Women Make Bid for Share in Exhibitions," *Globe and Mail*, 24 February 1939.
138 Miss Ina Hodgins, quoted in "Rural Women Make Bid for Share in Exhibitions."
139 Ibid.
140 "Rural Women Make Bid for Share in Exhibitions.".
141 "Recruiting Our Land Army," *Farmer's Advocate and Home Magazine* 76 (23 January 1941): 36, 40.
142 "New Power Vested in Farm Women as Executives of O.A.A.S.: Women from Rural Ontario Granted Equal Voting Status in Agricultural Association,"

Globe and Mail, 16 February 1940. Unfortunately, only a list of the proceedings instead of a complete report was provided for the women's division meeting at the OAAS Annual Convention in 1940, but newspaper reports highlight some of the key discussions.

143 Wilfred Walker, "President's Address," in *Ontario Department of Agriculture, Forty-First Annual Report of the Agricultural Societies and the Convention of the Association of Agricultural*, 12.
144 Ibid.
145 Mrs J.K. Kelly, "President's Address," in *Ontario Department of Agriculture, Forty-First Annual Report of the Agricultural Societies and the Convention of the Association of Agricultural Societies*, 34-5.
146 Miss Mary A. Clarke, "Objectives of Agricultural Societies with Special Reference to the Women's Division," in *Ontario Department of Agriculture, Forty-First Annual Report of the Agricultural Societies and the Convention of the Association of Agricultural Societies*, 37; and Mrs J.K. Kelly, quoted in "Farm Women Face Future Confident that Agriculture Will Not Fail in Its Duty," *Globe and Mail*, 13 February 1942.
147 Ibid.
148 J.A. Carroll, "Summary of Address by J.A. Carroll, Superintendent to Women's Division," in *Ontario Department of Agriculture, Forty-Third Annual Report of the Agricultural Societies and the Convention of the Association of Agricultural Societies*, 23.
149 Ibid.
150 Ibid., 23-4.
151 Mrs J.K. Kelly, "Fair Management," in *Ontario Department of Agriculture, 91st Annual Report of Ontario Agricultural Societies*, 48.
152 Ibid.
153 "Burford Fair Prize List, 1945," Ontario Fall Fair Catalogues, XA1 MS A073, University of Guelph Archival and Special Collections. Men served on the horse, dairy cattle, beef cattle, sheep, swine, poultry, fruit, grain, roots and vegetables, tobacco, pet and hobby show, buildings and grounds, gate, and announcer committees, while women served exclusively on the arts and crafts committee. Most committee members for dairy and provisions, ladies' work, victory garden, and children's work, were women, but at least one male also served on these committees. The horticultural committee had equal representation.

154 Buttel and Gillespie, "The Sexual Division of Farm Household Labor"; Sachs, *The Invisible Farmers*.
155 Good statistics exist for the US. See Jones and Rosenfeld, *American Farm Women*, and Salant, *Farm Women*.
156 Elva Fletcher, "They Grow Their Own Wines," *Country Guide: The Farm Magazine* (August 1979): 44–5.
157 Miss Ina Hodgins, "President's Address," in *Ontario Department of Agriculture, 97th Annual Report of Ontario Agricultural Societies*, 66.
158 Ibid.
159 Ibid., 67.
160 Mrs S.W. Rathwell, "How We Put Across the Small Fair," in *Ontario Department of Agriculture, 97th Annual Report of Ontario Agricultural Societies*, 78. Rathwell spoke about how the recently formed Cumberland Township Agricultural Society had open nominations for the election of officers and directors, but specific rules had been made to ensure eight of the twelve directors were men. Furthermore, female directors were confined to the Homecraft, Domestic Science, Horticulture, W.I., and Junior Work.
161 Mrs A.L. Dickson, quoted in "Woman Asks for Equality in Rural Fairs," *Globe and Mail*, 9 February 1956.
162 Aberfoyle Agricultural Society, *The Agricultural Society in Puslinch*, 32.
163 Fullerton, *Paisley Agricultural Society*, 10.
164 Mrs Allan Koehler, quoted in "Woman Asks for Equality in Rural Fairs."
165 "Prize List, Teeswater Fall Fair, 1959," Ontario Fall Fair Catalogues, XA1 MS A073, University of Guelph Archival & Special Collections.
166 "Women Active in Work of Fairs," *Globe and Mail*, 26 February 1959.
167 F.A. Lashley, "Report of the Superintendent," in *One Hundred and Seventh Annual Report of Ontario Agricultural Societies*, 10.
168 Ibid.
169 Many women credited 4-H with allowing them to learn important leadership skills, which they later applied when participating in agricultural societies. For more about the effect of 4-H on rural women, especially in the postwar period, see Chapter 5.
170 Interview with Jeanette Jameson, 22 July 2014.
171 Interview with Eleanor Wood, 22 July 2014.
172 Ibid.
173 Interview with Jeanette Jameson, 22 July 2014.

174 Ibid.
175 Interview with Eleanor Wood, 22 July 2014.
176 Interview with Glenda Benton, 28 October 2013.
177 Petrzalka and Mannon, "Keepin' This Little Town Going," 246.
178 Interview with Margaret Lovering, 19 August 2014.
179 Ibid.
180 Switzer, *Erin Fall Fair since 1850*, 53.
181 Ibid.
182 Interview with Myrtle Reid, 23 September 2013. Women often did the majority of the fundraising for agricultural societies. The superintendent for Ontario Agricultural Societies, F.A. Lashley, emphasized in his 1960 OAAS Convention address that the "women of Paris and Burford," were "to be commended for raising $3,000.00 and $1,000.00 respectively and turning this money over to the board to help finance the fair in general." F.A. Lashley, "Report of the Superintendent," in *One Hundred and Seventh Annual Report of Ontario Agricultural Societies*, 10.
183 Interview with June Switzer, 12 September 2013.
184 Interview with Glenda Benton, 28 October 2013.
185 Carbert, *Agrarian Feminism*; 145–6.
186 See Cebotavrev, "From Domesticity to the Public Sphere: Farm Women, 1945–86," 200–31.
187 Elva Fletcher, "New Voice on the Farm Front," *Country Guide: The Farm Magazine* (November 1979): 44.
188 Ibid.
189 The Canadian National Exhibition Association had disallowed women from attending directors' luncheons during the annual exhibition (a policy that began in the 1930s), and it was not until 1969 that "Prominent women who qualify" were invited, despite the fact that the CNE association had had women directors since 1959. L.C. Powell, the CNE general manager, said that the policy had been a result of limited space in the dining facilities, but Mrs W.A. Wood, named a director in 1967 to represent the National Council of Women, asserted that the "CNE cannot continue to ignore the accomplishment of women just because they are women." L.C. Powel and Mrs W.A. Wood, quoted in "CNE Decides to Allow Women to Attend Directors' Luncheons," *The Globe and Mail*, 26 April 1969.
190 Interview with June Switzer, 12 September 2013.

Notes to pages 54–60

191 Ibid.
192 *Annual Report for the year ending March 31, 1980 of the Minister of Agriculture and Food*, 21.
193 K.E. Lantz, "Address," *One Hundred and Eighteenth Annual Report of Ontario Agricultural Societies*, 24–5.
194 J.D. Tate, Sutton West, "President's Address," *One Hundred and Eighteenth Annual Report of Ontario Agricultural Societies*, 7.
195 President [name unknown], quoted in "First Hundred Years for Neustadt's Fall Fair: History and Prize List for 1970," A971.020.001, Bruce County Archives.
196 Interview with June Switzer, 12 September 2013.
197 Interview with Margaret Lovering, 19 August 2014.

Chapter Two

1 Walden, *Becoming Modern in Toronto*, 119.
2 See Cohen, *Women's Work*; Bradbury, *Working Families*, and Bradbury, "The Home as Workplace"; Campbell, *Respectable Citizens*; Hoffman and Taylor, *Much to Be Done*.
3 "Erin Agricultural Show," *Guelph Mercury*, 21 October 1864.
4 Single or widowed women were the first to be recognized by their full name or title, and some married women were identified by their title in the late nineteenth century, but others continued to exhibit under their husbands' names. Even in the 1970s, many newspapers and agricultural societies continued to identify women by their husbands' first names rather than their own. Only single women were routinely identified by their first and last names.
5 For more on women's restricted rights in agrarian communities in Canada, see Taylor, *Fashioning Farmers*; Kechnie, *Organizing Rural Women*; and Carbert, *Agrarian Feminism*. For an understanding of women's limited ownership rights see Cavanaugh, "The Limitation of the Pioneering Partnership," and Backhouse, *Petticoats and Prejudice*.
6 "Woman's Worth," *Farmer's Advocate and Home Magazine* 34 (1 June 1899): 325.
7 Mrs Signourney, cited in "The Two Sexes," *Ontario Farmer* 3 (June 1871): 191.
8 Jellison, *Entitled to Power*, 179.

9   Kubic, "Farm Women," 242.
10   "Ladies at the Fall Fairs," *Brampton Times*, 16 October 1868.
11   "How to Choose Your Wives," *Farmer's Advocate* 8 (January 1873): 10.
12   Minnie May, "A Few Words on Cooking," Minnie May's Department, *Farmer's Advocate* 12 (January 1877): 19.
13   Ibid.
14   The "good taste and industry of the blooming daughters and hearty matrons of Ontario's Albion" was lauded in the *Brampton Times* report of the 1868 Albion Fall Fair, while in 1870, the newspaper reported there was a general falling-off of exhibits at the county fair due to "Peel's fair daughters... [having] reserved their industry and talent to swell the triumph of the Provincial Exhibition." See "Albion Fall Fair," *Brampton Times*, 16 October 1868, and "Peel Fall Fair, Another Grand Success," *Brampton Times*, 30 September 1870.
15   "Chinguacousy Fall Fair," *Brampton Times*, 30 September 1870.
16   "Peel Fall Fair: Another Grand Success."
17   John F. Clark, "Staging Horticultural Products at a Fair," in *Ontario Department of Agriculture Thirty-First Annual Report of the Agricultural Societies and of the Convention of the Association of Fairs and Exhibitions*, 21.
18   Ibid.
19   Spike, "Agricultural Exhibitions in Nova Scotia."
20   Neal Knapp argues that late nineteenth- and early twentieth-century American livestock improvers began to question whether the aesthetic traits desired in the show ring increased an animal's economic worth; Knapp, "Transforming the Political Bodies of Livestock." W.E. Smallfield suggested at the 1923 Ontario Association of Fairs and Exhibitions Convention that the fine appearance of animals was not enough: "important as it is to have the fine looking animal, to please the eye of the spectators, the lesson will hardly be more than half taught if the placard of information is not there also ... the educational value would be greatly increased if some competent person ... gave a talk, elaborating on the points of superiority in the build and make-up of the pure-bred animal and emphasizing the profit, shown on the placard, of the good animal over the grade." It later became common practice for judges to provide verbal reasons for their placings at most cattle shows, but this practice was not yet standard in the early part of the twentieth century; W.E. Smallfield, "How Can Rural Fairs

Be Made More Attractive and of Greater Educational Value?" in *Ontario Department of Agriculture Twenty-Second Annual Report of the Agricultural Societies and of the Convention of the Association of Fairs and Exhibitions*, 24.

21 "Erin Fair," *Hillsburgh Beaver*, 22 October 1907.
22 "Hints for the Month," *Ontario Farmer* 3 (September 1871): 427.
23 "Minto Agricultural Show," *Elora Observer*, 23 October 1863.
24 "Peel Fall Fair: Another Grand Success."
25 "Aylmer's Great Fair," *Aylmer Sun*, 25 September 1902.
26 Ken W. MacTaggart, "Home Sciences Stressed at Fall Fair," *Globe and Mail*, 10 October 1944.
27 Although spring and early summer fairs existed, most fairs ran during the months between late August and late October.
28 "West Agricultural and Horticultural Society's Fair," *Daily Globe* [Toronto], 12 October 1876.
29 For a full discussion of the prevalence of apples at fairs, see discussion in section "Garden Vegetables and Fruits."
30 Mrs Alex Robinson, "Women's Division Reports; District 14," in *Ontario Department of Agriculture 97th Annual report of Ontario Agricultural Societies*, 71.
31 Arran-Tara Fall Fair Booklets, A959.034.001, Bruce County Archives, and Arran-Tara Agricultural Society Prize Lists, private collection of Ron Hammell, Dobbinton, Ontario. In 1956, exhibitors were instructed that bread had to be baked the day before judging, and by 1960 onward the prize list specified that all bread was to be twenty-four hours old.
32 "Brampton Fair," *Brampton Conservator*, 5 October 1883.
33 "Brampton Fair," *Brampton Conservator*, 12 October 1877.
34 Ibid.
35 "Co. of Peel Fall Fair," *Brampton Conservator*, 10 October 1884.
36 Ibid.
37 "Role of Fairs in War Told: Must Adapt Plans to Needs, Says McLoughry," *Globe and Mail*, 13 February 1943.
38 Miss May Needham, "Some Suggestions for Improving Fall Fair Women's Departments," in *Ontario Department of Agriculture, Thirty-Fourth Annual Report of the Agricultural Societies and the Convention of the Association of Fairs and Exhibitions*, 24.

39 "Rain Causes Delay at Tillsonburg Fair," *Globe* [Toronto], 3 September 1931; and "Role of Fairs in War Told."
40 K. Goodfellow, "Ladies' Exhibits at Fairs & Exhibitions," in *Ontario Department of Agriculture, Thirtieth Annual Report of the Agricultural Societies and of the Convention of the Association of Fairs and Exhibitions*, 47.
41 "The Provincial Exhibition," *Weekly Review* [Brampton], 7 October 1854.
42 "Springfield Fair Another Big Success," *Aylmer Express*, 29 September 1932.
43 Needham, "Some Suggestions for Improving Fall Fair Women's Departments," 24.
44 Miss Agnes Yuill, "Women's Division Reports; District No. 2," in *Ontario Department of Agriculture, 96th Annual report of Ontario Agricultural Societies*, 56.
45 Mrs J.H. Booth, "Women's Division Reports; District No. 15," in *Ontario Department of Agriculture, 96th Annual report of Ontario Agricultural Societies*, 60.
46 Mrs Guthrie Reid, "Women's Division Reports; District No. 10," in *Ontario Department of Agriculture, 96th Annual report of Ontario Agricultural Societies*, 59.
47 Mrs Allan Koehller, "Women's Division Reports; District No. 10," in *Ontario Department of Agriculture, 97th Annual report of Ontario Agricultural Societies*, 70.
48 "Wallacetown Fair," *Advance* [Dutton], 5 October 1916.
49 Ibid.
50 Hunter and Yates, *Thrift in America*, 13.
51 Taylor, *Fashioning Farmers*, chapters 4 and 5.
52 Historian Julie Guard notes that Canada was preoccupied with "food's price, distribution, and accessibility" during the Great Depression, and that food absorbed a third of the average household budget. Guard, *Radical Housewives*, 6. The importance of food is demonstrated by the many protests against scarcity during the period. See Liverant, *Buying Happiness*, 89–109.
53 Needham, "Some Suggestions for Improving Fall Fair Women's Departments," 25.
54 Ibid.
55 Walden, *Becoming Modern in Toronto*, 141–4.
56 "Prize List, Exeter Agricultural Society Fall Show ... 1934," Ontario Fall Fair

Catalogues, XA1 MS A073, University of Guelph Archival & Special Collections. The Exeter Fair was like other fairs during this period that requested only single jar samples, typically a quart or a pint, of each canned product. Butter was mostly shown in five-pound rolls or prints, and the largest amount required for one class was ten pounds. Factory-made dairy classes, however, did require larger amounts (one class for fifty pounds of butter and another for a fifty-pound cheese).

57 "Guelph Central Fair, Opening Day," *Daily Globe* [Toronto], 4 October 1876.
58 Ibid.
59 Ibid. George Elliot's 1876 novel *Daniel Deronda* consisted of two distinct parts, each with its own unique qualities. The newspaper reporters' reference to the popular novel would have conveyed to readers how different the various exhibits were.
60 Zeller, *Inventing Canada*, 79.
61 "Autumn Fair," *York Herald*, 19 October 1860.
62 "Toronto Township Fall Fair," *Brampton Times*, 23 October 1868.
63 "Bayham Agricultural Society," *Aylmer Sun*, 13 October 1892.
64 "Albion Fall Fair, *Brampton Times*, 16 October 1868, and "The Markham Fair," *Globe* [Toronto], 6 October 1894. The Albion Agricultural Society renovated the existing building on the fairgrounds. Although the reporter did not specify how the building had been changed, common renovations included resurfacing dirt floors, fixing leaky roofs and barn-board walls, installing new windows, and adding additional tables and display cases for exhibits. Markham Fair erected a large, new, exhibition hall made from wood and glass described as a "magnificent exhibition pavilion." It was reportedly 150 feet long and 40 feet wide, with two storeys, for a total of 11,000 square feet of floor space. It cost the agricultural society over $3,000 to build.
65 Booth, "Women's Division Reports; District No. 15," 60.
66 "County Agricultural Show," *Guelph Advertiser*, 31 October 1850.
67 Needham, "Some Suggestions for Improving Fall Fair Women's Departments," 25.
68 Ibid.
69 Ibid.
70 Jones, *Midways, Judges, and Smooth-Tongued Fakirs*, 101–2. Jones explains

that at early western-Canadian fairs, the judging skills were often very rudimentary and food exhibits were seldom tasted, or even smelled. Many exhibitors expected the judge to decide on appearance alone, so when a judge actually tasted the food, cases of cheating were discovered.

71 "Lovely Ladies Stoop to Fall Fair Follies," *Globe and Mail*, 14 February 1947.
72 "Central Agricultural Society Prize List of the First Annual Fall Exhibition, 1889," Ontario Fall Fair Catalogues, XA1 MS A073, University of Guelph Archival & Special Collections.
73 Ibid.
74 Often those who accused others of cheating were put under scrutiny and accused of being jealous of the final results. When John Smith wrote to the *Brampton Times* expressing his displeasure with a class placing, he recognized that "that when one finds fault with the decision of judges, he lays himself to the charge of undue interference; or when he complains of his being but an unsuccessful exhibitor, his opposition attributed to spring from disappointed ambition." John Smith, "A Grievance," *Brampton Times*, 23 October 1868. Some agricultural societies also deterred unwarranted protests by requiring monetary deposits before protests would be considered, and only if the accused parties were found guilty would the accusers have their money returned. Typically, the fair officers would discuss the merit of any protest before any action was taken.
75 Interview with June Switzer, 12 September 2013.
76 Ibid.
77 "Record Crowd Attend Ont.'s Preview of The Royal," *Erin Advocate*, 13 October 1960.
78 Cohen, *Women's Work*, 93–117. Cohen notes that dairying was one of the most productive and important aspects of women's farm work in Canada until the rise of cheese factories and creameries, which ultimately led to farm specialization and male control of the industry.
79 Ibid., 97, and Jensen, "Cloth, Butter and Boarders," 17.
80 Cohen, *Women's Work*, 104.
81 Ibid., 103–4.
82 "Canadian Cheese and Butter," *Farmer's Advocate* 3 (June 1868): 87.
83 The 1847 Eramosa Fair had one class for the "Best Butter fit for exportation," and one for "Best Butter for immediate use." "Eramosa Show," *Guelph & Galt Advertiser*, 8 October 1847.

84 Marling, "'She Brought Forth Butter in a Lordly Dish,'" 220.
85 "Butter Making," *Canadian Agriculturist*, May 1859, 110.
86 See Quaile, "Dairy Pin-Up Girls," 19–31.
87 Cohen, *Women's Work*, 96–7. At the 1847 Puslinch Fair, the "housewives" were praised for the excellent dairy exhibits and the "great credit" they bestowed on the township, and the next year the "wives and daughters" of Puslinch again elicited praise from the judges for making their community proud with the fine dairy products they exhibited. "Puslinch Agricultural Show," *Guelph & Galt Advertiser*, 22 October 1847, and "Puslinch Show," *Guelph & Galt Advertiser*, 5 October 1848.
88 "Eramosa Agricultural Show," *Guelph Advertiser*, 17 October 1850.
89 "County Agricultural Exhibition," *Guelph Advertiser*, 26 October 1854.
90 "Nichol Agricultural Show," *Guelph & Galt Advertiser*, 5 October 1848.
91 "Annual report for the South Huron Agricultural Society," 31 January 1879, Records of Agricultural Societies, Box 1, File 2, XA1 RHC A0386023, University of Guelph Archival & Special Collections.
92 "Guelph Central Fair," *Guelph Daily Mercury*, 2 October 2, 1884. A local report lamented that the "show of dairy produce would scarcely do credit to a township fair" because there were so few entries.
93 Cohen, *Women's Work*, 108.
94 Crowley, "Experience and Representation," 243.
95 "County Show – Upper Canada, Perth," *Canadian Agriculturist*, 16 December 1860, 654.
96 "Another Success: The West Elgin Fair Still Adds to Its Reputation," *Advance* [Dutton], 2 October 1890.
97 "The Markham Fair," *Globe* [Toronto], 6 October 1894.
98 Drummond, *Progress without Planning*, 36; Reaman, *A History of Agriculture in Ontario*, vol. 1, 86; and Cohen, *Women's Work*, 106–9.
99 Cohen, *Women's Work*, 107–9.
100 Ibid., 109.
101 "Erin Fall Show," *Guelph Weekly Mercury and Advertiser*, 24 October 1901.
102 "Brampton Fair, Delightful Weather," *Brampton Conservator*, 4 October 1878.
103 Ibid.
104 "Peel County Fall Fair," *Brampton Times*, 27 August 1869.
105 "The Toronto Fair," *Brampton Conservator*, 26 September 1884.

106 "Eramosa Show," *The Guelph Advertiser*, 6 October 1853. It was reported that at the 1879 Markham Fair, there was "an immense show of butter, and the judges, after two hours' laborious tasting, came to the conclusion that it is a first-class lot"; "Markham Fall Exhibition," *Daily Globe* [Toronto], 4 October 1879.

107 "West Riding and Vaughan Agricultural Society," *York Herald*, Friday, 30 October 1868.

108 "Most Successful Fair for Years in Dufferin," *Globe* [Toronto], 23 September 1912.

109 "Puslinch Fall Show," *Guelph Weekly Mercury and Advertiser*, 6 October 1892.

110 Reaman, *A History of Agriculture in Ontario*, vol. 2, 41.

111 Mrs A. Drysdale, ""Women's Division Reports; District No. 2," in *Ontario Department of Agriculture 97th Annual report of Ontario Agricultural Societies*, 68.

112 Mrs Kestle, "Women's Division Reports; District No. 2," in *Ontario Department of Agriculture 97th Annual report of Ontario Agricultural Societies*, 70.

113 Mrs W.C. Huckle, "Women's Division Reports; District No. 2," in *Ontario Department of Agriculture 97th Annual report of Ontario Agricultural Societies*, 70.

114 Reaman, *A History of Agriculture in Ontario*, vol. 2, 160. See also Adams, *The Transformation of Rural Life*, 189–90. Dairy specialization was also evidenced in the United States at this time, resulting in the loss of women's dairy income there too.

115 Henry Ives, "Prize Essay: The Farmer's Garden," *Farmer's Advocate and Home Magazine* 22 (April 1887): 109.

116 "Gardening as Women's Work," *Farmer's Advocate* 6 (June 1871): 86.

117 "Lady Farmers," *Farmer's Advocate* 7 (March 1872): 43.

118 "Economical Gardening for Women – An Example," *Ontario Farmer* 3 (1871 September): 440.

119 See the *British American Cultivator* and *Canadian Agriculturist* for monthly and annual reports of the Agricultural Association of Upper Canada and of township and county agricultural societies during the mid nineteenth century.

120 Trade catalogues, such as William W. Custead, *Catalogue of Fruit & Ornamental Trees, Flowering Shrubs, Garden Seeds and Greenhouse Plants, Bul-*

bous Roots & Flower Seeds Cultivated and for Sale at the Toronto Nursery (York: William Lyon Mackenzie, printer to the House of Assembly, 1827), and newspaper advertisements, such as those found in the *Colonial Advocate*, *Gleaner and Niagara*, *Gazette* [Toronto], and *Canadian Freeman*, provide evidence of the nursery stock available.

121 "Erin Agricultural Show: List of Premiums," *Guelph Advertiser*, 2 November 1854, and "Township of Erin," *Guelph Advertiser*, 8 November 1855.
122 Reaman, *A History of Agriculture in Ontario*, vol. 1, 30.
123 Hoffman and Taylor, *Much To Be Done*, 84; Bradbury, *Working Families*, 163; Gaffield, *Language, Schooling, and Cultural Conflict*, 67.
124 "Erin Agricultural Show," *Guelph Advertiser*, 2 November 1854.
125 "Erin Fall Show," *Guelph Weekly Mercury and Advertiser*, 24 October 1901.
126 "Erin Fall Show," *Guelph Weekly Mercury and Advertiser*, 26 October 1893, and "Erin Show, A Fine Exhibition and a Large Attendance," *Guelph Weekly Mercury and Advertiser*, 29 October 1896. Potatoes were especially popular at the Erin Fair, and local newspapers regularly commented on the strength of the potato competition and the township's reputation as a significant producer in the county. In the nineteenth century, potatoes were raised as a special crop in Ontario, and farms located on high, dry, and light soils were especially well-suited for their cultivation; Jones, *History of Agriculture in Ontario*, 317.
127 "West Riding of York and Township of Vaughan Agricultural Societies' Exhibition," *Brampton Conservator*, 24 October 1879.
128 "Peel Fall Fair! Another Grand Success," *Brampton Times*, 6 October 1871, and "Brampton Fair," *Brampton Conservator*, 11 October 1878.
129 Jones, *History of Agriculture in Ontario*, 319.
130 W.T. Macoun, "Canada's Horticulture Heritage and Harvest," *Farmer's Advocate and Home Magazine* 54 (11 December 1919): 22–7.
131 Jones, *History of Agriculture in Ontario*, 318.
132 Ontario Department of Agriculture, "Apples," *Bulletin XV*, Agricultural College, Guelph, 18 August 1887.
133 Reaman, *History of Agriculture in Ontario*, vol. 1, 99; Drummond, *Progress without Planning*, 37.
134 "Erin Fall Show," *Guelph Weekly Mercury*, 20 October 1892.
135 "Average Crowd Attended Erin Fair despite Inclement Weather," *Erin Advocate*, 15 October 1925.

136 Interview with Jessie Milton, 28 October 2013.
137 "Apples," *Canadian Agriculturist*, 16 December 1860, 662.
138 Interview with Jessie Milton, 28 October 2013.
139 "Fall Fairs on 1942 Model. Prize Lists and Events Will Reflect the Spirit of Our Times – Women Plan Educational Features," *Farmer's Advocate and Home Magazine* 77 (26 February 1942): 100.
140 "The Girls Who Saved the Apple Crop. All Over Canada Farm Women Are Doing Their Full Share – in the Essential Service of Food Production," *Farmer's Advocate and Home Magazine* (28 May 1942): 316, and "Youth on Its Mettle. Young Women of the Farm Rally to the Colours – Gardens Are Cultivated with Patriotic Fervour," *Farmer's Advocate and Home Magazine* (24 September 1942): 557.
141 Bentley, *Eating for Victory*, 114–41.
142 Lee, *Head, Heart, Hands, Health*, 187.
143 "Prize List, Collingwood Township Agricultural Society for the 1943 Fall Fair at Clarksburg," Ontario Fall Fair Catalogues Collection, XA1 MS A073, University of Guelph Archival & Special Collections. The Collingwood Township Agricultural Society sponsored the "Girl's Garden Brigade Exhibit" in 1943, which had three classes, including one for an exhibit from first-year members that included "1 pint or 1 quart of canned tomatoes, 1 record book, 1 specimen of 5 kinds of vegetables, 5 specimens of 1 kind of vegetable not included in the single samples." The second class was for entries from second-year members that included a poster and exhibit on the subject "Vegetables To Keep Us Fit." The third class was open to third-year members, and required a vegetable exhibit and poster on the subject "Our Well Stocked Cellar Conforms to Canada's Official Food Rules." The prizes ranged from 75 cents for worthy participation to $3.00 for the first-place exhibit.
144 For a study of food and food rationing during the Second World War, see Mosby, *Food Will Win the War*.
145 "New Seeds for 1964," *Farmer's Advocate* 99 (22 February 1964): 35.
146 Ibid.
147 "Family Jars," *Canadian Agriculturist* (April 1855): 120.
148 Gal, "Grassroots Consumption," 149–50.
149 Minnie May, "Minnie May's Department," *Farmer's Advocate and Home Magazine* 23 (August 1888): 247.

150 Ibid.
151 Root and de Rochemont, *Eating in America*, 156.
152 Levenstein, *Revolution at the Table: The Transformation of the American Diet*, 28.
153 Crowley, "Rural Labour," 34.
154 Ambrose, *For Home and Country*, 102.
155 At the 1918 C.N.E., the Consumers' Gas Company of Toronto sponsored the C.N.E. exhibit of food preservation using ring-top preserving jars: "Can What You Can't Eat" and "If U Can U Should Can." Consumer Gas Company Ltd. Fonds 1034, Item 712, City of Toronto Archives.
156 Taylor, *Fashioning Farmers*, 76.
157 Ibid., 76–82.
158 Neth, *Preserving the Family Farm*, 196.
159 Milkman, "Women's Work and Economic Crisis," 82.
160 Ibid.
161 Drummond, *Progress without Planning*, 39.
162 Interview with Jessie Milton, 28 October 2013.
163 Interview with Glenda Benton, 28 October 2013.
164 Ibid.
165 "Prize List of the Brooke and Alvinston Fall Fair, 1920," Ontario Fall Fair Catalogues, XA1 MS A073, University of Guelph Archival & Special Collections.
166 "Collingwood Township Agricultural Society Fall Fair Prize List, 1930"; "Collingwood Township Agricultural Society Fall Fair Prize List, 1935"; "Collingwood Township Agricultural Society Fall Fair Prize List, 1940"; "Collingwood Township Agricultural Society Fall Fair Prize List, 1945"; "Collingwood Township Agricultural Society Fall Fair Prize List, 1950." All in Ontario Fall Fair Catalogues, XA1 MS A073, University of Guelph Archival & Special Collections.
167 Neth, *Preserving the Family Farm*, 240.
168 van de Vorst, *Making Ends Meet*, 65; Comacchio, *The Infinite Bonds of Family*, 34.
169 Tye, *Baking as Biography*, 4.
170 "Woman of 84 Still Wins Prizes at Ridgetown Fair," *Globe and Mail*, 7 October 1938.

171 Bradbury, *Working Families*, 159, 167; Strasser, *Never Done*, 23–4; and Hoffman and Taylor, *Much To Be Done*, 104.
172 Cowan, *More Work for Mother*, 51.
173 Ibid.
174 R.J. Culverwell, "What To Eat, Drink, and Avoid – A Guide to Health and Long Life," *Canadian Agriculturist* 1 (September 1848): 249, and "Something about Bread-Making," *Ontario Farmer* 3 (June 1871): 190.
175 Diary of Matilda Bowers Eby, February 20, 1863, in Hoffman and Taylor, *Much to Be Done*, 54.
176 Minnie May, Minnie May's Department, *Farmer's Advocate* 12 (June 1877): 138.
177 "Guelph Township Agricultural Show," *Guelph Mercury*, 7 October 1864.
178 "Eramosa Agricultural Show," *Guelph Mercury*, 14 October 1864.
179 "County of Peel Fall Fair," *Brampton Times*, 27 August 1869.
180 "Brampton Fair, Delightful Weather – Immense Crowds – Magnificent Display, Receipts Far Above Any Former County Exhibition," *Brampton Conservator*, 4 October 1878.
181 "Erin Exhibition," *Guelph Mercury*, 24 October 1889; "Eramosa Show," *Guelph Mercury*, 10 October 1889; and "Central Agricultural Society Prize List of the First Annual Fall Exhibition, 1900." Ontario Fall Fair Catalogues, XA1 MS A073, University of Guelph Archival & Special Collections.
182 Root and de Rochemont, *Eating in America*, 225.
183 "Erin Fall Show," *Guelph Weekly Mercury and Advertiser*, 21 October 1897.
184 "Women and the Local Fair," *Farmer's Advocate and Home Magazine* 73 (24 February 1938): 100.
185 "Nassagaweya Fall Show," *Guelph Mercury and Advertiser*, 17 October 1889.
186 "The Markham Fair," *Globe* [Toronto], 6 October 1894.
187 "Purity Flour," *Farmer's Advocate and Home Magazine* 54 (October 2, 1919): 1753.
188 "West Durham Fair," *Globe* [Toronto], 18 September 1899.
189 "Judge's Decision Is Final," 1959, *St. Thomas Times-Journal* fonds, Elgin County Archives, C8 Sh3 B1 F11 1.
190 "Breadmaking," *Farmer's Advocate and Home Magazine* 55 (12 February 1920): 270.
191 "Central Agricultural Society Prize List of the Second Annual Fall Exhibition, 1890," Ontario Fall Fair Catalogues, XA1 MS A073, University of Guelph Archival & Special Collections.

Notes to pages 94–9

192 "Prize List of the Twelfth Annual Fall Exhibition to be Held in Walters Falls, 1900," Ontario Fall Fair Catalogues, XA1 MS A073, University of Guelph Archival & Special Collections.
193 "Prize List of the Twenty-Second Annual Fall Exhibition, to be Held at Walters Falls, 1910," Ontario Fall Fair Catalogues, XA1 MS A073, University of Guelph Archival & Special Collections.
194 "Prize List of the Pinkerton Exhibition, 1925," Pinkerton Fall Fair Prize Lists, A2011.502.001, Bruce County Archives.
195 Ibid.
196 Collingwood Township Agricultural Society Fall Fair Prize List, 1940," Ontario Fall Fair Catalogues, XA1 MS A073, University of Guelph Archival & Special Collections.
197 "100th Anniversary Puslinch Agricultural Society Prize List, 1940," Fergus Fall Fair Collection, MU 202, A1978.30, Wellington County Museum and Archives.
198 "Erin Fall Fair Prize List, 1940," Erin Agricultural Society collection, 1862–1987, A1989.97, Wellington County Museum and Archives. Organizers asked exhibitors to submit a grocer's sales slip showing the purchase of the flour within "a reasonable time prior to the date of the contest."
199 Ibid.
200 Ibid.
201 "Brussels Fall Fair Prize List, 1954," Ontario Fall Fair Catalogues, XA1 MS A073, University of Guelph Archival & Special Collections.
202 Ibid.
203 "Dorchester Fair Prize List, 1966," Ontario Fall Fair Catalogues, XA1 MS A073, University of Guelph Archival & Special Collections. The Dorchester Fair also had an additional thirteen classes sponsored by larger companies such as Standard Brands Limited (makers of Magic Baking Powder and Fleischmann's Yeast), J.M. Scheider Limited (Crispy-flake Shortening), and Fry-Cadbury (Fry's Cocoa), as well as six additional special prize classes sponsored by local businesses for baked goods.
204 Cowan, *More Work for Mother*, 71–2.
205 Mink, "Cooking in the Countryside," 12.
206 Ibid.
207 "Prize List of the Arran-Tara Agricultural Society Fall Fair, 1955," Arran-Tara Fall Fair Booklets, A959.034.001, Bruce County Archives, and "Prize List of the Arran-Tara Agricultural Society Fall Fair, 1960," Arran-Tara

Agricultural Society Prize Lists, private collection of Ron Hammell, Dobbinton, Ontario.
208 Walden, *Becoming Modern*, 80–118; Jones, *Midways, Judges, and Smooth-Tongued Fakirs*, 4.
209 Mrs Wm. Beattie, "Women's Division Reports; No. 9," *Ninety-Ninth Annual Report of Ontario Agricultural Societies*, 45.
210 "Record Crowd Attend Ont.'s Preview of The Royal," *Erin Advocate*, 13 October 1960.
211 See Chapter 5 for a discussion of the "Beauty Queen" and pageant culture of the postwar period.
212 "A Family Practice," 1968, *St. Thomas Times-Journal* fonds, Elgin County Archives, C8 Sh3 B1 F19 10.
213 "Prize List of the Arran-Tara Agricultual Society Fall Fair, 1960," Arran-Tara Agricultural Society Prize Lists, private collection of Ron Hammell, Dobbinton, Ontario, and "Erin Fall Fair Prize List, 1974," Erin Agricultural Society fonds, A1989.97, Wellington County Museum and Archives.
214 Elva Fletcher, "Spotlight on Homemade Breads," *Country Guide: The Farm Magazine* (February 1979): 68–9. The Ontario Ministry of Agriculture and Food (OMAF) economics branch promoted programs for home baking, cooking, and other domestic manufactures, and hundreds of women enrolled in these courses because they were pleased with the skills they learned and developed. The 200 women from Kent and Essex counties who enrolled in OMAF's "Spotlight on Bread" project were satisfied they gained valuable knowledge to improve their bread making, which was important because they believed "there's something special about homemade bread."
215 See Gal, "Grassroots Consumption."

Chapter Three

1 Bushman, *The Refinement of America*, 440–1.
2 Taylor, *Fashioning Farmers*, 82.
3 See trade catalogues such as Custead, *Catalogue of Fruit & Ornamental Trees*, and newspapers such as the *Colonial Advocate*, the *British American Cultivator*, the *Gleaner and Niagara*, the Toronto *Gazette*, and the *Canadian Freeman*.

4 Traill, *The Canadian Emigrant Housekeeper's Guide* and *The Backwoods of Canada*. Colin Coates discusses the desire of immigrants to replicate landscapes from their homeland in *The Metamorphoses of Landscape*.
5 Like other fairs of the period, the first Provincial Exhibition held in 1846 did not include flower classes.
6 Reaman, *A History of Agriculture in Ontario*, 96.
7 "Elora Horticultural Society," *Guelph Advertiser*, 25 September 1851. The other twenty-four classes were all for various kinds of vegetables and fruits, including beans, melons, carrots, parsnips, potatoes, onions, beets, kale, cabbage, tomatoes, celery, pumpkins, squash, citrons, cucumber, hops, apples, and plums.
8 "The Agricultural Association," in *Journal and Transactions of the Board of Agriculture of Upper Canada*, vol. 1, 25.
9 Victoria and Alberta Museum, "Mrs. Loudon & the Victorian Garden."
10 Loudon, *Gardening for Ladies*, xi.
11 von Baeyer, *Rhetoric and Roses*, 2.
12 "On Horticulture," *British American Cultivator* (April 1847): 126.
13 Ibid.
14 Ibid., 126–7.
15 "On Horticulture," *British American Cultivator* (April 1847): 127.
16 Ibid.
17 W.T.G., "Gardening in Canada," in *Canada Farmer* (1864), in Baeyer and Crawford, *Garden Voices*, 6.
18 Ibid.
19 "Home Adornment," *Farmer's Advocate* 3 (September 1868): 133.
20 "Letter to Minnie May," Minnie May's Department, *Farmer's Advocate* 9 (August 1874): 124.
21 "Illustrated Florist," *Farmer's Advocate* 10 (June 1875): 117.
22 Ibid.
23 "Flowers in the Country," *Farmer's Advocate* 11 (May 1876): 102.
24 "For What Is a Mother Responsible?" *British American Cultivator* (February 1846): 56.
25 Ibid.
26 "To the Ladies," *Farmer's Advocate* 7 (March 1872): 41.
27 "Pansies," *Farmer's Advocate* 4 (July 1869): 104; and "Ladies!" *Farmer's Advocate* 7 (May 1872): 73.

28 "Annual Show of the Provincial Agricultural Association," *Canadian Agriculturist* (1849): 258.
29 "County of Peel Fall Fair," *Brampton Times*, 9 October 1868.
30 "Township of Erin," *Guelph Advertiser*, 8 November 1855. Miss McKenzie was awarded a discretionary prize for her "Vase Flowers." In 1865, James Brown won a discretionary prize for a "Wreath of Flowers"; only men's names were listed in the winners' list that year, including for "Domestic Manufactures and Ladies' Work," so whether Mr or Mrs Brown constructed the floral wreath is unclear; "The Erin Fall Show," *Guelph Mercury*, 19 October 1865.
31 "Erin Exhibition," *Guelph Mercury*, 24 October 1889.
32 "Hespeler's First Show: Fine Weather and a Great Success," *Guelph Mercury*, 10 October 1889.
33 "Ladies at the Fall Fairs," *London Advertiser*, reprinted in *Brampton Times*, 16 October 1868.
34 For example, the same year that the *Brampton Times* reporter singled out women for their floral exhibits at the Peel County Fair in 1868, the printed winners' list revealed that both Judge Scott and Rev. R. Arnold won some of the top flower prizes. A judge and a reverend were educated men who were expected to have cultured tastes and could appreciate beauty. Judge Scott became a long-time exhibitor, eventually becoming a flower show judge.
35 "Central Exhibition: A Successful Show – The Receipts Fair," *Guelph Mercury*, 2 October 1890.
36 Ibid.
37 "County of Peel Fall Fair," *Brampton Times*, 9 October 1868.
38 Couckuyt, *The Aylmer Fair*, 50.
39 "Peel Fall Fair! Another Grand Success!" *Brampton Times*, 6 October 1871.
40 "Average Crowd Attended Erin Fair despite Inclement Weather," *Erin Advocate*, 15 October 1925. The reporter commented that the event was "a splendid opportunity for milady to show her new fur coat," and he noted that Miss Ella McQuarrie was "judged to be the finest looking young lady on the grounds."
41 "Toronto Township Fall Fair," *Weekly Review* [Brampton], 17 October 1857.
42 "Fall Fair at Markham," *York Herald*, 15 October 1869.
43 Ibid.

44 "Agricultural Exhibitions," *Farmer's Advocate* 3 (January 1868): 2.
45 George Douglas, "President's Address, Western Fair, 1880," East-Middlesex Agricultural Society Minute Book, 1873–1913, B5289-1, University of Western Ontario Archives.
46 Heaman, *The Inglorious Arts of Peace*; Greenhalgh, *Ephemeral Vistas*; McAleer and MacKenzie, *Exhibiting the Empire*; and Raizman and Robey, *Expanding Nationalisms at World's Fairs*.
47 For example, wheat was a commodity that had great economic importance, but also significant cultural significance to Canada and its place in the British Empire. See Tosaj, "Weaving the Imperial Breadbasket," 249–75.
48 George Douglas, "President's Address, Western Fair, 1880," East-Middlesex Agricultural Society Minute Book, 1873–1913, B5289-1, University of Western Ontario Archives.
49 Reaman, *A History of Agriculture in Ontario*, 149.
50 "The Farm," *Brampton Conservator*, 4 April 1879.
51 Harris and Mueller, "Making Science Beautiful," 110. Depopulation was an increasing concern to rural reformers in the early twentieth century, and the same types of projects were implemented to fight against migration and urban influence. See Young, "Conscription, Rural Depopulation, and the Farmers of Ontario," and Halpern, *And on That Farm He Had a Wife*, 88, 105.
52 Harris and Mueller, "Making Science Beautiful," 119. The Department of Agriculture, in conjunction with the Experimental Farms Branch, made beautification a priority at experimental farms across the province and the country. The Central Experimental Farm in Ottawa was designed to showcase such beautification.
53 "Farmer's Homes," *Farmer's Advocate* 6 (June 1871): 84.
54 Anon, "The Rights of Farmers' Girls," *Guelph Weekly Mercury*, 1 December 1881.
55 Schlereth, *Victorian America*, 136.
56 von Baeyer, *Rhetoric and Roses*, 38.
57 Ibid., 6.
58 Tessier, *Spencerville Fair*, 239.
59 "Flower Seeds," *Farmer's Advocate* 7 (February 1872): 24.
60 "The "Landscape" Lawn Mower," *Ontario Farmer* 3 (June 1871): 179.
61 von Baeyer, "The Home Gardener of the 1880s," 199.

62 Ibid., 200.
63 Mary McCulloch Diary, Friday, 15 April 1898. Peel Art Gallery, Museum, and Archives.
64 Ibid., Friday, 22 April 1898.
65 Ibid., Monday, 25 April; Friday, 29 April; Tuesday, 10 May; Friday, 13 May; Wednesday, 18 May; Thursday, 26 May; Wednesday, 8 June; Monday, 13 June; Tuesday, 14 June; Friday, 17 June; Tuesday, 28 June, and Wednesday, 4 October 1898. These dates represent entries specifically mentioning flowers; more general references to garden care and maintenance have not been included in this list, but it was a routine task throughout the spring, summer, and fall months.
66 Ibid., Monday, 25 April; Friday, 29 April.
67 Year: *1881*; Census Place: *Toronto Gore, Peel, Ontario*; Roll: *C_13252*; Page: *27*; Family No: *119*.
68 Year: *1881*; Census Place: *Brampton, Peel, Ontario*; Roll: *C_13252*; Page: *1*; Family No: *3*.
69 Year: *1881*; Census Place: *Chinguacousy, Peel, Ontario*; Roll: *C_13252*; Page: *18*; Family No: *92*.
70 Brampton Centennial Committee, *Brampton's 100th Anniversary*.
71 "Brampton Fair, Delightful Weather – Immense Crowds – Magnificent Display, Receipts Far Above Any Former County Exhibition," *Brampton Conservator*, 4 October 1878.
72 L. Woolverton, M.A., Grimsby, "Horticultural Specialties for the Canadian Farmer," and M. Pettit, Winona, Ont., "Ornamental and Profitable Tree-planting," *Farmer's Advocate and Home Magazine* 24 (April 1889): 116–17.
73 The exhibitors taking top prizes in these classes at the 1890 Central Fair in Guelph were all nurserymen. "Central Exhibition," *Guelph Mercury*, 2 October 1890.
74 "Nassagaweya Fall Show," *Guelph Mercury*, 17 October 1889. The first and second winners were as follows: "Selection of natural flowers, Mrs Joshua Norrish, Mrs Jas Ramsay; Selection of window plants, Mrs McLaughlan, Mrs W Anderson; Bouquet for table, Mrs Joshua Norrish, Mrs Jas Ramsay; Bouquet for hand, Mrs James Ramsay, Geo Strange."
75 "A Garden Talk," The Ingle Nook, *Farmer's Advocate and Home Magazine* 51 (23 March 1916): 521.
76 Reaman, *A History of Agriculture in Ontario*, vol. 2, 64–6.

Notes to pages 119–21

77 "Erin Fair: A Big Success," *Erin Advocate*, 19 October 1910.
78 "Fall Fair of Egremont Agricultural Society Prize List, 1910," Ontario Fall Fair Catalogues Collection, XA1 MS A073, University of Guelph Archival & Special Collections.
79 "Shedden Fair a Big Success," *Dutton Advance*, 20 October 1910.
80 "Prize List of the Arran-Tara Annual Fall Exhibition, 1910," Arran-Tara Fall Fair Booklets, A959.034.001, Bruce County Archives.
81 Reaman, *A History of Agriculture in Ontario*, 64–6; von Baeyer, *Roses and Rhetoric*, 165–7.
82 von Baeyer, *Roses and Rhetoric*, 166–7.
83 Ibid.
84 Ibid., 165.
85 "Our Gardens Owe Much to These Leaders," *Farmer's Advocate and Home Magazine* 73 (10 March 1938): 128. Preston was presented with the J.E. Carter medal for outstanding contributions in the origination of ornamental plants at the Ontario Horticultural Association's annual meeting in Toronto.
86 Junia, "A Garden Talk," The Ingle Nook, *Farmer's Advocate and Home Magazine* 51 (23 March 1916): 521.
87 Polly Primrose, Northumberland, Ont., "A Woman's Experience," *Farmer's Advocate and Home Magazine* 51 (27 April 1916): 767.
88 "Aylmer Fair Drew Big Attendance," *Aylmer Express*, 2 October 1919.
89 "Average Crowd Attended Erin Fair despite Inclement Weather."
90 "Language of Flowers," *Farmer's Advocate* 10 (March 1875): 46.
91 A.L. Potts, "Gleaned in Gardens," *Canadian Homes and Gardens* (April 1926): 96.
92 Ada L. Potts, "Gleaned in Gardens," *Canadian Homes and Gardens* (September 1926): 110.
93 "Prof. F.A. Waugh, "Useful Landscape Gardening on the Farm," *Farmer's Advocate and Home Magazine* 54 (11 December 1919): 2218.
94 Baskerville, *Sites of Power*, 193.
95 "Finest Exhibit and Largest Crowd Ever at Erin Fall Fair," *Erin Advocate*, 11 October 1934.
96 Levine, *The Unpredictable Past*, 266.
97 Ibid., 257–8.
98 Ibid., 258.

99 John F. Clark, "Staging Horticultural Products at a Fair," in *Ontario Department of Agriculture Thirty-First Annuals Report of the Agricultural Societies and of the Convention of the Association of Fairs and Exhibitions*, 19.
100 See Chapter 1 for a full analysis of shifts in women's roles in agricultural societies.
101 "Prize List of the Arran-Tara Annual Exhibition, 1910," and "Arran-Tara Agricultural Society Annual Fall Fair, 1933," Arran-Tara Fall Fair Booklets, A959.034.001, Bruce County Archives. Arran-Tara Fair had had a large flower show before many other fairs, but its classes grew from twenty-two before the First World War to thirty-four in 1933 once an all-female committee managed it.
102 "Prize List, Arran-Tara Agricultural Society Fall Exhibition, 1950," Arran-Tara Fall Fair Booklets, A959.034.001, Bruce County Archives.
103 "Seaforth Fall Fair Prize List, 1925," and "Seaforth Fall Fair Prize List, 1936," Ontario Fall Fair Catalogues Collection, XA1 MS A073, University of Guelph Archival & Special Collections.
104 "East Huron Fall Fair Prize List, 1936," Ontario Fall Fair Catalogues Collection, XA1 MS A073, University of Guelph Archival & Special Collections.
105 "Forest Agricultural Society Annual Exhibition Prize List, 1936," "Brooke and Alvinston Fall Fair Prize List, 1936," "1936 Ingersoll Fair Prize List," and "Drumbo Fair Prize List, 1936," Ontario Fall Fair Catalogues Collection, XA1 MS A073, University of Guelph Archival & Special Collections.
106 Rockwell and Grayson, *Flower Arrangement*.
107 Rockwell and Grayson were prolific writers, and many of their books were reprinted and revised over the years. Their work includes *Gardening Indoors*; *Flower Arrangement in Color*; *The Complete Book of Flower Arrangement*; *The Complete Book of Bulbs*; *The Complete Book of Annuals*; *The Rockwells' Complete Book of Roses*; *New Complete Book of Flower Arrangement*; *The Complete Book of Lilies*; *The Rockwells' Complete Guide to Successful Gardening*.
108 Rockwell and Grayson, *Flower Arrangement*, vii.
109 Ibid.
110 Rockwell and Grayson, *Flower Arrangement*, 3.
111 Ibid.
112 Beausaert, "Leisure, Consumption, and Cosmopolitanism," 215–47.
113 Ibid., 31.

114 Ibid., 38–45.
115 "Artistry with Summer's Bloom," *Canadian Home and Garden* (June 1940): 30.
116 Wilson, "Women and Gardening," [1933] in Baeyer and Crawford, *Garden Voices*, 13.
117 Ibid., 14–15.
118 Ibid., 15.
119 "Collingwood Township Agricultural Society Fall Fair Prize List, 1940." In 1945, one man joined three women to run the show, but by 1950 the committee was again all female. See "Collingwood Township Agricultural Society Fall Fair Prize List, 1945," and "Collingwood Township Agricultural Society Fall Fair Prize List, 1950," Ontario Fall Fair Catalogues, XA1 MS A073, University of Guelph Archival & Special Collections.
120 For example, see "Arran-Tara Agricultural Society Annual Exhibition Prize List, 1949," Arran-Tara Fall Fair Booklets, A959.034.001, Bruce County Archives; "Erin Fall Fair Prize List, 1949," Erin Agricultural Society fonds, A1989.97, Wellington County Museum and Archives; "Burford Fair Prize List, 1945;" "Shedden Fair Prize List, 1949," Ontario Fall Fair Catalogues Collection, XA1 MS A073, University of Guelph Archival & Special Collections.
121 Holson, "Private Gardening in Suburban Sourthern Ontario in the 1950s," 1.
122 Ibid.
123 Ethel Chapman, "Fairs and Their Trends in Family Living," in *One Hundredth Annual Report of Ontario Agricultural Societies*, 84.
124 Ibid.
125 Miss Margaret Dove, "Flower Arrangements," in *One Hundred and First Annual Report of Ontario Agricultural Societies*, 84.
126 Ibid.
127 Ibid., 85.
128 A.E. Hick, "The "New Look" in Flowers," *Your Home and Garden* 4, no. 1 (January 1950): 13.
129 Ibid.
130 "Erin Fall Fair Prize List, 1959," Erin Agricultural Society fonds, A1989.97, Wellington County Museum and Archives.
131 "Prize List of the Arran-Tara Agricultural Society Fall Fair, 1965," private collection of Ron Hammell, Dobbinton, Ontario.
132 "New Seeds for 1964," *Farmer's Advocate* 99 (22 February 1964), 35.

133 Laura Chisholm, "Living with Flowers," *Farmer's Advocate* 99 (12 September 1964): 28. "Glamourous Hostess" was a corsage class; "Barn Dance" was a composition using fruits, flowers, and weeds; "Tea time" was an arrangement in a cup and saucer; and "Housewife's Dream" was an arrangement that included a kitchen utensil in its composition.
134 "Erin Fall Fair Prize List, 1974," Erin Agricultural Society fonds, A1989.97, Wellington County Museum and Archives.
135 "Erin Fair Ladies Division, Home and Field Classes, 1976," Erin Agricultural Society Private Collection, Erin, Ontario.
136 "Erin Fall Fair Prize List, 1974."
137 "Erin Fair Ladies Division, Home and Field Classes, 1976.".
138 "Prize List of the Arran-Tara Agricultural Society Fall Fair, 1974," private collection of Ron Hammell, Dobbinton, Ontario.

Chapter Four

1 Hunter and Yates, "Introduction: The Question of Thrift," 13–14.
2 McCalla, *Planting the Province*, 100.
3 "Frontenac Cattle Show," *British Whig* [Kingston], 14 October 1835.
4 Heaman, *The Inglorious Arts of Peace*, 265; and Kelly, "'The Consumption of Rural Prosperity and Happiness," 580–1.
5 Comacchio, *The Infinite Bonds of Family*, 34; Hoffman and Taylor, eds., *Much to Be Done*, 56; and Bradbury, "The Home as Workplace," 448–9.
6 "Hints on Housekeeping," *Farmer's Advocate* 3 (April 1868): 60.
7 "The Managing Woman," *Farmer's Advocate* 3 (May 1868): 71.
8 Kelly, "The Consumption of Rural Prosperity and Happiness," 581.
9 "Toronto Township Fall Fair," *Weekly Review* [Brampton], 17 October 1857.
10 Hunter and Yates, "Introduction: The Question of Thrift," 13.
11 Ulrich, *The Age of Homespun*, 413.
12 Samson, *The Spirit of Industry and Improvement*, 12.
13 Ferry, *Uniting in Measures of Common Good*, 170.
14 "Puslinch Independent Fall Show," *Guelph Mercury*, 11 October 1866.
15 McCalla, *Consumers in the Bush*, 37–66, and McCalla, "Textile Purchases by Some Ordinary Upper Canadians," 4–27.
16 Inwood and Wagg, "The Survival of Handloom Weaving in Rural Canada circa 1870," 346–58.

Notes to pages 140–7

17 Craig, *Backwoods Consumers and Homespun Capitalists*, 183.
18 Ibid., 196.
19 "Erin Agricultural Show," *Guelph Mercury*, 21 October 1864; "Erin Exhibition," *Guelph Mercury*, 24 October 1889; and "Erin Fall Show," *Guelph Weekly Mercury and Advertiser*, 24 October 1901.
20 "Fall Fair at Markham," *York Herald*, 15 October 1869.
21 Inwood and Wagg, "The Survival of Handloom Weaving in Rural Canada circa 1870."
22 "Erin Fair," *Erin Advocate*, 19 October 1910.
23 Kelly, "The Consumption of Rural Prosperity and Happiness," 575–6.
24 Bradbury, "The Home as Workplace," 447.
25 Ibid., 448–9.
26 "For the Ladies," *The Farmers' Advocate* 2 (July 1867): 63.
27 Ulrich, *The Age of Homespun*, 415.
28 "Albion Fall Fair," *Brampton Times*, 16 October 1868.
29 "Eramosa Agricultural Society," *Guelph Mercury*, 18 January 1865.
30 "Peel County Fair," *Brampton Times*, 23 September 1870.
31 Ibid.
32 "Etobicoke Township Fair," *Daily Globe* [Toronto], 19 October 1881.
33 "The West Elgin Fair," *Advance* [Dutton], 18 September 1899.
34 "Beauty and Utility Combined," *Farmer's Advocate and Home Magazine* 34 (15 June 1899): 350.
35 Drummond, *Progress without Planning*, 6.
36 Ibid., 107.
37 Baskerville, *Sites of Power*, 159.
38 Bushman, *The Refinement of America*, xviii.
39 Handlin, *The American Home*, 458.
40 Marwick, *Beauty in History*, 220.
41 Minnie May, "Minnie May's Department," *Farmer's Advocate and Home Magazine* 23 (October 1888): 318.
42 "Education of Farmers' Daughters," *British American Cultivator* (March 1846): 89.
43 Ibid.
44 "Female Education," *Canadian Agriculturalist* 1 (January 1848): 27.
45 "Fanny Fern on Farmer's Wives," *Farmer's Advocate* 3 (February 1868): 19.
46 Schlereth, *Victorian America*, 120–1.

47 Kelly, "The Consumption of Rural Prosperity and Happiness," 592.
48 Cowan, *More Work for Mother*, 29–31.
49 Nurse, "Reaching Rural Ontario," 69. A study of the 1871 Peel County Fair winners' prize list reveals that across categories of women's work, women were more likely to be married than not. Still, 28 per cent of female exhibitors were single, a percentage that was six times greater than for men.
50 Minnie May, "Minnie May's Department," *Farmer's Advocate* 12 (May 1877): 115.
51 Volz, "The Modern Look of the Early-Twentieth-Century House," 33.
52 Minnie May, "Minnie May's Department," *Farmer's Advocate and Home Magazine* 21 (November 1886): 341.
53 "Peel Fall Exhibition," *Brampton Conservator*, 1 August 1879.
54 "Harriston Fall Show," *Mount Forest Confederate*, 9 October 1902.
55 "Erin Exhibition," *Guelph Mercury*, 24 October 1889.
56 Bowles, *Homespun Handicrafts*, 224.
57 Wardle, *Victorian Lace*, 31.
58 Hawkins, *Old Point Lace*, 11.
59 Ibid., 12–13.
60 Bushman, *The Refinement of America*, 326–7.
61 Parker, *The Subversive Stitch*, 158.
62 Ibid., 159.
63 Bushman, *The Refinement of America*, 327–8.
64 Ibid., 330.
65 Ulrich, *The Age of Homespun*, 111.
66 "Nichol and Pilkington Union Show," *Guelph Mercury*, 11 October 1866. Mrs Walker was signalled out for her "beautiful specimen of hair work."
67 Sheumaker, *Love Entwined*, xii.
68 "Mrs. Hamlin's Family History Wreath," *The Lost Art of Sentimental Hairwork*.
69 Grier, "The Decline of the Memory Palace," 56; 59.
70 Ulrich, *The Age of Homespun*, 117.
71 "The Art of Water-Color Landscape Painting," *Farmer's Advocate and Home Magazine* 21 (November 1886): 343.
72 See Weimann, *The Fair Women*; Palm, "Women Muralists, Modern Women and Feminine Spaces"; and Corn, *Women Building History*.

73 "The Arts' Department at the Recent Provincial Show," *Ontario Farmer* (October 1869): 318.
74 "The Arts Department of the Recent Provincial Exhibition," *Ontario Farmer* (November 1870): 344.
75 "The Guelph Central Fair, Opening Day, Number of Entries," *Daily Globe* [Toronto], 16 September 1874.
76 Lewis, *Myles Birkett Foster*.
77 Ibid.
78 Tippett, *By a Lady*, 14.
79 Ibid.
80 Ibid., 11.
81 "Erin Agricultural Show," *Guelph Advertiser*, 3 November 1853.
82 "Arthur Agricultural Society," *Guelph Mercury*, 5 October 1865.
83 Harbeson, *American Needlework*, 113–17.
84 Drummond, *Progress without Planning*, 110. For a discussion of these changes in the nineteenth century, see also Craig, *Backwoods Consumers and Homespun Capitalists*, McCalla, *Consumers in the Bush* and "Textile Purchases by Some Ordinary Upper Canadians," Bradbury, "The Home as Workplace," and Inwood and Wagg, "The Survival of Handloom Weaving in Rural Canada."
85 Strasser, *Never Done*, 126.
86 Council on Rural Development Canada, *Rural Women*, 87.
87 Donica Belisle describes how debates about what constituted appropriate fashion in early twentieth-century Canada were common and tied to ideas about respectability. Belisle, *Purchasing Power*, 125.
88 Egremont Agricultural Society Fall Fair Prize List, 1917, Ontario Fall Fair Catalogues, XA1 MS A073, University of Guelph Archival & Special Collections.
89 See Glassford and Shaw, *A Sisterhood of Suffering and Service*, Sangster, "Mobilizing Women for War," and Street, "Bankers and Bomb Makers."
90 Federated Women's Institutes of Ontario, *Ontario Women's Institute Story*, 53.
91 "The Knitting Brigade," *Farmer's Advocate and Home Magazine* 51 (8 June 1916): 1003.
92 Ibid.

93 "Prize List, Seaforth Fall Fair, 1925," Ontario Fall Fair Catalogues, XA1 MS A073, University of Guelph Archival & Special Collections.
94 Gal, "Grassroots Consumption," 225–8; and Strasser, *Never Done*, 126.
95 Kechnie, *Organizing Rural Women*, 31–2.
96 "The Fashions," *Farmer's Advocate and Home Magazine* 54 (13 November 1919): 2052.
97 "A Thrift Campaign Needed," *Farmer's Advocate and Home Magazine* 55 (29 January 1920): 157.
98 van de Vorst, *Making Ends Meet*, 67.
99 Erin Fall Fair Prize List 1934," Erin Agricultural Society fonds, A1989.97, Wellington County Museum and Archives.
100 "Seventy-Fourth Annual Exhibition of the West Elgin Fair and Beef Cattle Show Prize List, 1934," ECVF Box 48, File 39, Elgin County Archives.
101 "Prize List, North Bruce & Saugeen Agricultural Society Fall Fair to be held at Port Elgin, 1934," Box 29, Bruce County Archives.
102 Neth, *Preserving the Family Farm*, 31.
103 Ibid., 31–2.
104 1930 Prize List, Collingwood Township Agricultural Society," Ontario Fall Fair Catalogues, XA1 MS A073, University of Guelph Archival & Special Collections.
105 Ibid.
106 Ibid.
107 Watson, *Ontario Red Cross*, 32.
108 Ibid.
109 See Pierson, *"They're Still Women After All."*
110 "Erin Fall Fair Prize List, 1940," Erin Agricultural Society collection, 1862–1987, A1989.97, Wellington County Museum and Archives.
111 "Collingwood Township Agricultural Society Fall Fair Prize List, 1940," Ontario Fall Fair Catalogues, XA1 MS A073, University of Guelph Archival & Special Collections.
112 "Sombra Agricultural Society, 1940 Prize List," Ontario Fall Fair Catalogues, XA1 MS A073, University of Guelph Archival and Special Collections.
113 "Burford Fair Prize List, 1945," Ontario Fall Fair Catalogues, XA1 MS A073, University of Guelph Archival & Special Collections. The Red Cross asked

knitters to use Canadian wheeling wool, and this was required by many fairs. "Red Cross headquarters are asking that knitters realize the conditions overseas and that the coarse Canadian wheeling wool is what is wanted for men's socks. It has been learned from officers overseas that the men's socks do not last one really long day's marching, when made with a soft wool. The coarse wheeling is what farmers and miners use and the conditions of their work approach nearly to those of active service. Headquarters officials draw to the attention of knitters that country women who buy, knit and wash their men's socks, prefer this wool. It, like other wools issued, has been thoroughly tested." "Wheeling Wool Is Best for War Socks," *Ottawa Citizen*, 29 February 1940.

114 "Collingwood Township Agricultural Society Prize List, 1945," Ontario Fall Fair Catalogues, XA1 MS A073, University of Guelph Archival & Special Collections.

115 Hearthstone Circle, "Philosophy of Sock-Knitting," *Farmer's Advocate and Canadian Countryman* 89 (11 September 1954): 14; "Sew It Yourself," *Farmer's Advocate and Canadian Countryman* 89 (27 November 1954): 12, 22. The *Farmer's Advocate* continued to publish instructions for home manufactures throughout this period.

116 "Prize List of the Arran-Tara Agricultural Society Fall Fair, 1950," Arran-Tara Fall Fair Booklets, A959.034.001, Bruce County Archives.

117 "Prize List of the Arran-Tara Agricultural Society Fall Fair, 1960," Arran-Tara Agricultural Society Prize Lists, private collection of Ron Hammell, Dobbinton, Ontario; and "Prize List of the Arran-Tara Agricultural Society Fall Fair, 1970," Arran-Tara Agricultural Society Prize Lists, private collection of Ron Hammell, Dobbinton, Ontario. The number of regular classes in "Ladies Work" declined from sixty-six in 1960 to fifty-nine in 1970.

118 Council on Rural Development Canada, *Rural Women*, 87.

119 "Poncho Appeal," *Country Guide: The Farm Magazine* (March 1979): 78.

120 "Erin Fall Fair Prize List, 1974," Erin Agricultural Society fonds, A1989.97, Wellington County Museum and Archives.

121 Interview with Gail Bartlett, 19 August 2014.

122 See McLeod, *In Good Hands*.

123 Ibid., 297.

124 Grier, "The Decline of the Memory Palace," 68–9.

125 "Erin Fair Was Best in Years," *Erin Advocate*, 16 October 1924; and "Average Crowd Attended Erin Fair despite Inclement Weather," *Erin Advocate*, 15 October 1925.
126 Handlin, *The American Home*, 446–8.
127 Miss Lillian M. Rutherford, "Address of President," in *Ontario Department of Agriculture, Forty-Third Annual Report of the Agricultural Societies and the Convention of the Association of Agricultural Societies*, 21.
128 Hamilton Spectator, "Waterdown Fair," 1955, Local History & Archives, Hamilton Public Library, 32022191097548.
129 "Tatting Comes Back," *Farmer's Advocate* 99 (25 April 1964), 28.
130 "Peel Fall Fair! Another Grand Success!" *Brampton Times*, 6 October 1871. J.W. Cole, the local Brampton photographer, was the only competitor in the 1871 classes for "Collection of ambrotypes" and "Photographs." Ambrotypes were a type of collodion positive in which Canadian balsam was used to seal the collodion plate to the cover glass. This process was patented by American photographer James Ambrose Cutting in 1854, and commonly used in North America until the 1880s. Harding, "How to Spot a Collodion Positive."
131 "Kodak on the Farm," *Farmer's Advocate and Home Magazine* 51 (2 March 1916): 366.
132 Minnie May, "Minnie May's Department," *Farmer's Advocate and Home Magazine* 34 (16 October 1899): 586.
133 "The Guelph Central," *Guelph Weekly Mercury*, 29 September 1892.
134 "Eramosa Show," *Guelph Mercury*, October 10, 1889.
135 Wilson, "Shaking Out the Quilt and Quilting Bee."
136 Tessier, *Spencerville Fair*, 243.
137 "Country Crafts," *Country Guide: The Farm Magazine* (January 1979): 47; "Country Crafts," *Country Guide: The Farm Magazine* (February 1979): 69; "Country Crafts," *Country Guide: The Farm Magazine* (August 1979): 44.
138 "Record Crowd Attend Ont.'s Preview of The Royal," *Erin Advocate*, 15 October 1960.
139 Provenance descriptions for Sarah Jane (Wheeler) Jackson, "Sock," circa 1900, 2004.14.5.02, and Sarah Jane (Wheeler) Jackson, "Nightgown", circa 1900, 1954.97.1, Wellington County Museum and Archives.
140 Registrations of Marriages, Series MS932-8; Reel 8, Archives of Ontario.
141 "Account Ledger: 1889–1899," Erin Agricultural Society fonds, A1989.97,

Notes to pages 168–75

Reel 6; and provenance description for Sarah Jane (Wheeler) Jackson, "Nightgown", circa 1900, 1954.97.1, Wellington County Museum and Archives.

142 Museum staff believe that the embroidered trim on the flannel nightgown is homemade.

143 Cairns and Silverman, *Treasures: The Stories Women Tell about the Things They Keep*, xii– xiii.

144 Archives of Ontario; Series: MS932; Reel: 91.

145 Provenance description for Clara (Jackson) Robertson, "Shirt, Work," circa 1925, 1994.34.1, Wellington County Museum and Archives.

146 Nasby, "Historical Views of Guelph."

147 The specific amount and type of taxes paid was not recorded. Providence description for "Handiworked laces of Fanny Calvert, 68 Queen St., Guelph, 1902–1910," Regional History Collection, XR1 MS A 743, University of Guelph Archival & Special Collections.

148 [Untitled documents], Fanny Colwill Calvert files, MacDonald Stewart Art Centre, UG 997.006.001 – UG 997.006.006.

149 Kaethler and Shantz, *Quilts of Waterloo County*, 44.

150 *Historical Atlas of the County of Wellington, Ontario*, 59.

151 Ibid.; Archives of Ontario; Series: *MS932*; Reel: *103*; Archives of Ontario; Series: *MS932*; Reel: *120*; Year: *1901*; Census Place: *Eramosa, Wellington (south/sud), Ontario*; Page: *3*; Family No: *19*. William's sons Ralph and Cleveland helped show at the fairs, especially in the years leading up to the turn of the century. In 1900, William's oldest son Ralph married Sarah Emma Wilson on 22 March and moved to his own farm near Arkell. Cleveland, William's second eldest son, married Mary Ethel McFarlane on 21 June 1905 and became the manager for his father's large operation. While his oldest daughter Ida had already married and left the farm, his younger daughters Bernice, Irene, and Edna remained home along with his youngest son George.

152 The prize ribbons used for the quilt were won at such prestigious shows as Provincial Fat Stock Show (later renamed the Provincial Fat Stock and Dairy Show), the Ontario Provincial Winter Fair held in Guelph, the Toronto Industrial Exhibition, and the Pan-American Exposition held in Buffalo, New York. They appear to have been only a portion of the ribbons William was awarded over the years, because he also exhibited in Chicago,

Quebec City, Montreal, Ottawa, and Kingston. *Historical Atlas of the County of Wellington, Ontario*, 59.

153 Year: *1901*; Census Place: *Eramosa, Wellington (south/sud), Ontario*; Page: *3*; Family No: *19*.

154 Crazy quilts appealed to many needleworkers because the irregular-shaped scraps of fabric allowed women to incorporate remnants and used material more easily, while also showcasing a variety of embroidery stitches and embellishments. Elizabeth Donaldson's skill was evident in the brightly coloured thread which highlighted each stitch. Embroidery was a talent expected of all "refined" ladies. For more on crazy quilting, see Walker, *Ontario's Heritage Quilts*, 63.

155 Providence description for "Quilt, Crazy," 1910–1920, made by Elizabeth Ann (Huxley) Donaldson, 1977.37.1, Wellington County Museum and Archives; and Year: *1891*; Census Place: *Erin, Wellington South, Ontario*; Roll: *T-6377*; Family No: *182;* Year: *1901*; Census Place: *Erin, Wellington (south/sud), Ontario*; Page: *9;* Family No: *96;* Year: *1911;* Census Place: *11 - Erin Township, Hillsburg Police Village, Wellington South, Ontario;* Page: *3; Family No: 29; Reference Number: RG 31; Folder Number: 97; Census Place: Erin (Township), Wellington South, Ontario; Page Number: 22.*

156 The quilt also included two souvenir ribbons for Queen Victoria's birthday celebrations in 1898, a member's ribbon for the Georgetown Driving Park in 1891, a Hillsburg Turf Association Dominion Day Celebration ribbon for 1894, and an Elora Citizen's Demonstration guest ribbon, dated 6 August 1894.

157 Janet Floyd, "Back to Memory Land? Quilts and the Problem of History," *Women's Studies* 37:38, no. 56 (2008): 41.

158 Archives of Ontario; Series: *MS932*; Reel: *664*; and Providence description for Christena Nelena (McMillan) Neal, "Quilt," circa 1950-60, 2010.36.1., Wellington County Museum and Archives.

159 Woodward, "Looking Good: Feeling Right," 21–2.

160 "The Guelph Central Fair, Opening Day, Number of Entries," *Daily Globe* [Toronto], 16 September 1874.

161 T. Manning, "Deaths – Nancy Strickland." *Christian Guardian*, 12 May 1886.

162 Ibid.

163 "Canada," *Huron Expositor*, 26 March 1886.

164 *Census of 1851 (Canada East, Canada West, New Brunswick, and Nova Scotia).*

Notes to pages 181–6

Ottawa, Canada. Place: *Whitby, Ontario County, Canada West (Ontario)*; Schedule: *A*; Roll: *C_11742*; Page: *21*; Line: *17*, Library and Archives Canada.
165 Kelly, "The Consumption of Rural Prosperity," 589.
166 Interview with Gail Bartlett, 19 August 2014.
167 Ibid.

Chapter Five

1 See Cohen, *Women's Work*; Crowley, "Rural Labour"; and Errington, *Wives and Mothers, Schoolmistresses and Scullery Maids*.
2 Wilkie, *Livestock/Deadstock*, 46.
3 Cohen, *Women's Work*, 57.
4 For a discussion of the importance of improved farm animals and scientific breeding, see Margaret Derry's extensive research, including *Ontario's Cattle Kingdom*; *Bred for Perfection*; *Horses in Society*; and *Art and Science in Breeding*.
5 Vamplew and Kay, *Encyclopedia of British Horseracing*, 342.
6 Larsen, *Gender, Work, and Harness Racing*, 1.
7 Ibid., 2.
8 See May, *Canada's International Equestrians*.
9 Irvine and Vermilya, "Gendered Work in a Feminized Profession," 56–82. Women today are still broadly characterized in this way. Irvine and Vermilya find that veterinary medicine maintains institutionalized inequality and hegemonic masculinity, despite women constituting half of its practitioners and almost 80 per cent of students. Women are seen as less capable and professional because of on gendered ideas of women's inherent traits and abilities.
10 "Hints to Cattle Breeders," *Farmer's Advocate* 6 (August 1871): 119.
11 Ibid.
12 Morgan, *Public Men and Virtuous Women*, 70.
13 Campbell and Bennett, "Harvey, Eliza Maria (Jones)."
14 Ibid.
15 MacEwan and Ewen, *The Science and Practice of Canadian Animal Husbandry*, 52.
16 "Belvedere: The Home of Mrs. E.M. Jones and Her Jerseys," *Farmer's Advocate and Home Magazine* 34 (20 December 1899): 664.

17 Ibid.
18 Ibid., 665.
19 Ibid.
20 Gidney and Millar, *Professional Gentlemen*, 328.
21 Cohen, *Women's Work*, 108; 116.
22 "Women and Poultry," *Farmer's Advocate and Home Magazine* 34 (April 1899): 178.
23 Ibid.
24 Derry, *Art and Science in Breeding*, 208.
25 Ibid., 173.
26 McMurray, "Women's Work in Agriculture," 253–4. McMurray emphasizes the collaboration that took place between men and women in order to produce milk. She notes that women participated in decisions regarding dairy operations, but she also notes that "men's and women's duties were clearly defined" and that "women could not always control the circumstances of their work."
27 See chapter 2 for a discussion of how women were initially not recognized for their exhibits at fairs.
28 "Home District Agricultural Society," *The Royal Standard and Toronto Daily Advertiser*, 9 November 1836.
29 "Home District Agricultural Spring Fair, and Horse and Cattle Show," [Toronto] *Patriot*, 28 April 1840.
30 "The Spring Fair and Cattle Show," *British Colonist*, 19 May 1841.
31 "Eramosa Show," *Guelph and Galt Advertiser*, 8 October 1847.
32 "Puslinch Agricultural Society," *Guelph Advertiser*, 20 October 1853.
33 "Central Riding Exhibition," *Guelph Weekly Mercury and Advertiser*, 17 October 1889.
34 Miller, "Introduction," ix–xv.
35 Mary O'Brien, journal entry for 26 April 1830, in *The Journals of Mary O'Brien*, 105.
36 Mary O'Brien, journal entry for 15 May 1830, in *The Journals of Mary O'Brien*, 105. Mary notes that her brother William had strained his back and could not attend the agricultural society meeting, but that her brother Richard and husband Edward attended. Mary also notes that her sisters-in-law "exclaimed at this most unbridegroom-like proceeding" when her husband left his bride so soon after the wedding, but Mary explains that "as he

had been chiefly instrumental in promoting it, I encouraged him to go." In the editor's endnotes, Miller explains that this meeting was to set up the executive of the Home Agricultural Society, which had been postponed in April.

37 Ancestry.com, *Canada, Find A Grave Index, 1600s-Current* [database on-line], Provo, UT, USA: Ancestry.com Operations, Inc., 2012.

38 Year: *1851*; Census Place: *Brock, Ontario County, Canada West (Ontario)*; Schedule: *A*; Roll: *C_11743*; Page: *9*; Line: *2*; Year: *1871*; Census Place: *Puslinch, Wellington South, Ontario*; Roll: *C-9945*; Page: *35*; Family No: *115*; Year: *1881*; Census Place: *Arthur, Wellington North, Ontario*; Roll: *C_13260*; Page: *80*; Family No: *325*; Year: *1891*; Census Place: *Arthur, Wellington North, Ontario*; Roll: *T-6376*; Family No: *270*; *Historical Atlas of the County of Wellington, Ontario*; and "Badenoch: 1832–1967," Puslinch Township History from The Clarks of Tomfad.

39 "Puslinch Agricultural Society," *Guelph Advertiser*, 20 October 1853, and "Puslinch Show," *Guelph Advertiser*, 19 October 1854.

40 Archives of Ontario; Series: *MS935*; Reel: *35*.

41 Year: *1891*; Census Place: *Elora, Wellington Centre, Ontario*; Roll: *T-6376*; Family No: *210*.

42 Year: *1881*; Census Place: *Elora, Wellington Centre, Ontario*; Roll: *C_13259*; Page: *3*; Family No: *11.*; George was twenty-eight years old and listed as a harness maker, while her twenty-four-year-old son William was a law student, and her fourteen-year-old son was still in school. Isabella, the eldest and only daughter, remained single throughout her lifetime and lived with her mother as her companion and dependent after the death of her father. Year: *1901*; Census Place: *Elora (Village), Wellington (centre), Ontario*; Page: *7*; Family No: *87*; Year: *1911*; Census Place: *49 - Elora, Wellington South, Ontario*; Page: *6*; Family No: *70*.

43 Chambers, *Married Women and Property Law in Victorian Ontario*, 19.

44 Ibid., 19–20.

45 See Morgan, *Public Men and Virtuous Women*. The scholarly literature on women in the public sphere is well developed in the United States. See Lasser and Robertson, *Antebellum Women*; Giesberg, *Civil War Sisterhood*; Ryan, *Civic Wars* and *Women in Public*; Varon, *We Mean to Be Counted*; Attie, *Patriotic Toil*; Hoffert, *When Hens Crow*; Yellin and Van Horne, *The Abolitionist Sisterhood*; Yee, *Black Women Abolitionists*; Hewitt, *Women's*

Activism and Social Change; Berg, *The Remembered Gate*; Cott, *The Bonds of Womanhood*; Melder, *The Beginnings of Sisterhood*.
46 Gidney and Millar, *Professional Gentlemen*, 328–32.
47 See Chapters 2, 3, and 4.
48 "Take the Children to the Fair," *Brampton Conservator*, 17 September 1880.
49 McMurray, "Women's Work in Agriculture," 253.
50 John McLoughlin, correspondence to HBC officials, cited in Carlson, "The Cattle Battle," 35–6.
51 Young, "The Reins in Their Hands," 4.
52 Ibid., 3–7.
53 "Attractions at Fairs," *Farmer's Advocate* 4 (December 1869): 181; reprint from *Moore's Rural New Yorker*.
54 "Woman on the Turf," *Farmer's Advocate* 4 (December 1869): 183.
55 Ibid.
56 Derry, *Horses in Society*, 33.
57 "Mono Mills, Oct. 31, 1868," Beeton Agricultural Society Minute Book entry, Electoral Division of Cardwell Agricultural Society ... Minute Book, accession 988-3, R2B S5 Sh4, Simcoe County Archives.
58 "Cooksville Fair," *Brampton Conservator*, October 17 October 1879.
59 "The County Exhibition," *Brampton Conservator*, 7 October 1881.
60 "Centre Wellington," *Globe and Mail*, 8 October 1881.
61 "Fiftieth Anniversary: West Elgin Fair Celebrates Its Golden Jubilee," *Advance* [Dutton], 6 October 1910.
62 It was reported that at the 1886 Toronto Industrial Exhibition, the "Lady riders and drivers will attract even without the horses ... and when a horse rears up and falls back flat on a lady and kills her, as was the case in Toronto, the excitement is too great for many, but still the most attractive to some, despite the accident." "The Agricultural Exhibitions of 1886," *Farmer's Advocate* 21 (October 1886): 290.
63 "Burwick Fair," *Brampton Conservator*, 22 October 1880.
64 "The Central Exhibition," *Guelph Daily Mercury and Advertiser*, 29 September 1884.
65 "The Wallacetown Fair," *Advance* [Dutton], 26 September 1889. Other classes in this category of competition included "trotters and pacers, purse $50; farmers' trot, purse, $15; boys' running race, purse $6; running race, purse $20; lady drivers, purse $9."

66 "Fiftieth Anniversary: West Elgin Fair Celebrates Its Golden Jubilee."
67 Ibid.
68 "Springfield Had Most Successful Fair," *Aylmer Express*, 29 September 1921.
69 "Better than Ever: The Celebrated Wallacetown Fair Scores Another Success," *Advance* [Dutton], 3 October 1912; "Special Prizes for the Fair," *Advance* [Dutton], 17 September 1914; "Wallacetown Fair," *Advance* [Dutton], October 5, 1916; "The Shedden Fair," *Advance*, [Dutton], 25 September 1919; "Springfield Had Most Successful Fair."
70 "Fiftieth Anniversary: West Elgin Fair Celebrates Its Golden Jubilee."
71 J. Lockie Wilson, "Rules for Judging Lady Drivers," Fairs & Exhibitions Minute Book 1909–1930, B228850/RG 16-52/1, Archives of Ontario.
72 Walden, *Becoming Modern in Toronto*, 184–5. Keith Walden discusses how female performers who transgressed gender norms often had to be described in feminine terms in order to reassure the public that they remained tied to conventional womanhood and the domestic sphere.
73 "Prize List of the Hanover Fair, 1938." Reference Box, Sports and Fairs B, Box 29, Bruce County Archives.
74 "Arran-Tara Fall Show," [1899] *Tara Leader*, cited in Miller, *The Fairest of Them All*, 23.
75 See page ooo for a deeper discussion of youth showing livestock.
76 "Erin Fair Was Best in Years," *Erin Advocate*, 16 October 1924.
77 "Erin Fair Broke All Records," *Erin Advocate*, 18 October 1938.
78 Rural school fairs began in Ontario in 1909, when three schools located in North Dumfries Township in Waterloo County held a fair. F.C. Hart, a provincial agricultural representative, had arranged for vegetable seeds to be purchased and distributed to school children the year before. By 1912, the Ontario Department of Agriculture supported rural school fairs and assigned representatives to manage them, because they hoped to stimulate children's interest in agricultural pursuits and introduce them to the idea of competition as a means of determining merit. Most rural school fairs were separate from fairs sponsored by agricultural societies, although some agricultural societies did host school competitions as well. I will use the term "rural school fair" to identify school fairs; the simple use of "fair" will continue to imply a local fair sponsored by an agricultural society. Lee, *Head, Heart, Hands, Health*, 173; Cormack, *Learn to Do by Doing*, 5; Reaman, *A History of Agricultural in Ontario*, vol. 2, 100; Madill, *History of Agricultural Education in Ontario*, 191; and Scott, *A Fair Share*, 100.

79 Madill, *History of Agricultural Education in Ontario*, 187–8. Madill reports that prize lists usually included classes for "grain, roots, corn, flowers, poultry, essays, writing, collections of weeds, of woods, and of leaves, cooking, sewing, manual training, and live stock." He also noted that contests for "oratory, singing, sewing, stock-judging, and driving," as well for "races of various kinds," were held because of spectator interest. Older boys competed in "poultry plucking, weed naming, whittling, carpenter work," while older girls exhibited "preserved fruit and loaves of bread."
80 "Vaughan Township Eighth Annual Rural School Fair Prize List, 1920." Rural School Fairs Programs, M002.3 Box 4, File 8, City of Vaughan Archives.
81 Ibid.
82 Lee, *Head, Heart, Hands, Health*, 182.
83 In his work on 4-H organizations in the United States and abroad, Gabriel Rosenberg emphasizes the "gendered bodies of rural youth" and how state-sponsored 4-H organizations were used to educate rural youth on appropriate gender roles. Rosenberg notes that most 4-H projects in the first half of the twentieth century were segregated by sex. While this is largely true in Canada as well, important exceptions did exist. Rosenberg, *The 4-H Harvest*, 14.
84 "Erin Fall Fair Prize List, 1917," Erin Agricultural Society Collection, 1862–1987, A1989.97, Wellington County Museum and Archives.
85 Ibid.
86 "Successful Calf Club in Grenville County," *Farmer's Advocate and Home Magazine* 54 (30 October 1919): 1956.
87 "Big Crowds Attended the Aylmer & East Elgin County Fair," *Aylmer Express*, 29 September 1921.
88 "Elgin County Fair and Cattle Show," *Aylmer Express*, 17 September 1931. The majority of prizes were won by Betty Armour and Dorothy and Leslie Davis.
89 Lee, *Head, Heart, Hands, Health*, 65.
90 "The Biggest Crowd in Aylmer's History Expected at the Fair," *Aylmer Express*, 22 September 1938.
91 "Belmont Fair Today," *Aylmer Express*, 26 September 1940.
92 Lee, *Head, Heart, Hands, Health*, 80.
93 Only boys are shown showing cattle and horses in the photographs in the

Notes to pages 197–201

      Archives of Ontario 4-H Collection that predate 1940. 4-H Collection, RG 16-275-2, Container 1, Archives of Ontario.
94 Agnes Foster, "Agnes Foster: One Leader's Story," in Lee, *Head, Heart, Hands, Health*, 151.
95 "Young Girl Cynosure of All Eyes as She Shows Her Heifer Calf," [unidentified newspaper] cited in Heald, *The Ottawa Winter Fair*, 108–9.
96 Ibid.
97 Foster, "Agnes Foster: One Leader's Story," 151, and 4-H Collection, RG 16-275-2, Container 1, Archives of Ontario.
98 "Erin Fall Fair Prize List, 1935." Erin Agricultural Society collection, 1862–1987, A1989.97, Wellington County Museum and Archives.
99 Wilson, "A Manly Art," 171, 186.
100 Wessel and Wessel, *4-H: An American Idea 1900–1980*, 14. Wessel and Wessel explain that the "principal emphasis in boys' club work until 1910 had been on finding a means of conveying new agricultural techniques from experiment stations and land-grant colleges to farm operators. The basic purpose was to improve agricultural techniques and increase production or shift production to other crops. Girls' clubs – confined to canning, sewing, baking and the like – had no such technological goal."
101 See Carbert, *Agrarian Feminism*; and Kechnie, *Organizing Rural Women*.
102 Kechnie, *Organizing Rural Women*, 97.
103 Lee, *Head, Heart, Hands, Health*, 69.
104 Mosby, *Food Will Win the War*, 119.
105 "Recruiting Our Land Army," *Farmer's Advocate and Home Magazine* 76 (23 January 1941): 36, 40; "The Girls Who Saved the Apple Crop. All Over Canada Farm Women Are Doing Their Full Share – in the Essential Service of Food Production," *Farmer's Advocate and Home Magazine* (28 May 1942): 316; and "Youth on Its Mettle. Young Women of the Farm Rally to the Colours – Gardens Are Cultivated with Patriotic Fervour," *Farmer's Advocate and Home Magazine* (24 September 1942): 557.
106 Lee, *Head, Heart, Hands, Health*, 195.
107 Ibid., 191, 195.
108 Mae Todd, "Junior Association Teaches Way of Co-Operative Living," in *Bruce County Year Book*, 37.
109 Ibid.
110 Smith, *The Elements of Live Stock Judging*, 1–6.

111 4-H Interclub Competitions, [1955?], Print, black and white, sound; 16 mm, RG 16-35-0-33-1, Archives of Ontario.
112 Ready, *A Manual in Canadian Agriculture*, 143. Ready, formerly a supervisor of agricultural education in British Columbia, included in the dairying chapter of his book, *A Manual in Canadian Agriculture*, a picture of a young girl rehearsing judging with her male coach entitled "Girls Make Excellent Judges." Unfortunately, no explanation for this contention was given, but other contemporary assessments of female club members suggested that girls' eyesight and patience were useful qualifications for livestock judging. See Cormack, *Learn to Do by Doing*, 30.
113 "Halton Girl Competes with 180 to Win Queen's Guineas," *Canadian Champion* [Milton], 25 November 1954.
114 "Alymer Fair – Event Winners," *St. Thomas Times-Journal*, 19 August 1958.
115 "Rodney Fair – Thousands Attend," *St. Thomas Times-Journal*, 24 September 1959.
116 "Awarded Grand Championship," *Canadian Champion* [Milton], 26 November 1953.
117 While the heavy horse show continued to be dominated by men, youth classes did exist that included females. At the 1947 Arran-Tara Agricultural Society's Annual Exhibition, the special classes for the heavy horses included one for "Boy Driver – 15 years or under." No such class existed for girls and women drivers in the heavy horse competition. Still, that same year there was an award for the best halter-broken foal exhibited by a showman under the age of fifteen. The next year, in 1948, it was clarified in the prize list that the competition was open to "boy or girl under 16 years." "Arran-Tara Agricultural Society's Fall Exhibition Prize List, 1947," Private Collection.
118 "Erin Fall Fair Prize List, 1949," Erin Agricultural Society collection, 1862–1987, A1989.97, Wellington County Museum and Archives.
119 "Record Exhibit and Crowd at Erin Fair," *Erin Advocate*, 13 October 1949.
120 "Fall Fair Time in Ontario," *Woodbridge News*, 26 October 1950.
121 May, *Canada's International Equestrians*, 64.
122 Ibid., 71.
123 Ibid., 111.
124 Ibid., 112.

125 The Extension Branch Ontario Department of Agriculture and Food published a number of such works: *4-H Beef Junior Manual, Production and Marketing*; *First Aid*; *Human Factors in Safety*; *Farm Fire Safety*; *Pesticide Safety*; *Common Machine Hazards*; 4-H Collection, RG 16-275-1, Archives of Ontario.
126 "Beauty and the Beast at Erin," 1952; newspaper clipping from 1952 *Guelph Mercury* found in the Erin Agricultural Society Private Photo Collection, Erin, Ontario.
127 Interview with June Switzer, 12 September 2013.
128 Interview with Phyllis MacMaster, 19 Augus, 2015.
129 Ibid.
130 Wilkie, *Livestock/Deadstock*, 59–63.
131 For a full discussion of more recent shifts in women's roles in agricultural industries, see Devine, *On Behalf of the Family Farm*; Sachs, *Gendered Fields*; Carbert, *Agrarian Feminism*; and Knowles and Haney, *Women and Farming*.
132 Devine, *On Behalf of the Family Farm*, 139.

Chapter Six

1 MacDonald, *Splendor in the Fall*, 88.
2 Smith and Boyd, "Talking Turkey," 116–44.
3 Wilson, "Reciprocal Work Bees and the Meaning of Neighbourhood," 451–2.
4 "Fergus Fair," *Guelph & Gault Advertiser*, 1 October 1847.
5 "'Burwick Fair': The Annual Autumnal Exhibition of West Riding of York and Township of Vaughan Agricultural Societies," *Brampton Conservator*, 22 October 1880.
6 "Toronto Township Fall Fair," *Weekly Review* [Brampton], 17 October 1857.
7 "Etobicoke Annual Fair and Cattle Show," *Globe* [Toronto], 15 October 1868.
8 "The Woodbridge Fair," *Globe* [Toronto], 20 October 1892.
9 "August 27, 1913," Shedden Women's Institute Minute Book, 1913–1919, Elgin County Archives, C9 Sh4 B1 F1.
10 See Chapter 1 for a discussion of lady directors and women's sections of agricultural societies.

11 For an excellent discussion of the centrality of food to women's organizations such as the Women's Institute, see Ambrose, "Forever Lunching," 174–85.
12 "Wallacetown Fair," *Advance* [Dutton], 8 October 1914.
13 "Wallacetown Fair," *Advance* [Dutton], 5 October 1916.
14 MacDonald, *Splendor in the Fall*, 88.
15 "Home Sciences Stressed at Fall Fair," *Globe and Mail*, 10 October 1944.
16 "August 26, 1925," Shedden Women's Institute Minute Book, 1922–1926, Elgin County Archives, C9 Sh4 B1 F2.
17 "25,000 People Present at Aylmer's Three Day Fair: Smash All Previous Records – Weather Fine," *Aylmer Express*, 29 September 1938.
18 "Thousands Attend Aylmer Fair despite Cold Winds and Threatening Rains," *Aylmer Express*, 28 September 1939.
19 "Fall Fair Time in Ontario," *Woodbridge News*, 12 October 1950.
20 [Untitled photograph caption] *Woodbridge News*, 13 October 1949.
21 Ibid.
22 Tye, *Baking as Biography*, 150–1.
23 Ibid., 150.
24 Petrzelka and Mannon, "Keepin' this Little Town Going," 253.
25 "County of Peel Fall Fair," *Brampton Times*, 27 August 1869.
26 "Big Crowds Attended the Aylmer & East Elgin County Fair," *Aylmer Express*, 29 September 1921.
27 Scott, *Country Fairs in Canada*, 58–9. Midways were the more organized and corporatized version of the early sideshow entertainment found at some fairs in the nineteenth century. Modern midways with mechanized rides appeared after 1900.
28 "Aylmer Fair Drew Big Attendance," *The Aylmer Express* (2 October 1919).
29 "Come to Elgin County Fair Next Week," *Aylmer Express* (1 September 1932).
30 "Thousands Attend Aylmer Fair despite Cold Winds and Threatening Rains." The *Woodbridge News* also thanked "the large contingent of women from the Presbyterian Ladies' Aid group at the 1949 Woodbridge Fair who fed the crowds of people in attendance" [No title - Picture]. *Woodbridge News*, 13 October 1949."
31 "Big Crowds Attended the Aylmer & East Elgin County Fair."
32 Howard, "The Mother's Council of Vancouver," 261.

33 "Thousands Attend Aylmer Fair despite Cold Winds and Threatening Rains."
34 Ibid.
35 "Red Cross Well Patronized at Aylmer Fair," *Aylmer Express*, 28 September 1939.
36 "Thousands Attend Aylmer Free Fair," *Aylmer Express*, 25 September 1941.
37 "County and Township Fairs," *Globe* [Toronto], 8 October 1881.
38 Cork, *A History of the Mount Forest Agricultural Society*, 122–3.
39 "Excellent Exhibits in All Classes and Large Crowd at Fair," *Erin Advocate*, 17 October 1940.
40 *Eaton's Fall and Winter 1948–49* (Toronto: T. Eaton Co., 1948–49), 293. The wool yarn advertised in 1948 in Eaton's catalogue cost between 18 and 35 cents for a one-ounce skein.
41 Lewis, *From Traveling Show to Vaudeville*.
42 Tenneriello, *Spectacle Culture and American Identity*, 1.
43 Rydell, *All the World's a Fair*, 6.
44 "Idle Exhibitions," *York Herald*, 24 June 1859.
45 County Member, letter to the editor, "More about the Central Exhibition," *Guelph Mercury and Advertiser*, 26 September 1889.
46 "Agricultural Exhibitions," *Farmer's Advocate* 3 (January 1868): 2.
47 H.B. Cowan, "The Agricultural Societies of Ontario," *O.A.C. Review* 17, no. 7 (April 1905): 381–4.
48 "Simcoe's Model Fair," *The Farmer's Advocate*, 2 November 1903.
49 Jones, *Midways, Judges, and Smooth-Tongued Fakirs*, 97.
50 "'Burwick Fair': The Annual Autumnal Exhibition of West Riding of York and Township of Vaughan Agricultural Societies," *Brampton Conservator*, 22 October 1880.
51 "Eramosa Fall Show," *Guelph Mercury and Advertiser*, 18 October 1891.
52 "Minto Agricultural Society," *Elora Observer*, 30 September 1870.
53 "Markham Fair," *Liberal* [Richmond Hill], 14 October 1881.
54 "Markham Fair," *Globe* [Toronto], 8 October 1881.
55 OAAS Women's Section past-president Ethel Brant Monture wrote about the life of Oronhyatekha in her book, *Canadian Portraits*, 131–58.
56 Ibid., 133.
57 For more on First Nations in exhibitions and fairs, see "Chapter 10: Making a Spectacle: Exhibitions of the First Nations," in Heaman, *The Inglorious*

Arts of Peace, 285–310. For the legacy of nineteenth-century white people's fascination with "the other," and how "playing Indian" became a cultural phenomenon see Wall, *The Nurture of Nature*.

58 "Centre Wellington Fair," *Guelph Weekly Mercury and Advertiser*, 15 October 1891.
59 See Beausaert, "Foreigners in Town."
60 Rydell, *All the World's a Fair*, 6.
61 "Aylmer's Great Fair," *Aylmer Sun*, 25 September 1902.
62 "Norfolk County Fair [Advertisement]," *British Canadian* [Simcoe], 7 October 1914.
63 "Norfolk County Fair," *British Canadian* [Simcoe], 21 October 1914.
64 Ibid.
65 Summerfield, "Patriotism and Empire," 17–48.
66 MacDonald, *Splendor in the Fall*, 58.
67 "Norfolk County Fair, Hon. John S. Martin at Opening," *British Canadian* [Simcoe], 26 September 1923.
68 "Erin Fair Was Best in Years," *Erin Advocate*, 16 October 1924.
69 "Erin Fair Broke All Records," *Erin Advocate*, 18 October 1938.
70 Ibid.
71 "Norfolk County Fair [Advertisement]," *Norfolk Observer*, 7 October 1935.
72 "The Biggest Crowd in Aylmer's History Expected at the Fair," *Aylmer Express*, 22 September 1938.
73 "25,000 People Present at Aylmer's Three Day Fair"
74 "Thousands Attend Aylmer Free Fair."
75 "Fall Fair Time in Ontario," *Woodbridge News*, 29 September 1949.
76 Walden, *Becoming Modern in Toronto*, 184–5.
77 Ibid., 73.
78 Ed Youngman, "Ed Youngman's Column," *Canadian Statesman* [Bowmanville], 18 September 1963.
79 "Wallacetown Fair," *Advance* [Dutton], 5 October 1916.
80 "Fair," *Hamilton Spectator*, 16 September 1946.
81 "Fall Fair Time in Ontario," *Woodbridge News*, 5 October 1950.
82 "Young at 94," *Stouffville Tribune*, 5 October 1978.
83 Malcolm Calder, quoted in "Directors' Meeting Minutes, May 4, 1934," OAAS Minutes, B228850, RG 16-52, Archives of Ontario.

Notes to pages 224–7

84 "Director's Meeting Minutes, May 3, 1935," OAAS Minutes, B228850, RG 16-52, Archives of Ontario.
85 "Erin Fall Show: A Fine Exhibition and a Great Crowd," *Guelph Weekly Mercury and Advertiser*, 21 October 1897.
86 "Erin Fair Was Best in Years." First place was won by Mildred Miller, second by Leslie Jessop, third place by Lila Miller and Wilma Walker. Mildred Miller and Wilma Walker also won first and second place in 1925. "Average Crowd Attended Erin Fair despite Inclement Weather," *Erin Advocate*, 15 October 1925.
87 "Good Crowd Present at Georgetown Fair," *Globe* [Toronto], 25 September 1928.
88 Woolum, *Outstanding Women Athletes*, 3–4.
89 "Rowing for Girls," *Farmer's Advocate and Home Magazine* 21 (September 1886): 279.
90 Woolum, *Outstanding Women Athletes*, 5–11.
91 "Weston Fall Fair," *Times & Guide* [Weston], 14 October 1910.
92 "Springfield Fair Next Week," *Aylmer Express*, 11 September 1941.
93 "Newmarket Fair Well Attended," *Globe* [Toronto], 26 October 1911.
94 Fidler, *The Origins and History of the All-American Girls Professional Baseball League*, 15.
95 Ibid., 15; 36.
96 Eliza Berman, "Meet the Real Women Who Inspired A League of Their Own," *Time Magazine* (8 April 2015), http://time.com/3760024/women-professional-baseball/ (accessed 29 April 2016).
97 "Good Crowd Present at Georgetown Fair."
98 "Splendid Exhibits at Elgin County Fair and Dairy Cattle Show," *Aylmer Express*, 10 September 1931.
98 "100th Anniversary Puslinch Agricultural Society Prize List, 1940," Fergus Fall Fair Collection, MU 202, A1978.30, Wellington County Museum and Archives.
100 "Fall Fair Time in Ontario," *Woodbridge News*, 6 October 1949.
101 "Woman's Crowning Glory Is Feature of Fall Fair," *Globe* [Toronto], 8 October 1925.
102 "A Novel Prize at Cooksville Fair: Cake for Best-Looking Young Lady – Fine Exhibit of Fruit and Stock," *Globe* [Toronto], 7 October 1911.

103 Cohen, Wilk, and Stoeltje, *Beauty Queens on the Global Stage*, 4.
104 Walden, *Becoming Modern in Toronto*, 155.
105 Ballerino, Wilk, and Stoeltje, *Beauty Queens on the Global Stage*, 4–5.
106 "Miss Canada Pageant To Be Held in Toronto," *Globe and Mail*, 28 May 1949.
107 Ibid.
108 MacDonald, *Splendor in the Fall*, 73.
109 Daniel R. Pearce, "She's a Real Beauty," *Simcoe Reformer*, 23 July 2013. http://www.simcoereformer.ca/2013/07/23/shes-a-real-beauty (accessed 12 May 2016).
110 See Rowe, *Imagining Caribbean Womanhood*; Yano, *Crowning the Nice Girl*; King-O'Riain, *Pure Beauty*; Watson and Martin, *There She Is, Miss America*; Banet-Weiser, *The Most Beautiful Girl in the World*; and Cohen, Wilk, and Stoeltje, *Beauty Queens on the Global Stage*.
111 The prize for first place was $25, second place $15, and third place $10. "Beauty Contest at Orono Fair," *Canadian Statesman* [Bowmanville], 18 August 1965.
112 "Seek Esquesing Entrants for Miss Acton Contest," *Georgetown Herald*, 2 September 1965.
113 Jennifer Barr, "Profiles," *Acton Free Press*, 5 September 1979.
114 "Seek Esquesing Entrants for Miss Acton Contest."
115 King-O'Riain, *Pure Beauty*, 116.
116 "TV, Radio Personalities Act as Judges Now 19 Vying for Fair Queen Crown," *Acton Free Press*, 5 September 1963.
117 "Seek Esquesing Entrants for Miss Acton Contest."
118 Ibid.
119 "Charmaine Bigelow Wins Acton Fair Queen Title," *Georgetown Herald*, 22 September 1976.
120 Ibid.
121 Charmaine Bigelow, quoted in "Charmaine Bigelow Wins Acton Fair Queen Title."
122 "The Fair(est) of All," *Stouffville Tribune*, 20 September 1979.
123 Ibid.
124 "Shedden Fair – Fair Queen," *St. Thomas Times-Journal*, 30 June 1973. St. Thomas Times-Journals fonds, Elgin County Archives, C8 Sh3 B1 F19 19.
125 "Counties to Pick Prettiest Dairy Maids for C.N.E.," *Stouffville Tribune*,

18 July 1957; and "Local Girls Have Chance in Dairy Princess Contest," *Newmarket Era and Express*, 25 July 1957. The first CNE Dairy Queen contest was held in 1956. More than fifty contestants competed, but they were mainly from the greater Toronto area. Other regions in the province also wanted the opportunity to "show off their milkmaids," and therefore they coordinated to create county competitions sponsored by the Ontario Dairy Producers Co-ordinating Board, the Canadian National Exhibition, the *Toronto Telegram*, and local committees. The competition was open to female residents from sixteen to twenty-nine years of age, married or single, and the winners of each county competition competed at the provincial competition at the CNE, where valuable cash and merchandise could be won. Women were judged on their "on appearance, self expression [sic], and the ability and efficiency in operating a milking machine in milking a cow."

126 "Choose Hazel Reid Dairy Princess at Annual Hornby Garden Party," *Canadian Champion* [Milton], 24 July 1958, and "Queen Leaves for U.K.," *Canadian Statesman* [Bowmanville], 30 October 1958.

127 "Girls Await Dairy Queen Decision," *The Globe and Mail*, 3 September 1964.

128 Tessier, *Spencerville Fair*, 235. At the Spencerville Fair, the first baby shows were conducted in front of the grandstand, but they were moved to the Town Hall auditorium in the postwar years.

129 Selden, "Transforming Better Babies into Fitter Families," 206–8.

130 Taylor, *Fashioning Farmers*, 78.

131 Comacchio, *Nations Are Built of Babies*, 3–4.

132 McPherson, Morgan, and Forestell, "Introduction: Conceptualizing Canada's Gendered Pasts," 9.

133 Comacchio, *Nations Are Built of Babies*, 4.

134 Exhibits related to infant care were popular. At the 1938 Western Fair in London, two of the many health exhibits on display included one from the Memorial Hospital for Sick Children that stressed the importance of proper feeding for infants and the Queen Alexandra Sanitorium's display about how to avoid infections and disease. See "Good Health Plays the Leading Role," *Farmer's Advocate and Home Magazine* 73 (22 September 1938): 580. At the 1949 Lion's Head Fair, the Bruce County Health Unit's display was reported to be a "great crowd-drawer" because of Miss Turnam,

one of the nurses at the Health Exhibit, who was on hand to answer fairgoers' questions about infant and preschool health. "Fall Fair Time in Ontario," *Woodbridge News*, 6 October 1949.
135 Selden, "Transforming Better Babies," 210.
136 "For What Is a Mother Responsible?" *British American Cultivator* (February 1846): 56.
137 Ibid.
138 "Attractions at Fairs," *Farmer's Advocate* 4 (December 1869): 181; reprint from *Moore's Rural New Yorker*.
139 Susan from Streetsville, letter to the editor, "Cooksville Fall Fair Prizes," *Brampton Conservator*, 5 April 1878.
140 Geo. Savage, letter to the editor, "Cooksville Fair Prizes," *Brampton Conservator*, 29 March 1878.
141 "Cooksville Fair," *Brampton Conservator*, 18 October 1878.
142 "Norfolk County Fair," *British Canadian* [Simcoe], 21 October 1914.
143 "Wallacetown Fair," *Advance* [Dutton], 5 October 1916.
144 "25,000 People Present at Aylmer's Three Day Fair."
145 Report from the *Mount Forest Confederate*, quoted in Cork, *A History of the Mount Forest Agricultural Society*, 107.
146 Ibid., 107–8.
147 MacDonald, *Splendor in the Fall*, 21.
148 "Springfield Fair Another Big Success," *Aylmer Express*, 29 September 1931.
149 Ibid.
150 "Wallacetown Fair," *Advance* [Dutton], 8 October 1914.
151 "Aylmer Fair Drew Big Attendance."
152 Comacchio, *Nations Are Built of Babies*, 3.
153 "Fall Time in Ontario," *Woodbridge News*, 3 November 1949.
154 "Fall Fair Time in Ontario," *Woodbridge News*, 28 September 1950.
155 Isabella Bardoel, "Rockton World's Fair Drooling Babies Steal the Show," *Globe and Mail*, 10 October 1978.
156 Jeanette Jamieson, quoted in "Rockton World's Fair Drooling Babies Steal the Show."
157 Mary Lee Rainy, quoted in "Rockton World's Fair Drooling Babies Steal the Show.
158 Mizener, "Furrows and Fairgrounds," 124.
159 "County of Peel Agricultural Exhibition," *Brampton Times*, 1 October 1869.

Notes to pages 240–9

160 "Peel County Fair," *Brampton Times*, 23 October 1870.
161 "The West Elgin Fair," *Advance* [Dutton], 3 October 1901.
162 John Ferguson diary entry, Monday, 11 October 1869, John H. Ferguson fonds, [Microfilm] MS 297, Reels 1-2, Archives of Ontario.
163 Nurse, "Reaching Rural Ontario," 69.
164 "Toronto Township Fall Fair," *Weekly Review* [Brampton], 17 October 1857.
165 "Fall Fair at Markham," *York Herald*, 15 October 1869.
166 Frances Tweedie Milne diary entries, Friday, 21 to 28 September, 1866, Frances Milne fonds, F 763, Archives of Ontario.
167 Switzer, *Erin Fall Fair*, 36.
168 Kathleen Rex, "Caledon Fair Way of Life for Newly Wed Pair in 80s," *Globe and Mail*, 8 September 1979.
169 "Toronto Township Fall Fair."

Conclusion

1 Statistics Canada, "Canada Goes Urban."
2 Whatmore, Marsden, and Lowe, "Introduction: Feminist Perspectives in Rural Studies," 1.
3 Ibid., 2.
4 For a useful discussion of important foundational works and current scholarly research contributing to feminist history, see Carstairs and Janovicek, *Feminist History in Canada*.
5 Barker Devine, *On Behalf of the Family Farm*, 140–2.
6 Ibid., 142; see also Andrews, *The Acceptable Face of Feminism*, and Halpern, *And on That Farm He Had a Wife*.
7 Tye, *Baking as Biography*, 150.
8 Walden, *Becoming Modern in Toronto*, 119.
9 Heaman, *The Inglorious Arts of Peace*; Walden, *Becoming Modern in Toronto*; Boisseau and Markwyn, *Gendering the Fair*; Greenhalgh, *Ephemeral Vista*; Weimann, *The Fair Women*; Palm, "Women Muralists, Modern Women and Feminine Spaces"; Corn, *Women Building History*; Bland, "Women and World's Fairs"; and Townsend Cummins, "From the Midway to the Hall of State at Fair Park."
10 Pini, Brandth, and Little, "Introduction," 3.
11 The *Guelph Tribune* published a photograph of local resident Rick Westgarth

surrounded by his fall fair prizes with the caption: "South end resident Rick Westgarth poses with just a few of the ribbons he has picked up this summer for the baked and canned goods he's entered in country fairs. Competing in fairs has become a year-round hobby for the retired teacher, who spends his winters working on homecrafts and preserving and then bakes up a storm just before fair weekend. This season, he will compete in seven fairs, finishing in Erin on Thanksgiving weekend." "Earning His Fair Share," *Guelph Tribute*, 18 September 2014.

12 Andrews, *The Acceptable Face of Feminism*; and Pini, Brandth, and Little, *Feminisms and Ruralities*.

Bibliography

Archival Sources

Archives of Ontario
Bruce County Archives
City of Toronto Archives
City of Vaughan Archives
Elgin County Archives
Hamilton Public Library
MacDonald Stewart Art Centre
Peel Art Gallery, Museum, and Archives
Peterborough Centennial Museum
Port Hope Archives
Simcoe County Archives
Thunder Bay Public Library
Trent University Archives
University of Guelph Archival and Special Collections
Wellington County Museum and Archives
Western University Archives

Private Collections

Arran-Tara Agricultural Society Prize List Collection. Ron Hammell, Dobbinton, Ontario.

Erin Agricultural Society Photo Collection, Erin Agricultural Society, Erin, Ontario.

Interviews

Gail Bartlett, 19 August 2014.
Glenda Benton, 28 October 2013.
Martha Cranston, 19 August 2014
Ruth Gunby, 22 July 2014.
Jeanette Jameson, 22 July 2014.
Margaret Lovering, 19 August 2014.
Phyllis MacMaster, 19 August 2015.
Jessie Milton, 28 October 2013.
Myrtle Reid, 23 September 2013.
June Switzer, 12 September 2013
Eleanor Wood, 22 July 2014.

Government Publications

1961 Census of Agriculture. Ottawa: Statistics Canada, 1962.
1971 Census of Canada. Vol. 1. Part 1. Ottawa: Statistics Canada, 1972.
Annual Report of the Ontario School of Agriculture and Experimental Farm, for the Year Ending 30th September 1875. Printed by Order of the Legislative Assembly. Toronto: Hunter, Rose and Co., 1875.
Annual Report for the year ending March 31, 1980 of the Minister of Agriculture and Food. Printed by order of the Legislative Assembly of Ontario, 1980.
Census of Canada 1665–1871: Statistics of Canada. Ottawa: I.B. Taylor, 1876.
Census of Canada 1870–71. Ottawa: I.B. Taylor, 1873.
Census of Canada 1880–1. Ottawa: Maclean, Roger, 1883.
Census of Canada 1890–1. Ottawa: S.E. Dawson, 1893.
Census of Canada 1901. Ottawa: S.E. Dawson, 1904.
Census of Canada 1911. Volume IV – Agriculture. Ottawa: J. de L. Taché, 1914.
Census of Canada 1921. Volume V – Agriculture. Ottawa: F.A. Acland, 1925.
Council on Rural Development Canada. *Rural Women: Their Work, Their Needs, and Their Role in Rural Development.* Ottawa: Ministry of Supply and Services, 1979.

Bibliography

Journal and Transactions of the Board of Agriculture of Upper Canada. Vol. 1. Toronto: Board of Agriculture, 1856.

Ontario Department of Agriculture. "Apples." *Bulletin XV*, Agricultural College, Guelph, 18 August 1887.

One Hundredth Annual Report of Ontario Agricultural Societies, 1953. Printed by authority of Hon. F.S. Thomas, Minister of Agriculture for Ontario.

One Hundred and First Annual Report of Ontario Agricultural Societies, 1954. Printed by authority of Hon. F.S. Thomas, Minister of Agriculture for Ontario.

One Hundred and Seventh Annual Report of Ontario Agricultural Societies, 1960. Toronto: Ministry of Agriculture, 1961.

One Hundred and Twelfth Annual Report of Ontario Agricultural Societies, 1965. Toronto: Ontario Department of Agriculture and Food, 1966.

One Hundred and Eighteenth Annual Report of Ontario Agricultural Societies, 1971. Toronto: Ontario Ministry of Agriculture and Food, 1972.

Ontario Department of Agriculture Twentieth Annual Report of the Agricultural Societies and of the Convention of the Association of Fairs and Exhibitions for the Year 1920. Toronto: A.T. Wilgress, Printer to the King's Most Excellent Majesty, 1920.

Ontario Department of Agriculture Twenty-Second Annual Report of the Agricultural Societies and of the Convention of the Association of Fairs and Exhibitions for the Year 1922. Toronto: Clarkson W. James, Printer to the King's Most Excellent Majesty, 1922.

Ontario Department of Agriculture Thirtieth Annual Report of the Agricultural Societies and of the Convention of the Association of Fairs and Exhibitions for the Year 1930. Toronto: Herbert H. Ball, Printer to the King's Most Excellent Majesty, 1930.

Ontario Department of Agriculture Thirty-First Annual Report of the Agricultural Societies and of the Convention of the Association of Fairs and Exhibitions for the Year 1931. Toronto: Herbert H. Ball, Printer to the King's Most Excellent Majesty, 1931.

Ontario Department of Agriculture Thirty-Fourth Annual Report of the Agricultural Societies and the Convention of the Association of Fairs and Exhibitions for the Year 1934. Toronto: Herbert H. Ball, Printer to the King's Most Excellent Majesty, 1934.

Ontario Department of Agriculture Thirty-Eighth Annual Report of the Agricultural Societies and the Convention of the Association of Agricultural Societies for 1938. Toronto: T.E. Bowman, Printer to the King's Most Excellent Majesty, 1938.

Ontario Department of Agriculture Forty-First Annual Report of the Agricultural Societies and the Convention of the Association of Agricultural Societies for 1941. Toronto: T.E. Bowman, Printer to the King's Most Excellent Majesty, 1941.

Ontario Department of Agriculture Forty-Third Annual Report of the Agricultural Societies and Convention of Association of Agricultural Societies for 1943. Toronto: T.E. Bowman, Printer to the King's Most Excellent Majesty, 1943.

Ontario Department of Agriculture 91st Annual Report of Ontario Agricultural Societies. Toronto: T.E. Bowman, Printer to the King's Most Excellent Majesty, 1945.

Ontario Department of Agriculture 96th Annual Report of Ontario Agricultural Societies, 1949, also Report of Ontario Association of Agricultural Societies. Toronto: Printer to the King's Most Excellent Majesty, 1950.

Ontario Department of Agriculture 97th Annual Report of Ontario Agricultural Societies, 1950, Also Report of Ontario Association of Agricultural Societies. Toronto: Printer to the King's Most Excellent Majesty, 1951.

Ninety-Ninth Annual Report of Ontario Agricultural Societies, 1952. Printed by authority of Hon. F.S. Thomas, Minister of Agriculture for Ontario.

Report of the Commissioner of Agriculture and Arts of the Province of Ontario for the Year 1868. Toronto: Hunter, Rose and Co., 1869.

Statistics Canada. "Canada Goes Urban." Statistics Canada, https://www150.statcan.gc.ca/n1/pub/11-630-x/11-630-x2015004-eng.htm. Accessed 16 November 2020.

Newspapers and Periodicals

Acton Free Press
Advance [Dutton]
Aylmer Express
Aylmer Sun
Brampton Conservator
Brampton Times
British American Cultivator
British Canadian [Simcoe]
British Whig
Canadian Agriculturalist
Canadian Champion [Milton]

Bibliography

Canadian Freeman
Canadian Homes and Gardens
Canadian Statesman [Bowmanville]
Christian Guardian
Colonial Advocate
Country Guide: The Farm Magazine
Daily Globe [Toronto]
Elora Observer
Erin Advocate
Farmer's Advocate
Farmer's Advocate and Canadian Countryman
Farmer's Advocate and Home Magazine
Gazette [Toronto]
Georgetown Herald
Gleaner and Niagara
Globe [Toronto]
Globe and Mail
Guelph Advertiser
Guelph & Galt Advertiser
Guelph Daily Mercury
Guelph Mercury
Guelph Tribune
Guelph Weekly Mercury and Advertiser
Hamilton Spectator
Hillsburgh Beaver
Huron Expositor
Liberal [Richmond Hill]
Mount Forest Confederate
Newmarket Era and Express
Norfolk Observer
O.A.C. Review
Ontario Farmer
Ottawa Citizen
Simcoe Reformer
St. Thomas Times-Journal

Stouffville Tribune
Time Magazine
Times & Guide [Weston]
Weekly Review [Brampton]
Woodbridge News
York Herald
Your Home and Garden

Trade Catalogues

Custead, William W. *Catalogue of Fruit & Ornamental Trees, Flowering Shrubs, Garden Seeds and Greenhouse Plants, Bulbous Roots & Flower Seeds Cultivated and for Sale at the Toronto Nursery*. York: William Lyon Mackenzie, printer to the House of Assembly, 1827.
Eaton's Fall and Winter 1948–49. Toronto. T. Eaton Co., 1948–49.

Secondary Sources

Aberfoyle Agricultural Society. *The Agricultural Society in Puslinch, 1840–1990*. Brantford, ON: Beck's Printing Services, 1990.
Abrahams, Naomi. "Negotiating Power, Identity, Family, and Community: Women's Community Participation." *Gender and Society* 10, no. 6 (December 1996): 768–96.
Adams, Jane. *The Transformation of Rural Life: Southern Illinois, 1890–1990*. Chapel Hill: The University of North Carolina Press, 1994.
Ambrose, Linda M. *A Great Rural Sisterhood: Madge Robertson Watt & the ACWW*. Toronto: University of Toronto Press, 2015.
– "'Better and Happier Men and Women': The Agricultural Instruction Act, 1913–1924." *Historical Studies in Education* 16, 2 (2004): 257–85.
– *For Home and Country: The Centennial History of the Women's Institutes in Ontario*. Guelph: Federated Women's Institutes of Ontario, 1996.
– "'Forever Lunching': Food, Power, and Politics in Rural Ontario Women's Organizations." In *Women in Agriculture: Professionalizing Rural Life in North America and Europe, 1880–1965*, ed. Linda A. Ambrose and Joan M. Jensen, 174–85. Iowa City: University of Iowa Press, 2017.

Bibliography

Ambrose, Linda, and Margaret Kechnie. "Social Control or Social Feminism? Two Views of the Ontario Women's Institutes." *Agricultural History* 73, no. 2 (Spring 1999): 222–37.

Andrews, Maggie. *The Acceptable Face of Feminism: The Women's Institute as a Social Movement*. London: Lawrence and Wishart, 1997.

Attie, Jeanie. *Patriotic Toil: Northern Women and the American Civil War*. Ithaca, NY: Cornell University Press, 1998.

Backhouse, Constance. *Petticoats and Prejudice: Women and Law in Nineteenth-Century Canada*. Toronto: Women's Press, 1991.

"Badenoch: 1832–1967," Puslinch Township History from The Clarks of Tomfad, http://www.clarksoftomfad.ca/BadenochCentennial1832-1967.htm (accessed 17 May 2015).

Badgley, Kerry. *Bringing in the Common Love of Good: The United Farmers of Ontario, 1914–1926*. Toronto: University of Toronto Press, 2000.

Banet-Weiser, Sarah. *The Most Beautiful Girl in the World: Beauty Pageants and National Identity*. Berkeley: University of California Press, 1999.

Baskerville, Peter A. *Sites of Power: A Concise History of Ontario*. Don Mills, ON: Oxford University Press, 2005.

Beausaert, Rebecca. "'Foreigners in Town': Leisure, Consumption, and Cosmopolitanism in Late-Nineteenth and Early-Twentieth Century Tillsonburg, Ontario." *Journal of the Canadian Historical Association* 23, no. 1 (2012): 215–47.

Belisle, Donica. *Purchasing Power: Women and the Rise of Canadian Consumer Culture*. Toronto: University of Toronto Press, 2020.

Bentley, Amy. *Eating for Victory: Food Rationing and the Politics of Domesticity*. Chicago: University of Illinois Press, 1998.

Berg, Barbara. *The Remembered Gate: The Origins of American Feminism, the Woman and the City, 1800–1860*. New York: Oxford University Press, 1978.

Black, Naomi. *Social Feminism*. Ithaca, NY: Cornell University Press, 1989.

Bland, Sidney R. "Women and World's Fairs: The Charleston Story." *The South Carolina Historical Magazine* 94, no. 3 (July 1993): 166–84.

Boisseau, T.J. "White Queens at the Chicago World's Fair, 1893: New Womanhood in the Service of Class, Race, and Nation." *Gender & History* 12, 1 (April 2000): 33–81.

Boisseau, T.J., and Abigail M. Markwyn, eds. *Gendering the Fair: Histories of Women and Gender at World's Fairs*. Urbana: University of Illinois Press, 2010.

Borish, Linda J. "'A Fair, Without *the* Fair, Is No Fair at All': Women at the New England Agricultural Fair in the Mid-Nineteenth Century." *Journal of Sport History* 24, no. 2 (Summer 1997): 155–76.

Bowers, William. *The Country Life Movement, 1900-1920*. Port Washington, NY: Kennikat Press, 1974.

Bowles, Ella Shannon. *Homespun Handicrafts*. New York: Benjamin Blom, 1972.

Bradbury, Bettina. "The Home as Workplace." In *Labouring Lives: Work and Workers in Nineteenth-Century Ontario*, ed. Paul Craven, 412–76. Toronto: University of Toronto Press, 1995.

– *Working Families: Age, Gender, and Daily Survival in Industrializing Montreal*. Toronto: Oxford University Press, 1993.

Brampton Centennial Committee. *Brampton's 100th Anniversary as an Incorporated Town, 1873–1973*. Brampton, ON: Corporation of the Town of Brampton and the Brampton Centennial Committee, 1973.

Bruce County Year Book. Mildmay, ON: Bruce County Federation of Agriculture, 1952.

Bushman, Richard. *The Refinement of America: Persons, Houses, Cities*. New York: Alfred A. Knopf, 1992.

Buttel, Frederick H., and Gilbert W. Gillespie, Jr. "The Sexual Division of Farm Household Labor: An Exploratory Study of the Structure of On-Farm and Off-Farm Labor Allocation among Farm Men and Women." *Rural Sociology* 49, no. 2 (1984): 183–209.

Bye, Cristine Georgina. "'I Like to Hoe My Own Row': A Saskatchewan Farm Woman's Notions about Work and Womanhood during the Great Depression." *Frontiers: A Journal of Women's Studies* 26, no. 3 (2005).

– "'I Think So Much of Edward': Family, Favouritism, and Gender on a Prairie Farm in the 1930s." In *Unsettled Pasts: Reconceiving the West through Women's History*, ed. Sarah Carter et al., 205–37. Calgary, AB: University of Calgary Press, 2005.

Cairns, Kathleen, and Eliane Leslau Silverman. *Treasures: The Stories Women Tell about the Things They Keep*. Calgary, AB: University of Calgary Press, 2004.

Campbell, Lara. *Respectable Citizens: Gender, Family, and Unemployment in Ontario's Great Depression*. Toronto: University of Toronto Press, 2009.

Campbell, S. Lynn, and Susan L. Bennett. "Harvey, Eliza Maria (Jones)." Dictionary of Canadian Biography. http://www.biographi.ca/en/bio/harvey_eliza_maria_13E.html (accessed 6 July 2015).

Bibliography

Carbert, Louise I. *Agrarian Feminism: The Politics of Ontario Farm Women.* Toronto: University of Toronto Press, 1995.

– *Rural Women's Leadership in Atlantic Canada: First-hand Perspectives on Local Public Life and Participation in Electoral Politics.* Toronto: University of Toronto, 2006.

Carlson, Laurie Winn. "The Cattle Battle: Livestock Ownership and the Settlement of the Pacific Northwest." *Columbia: The Magazine of Northwest History* 18, no. 4 (Winter 2004/2005): 34–40.

Carstairs, Catherine, and Nancy Janovicek, eds. *Feminist History in Canada: New Essays on Women, Gender, Work, and Nation.* Vancouver: UBC Press, 2013.

Cavanaugh, Catherine. "The Limitation of the Pioneering Partnership: The Alberta Campaign for Homestead Dower, 1909–1925." *Canadian Historical Review* 74, no. 2 (June 1993): 198–225.

Cebotavrev, E.A. (Nora). "From Domesticity to the Public Sphere: Farm Women, 1945–86." In *A Diversity of Women: Women in Ontario since 1945*, ed. Joy Parr, 200–31. Toronto: University of Toronto Press, 1995.

Chambers, Lori. *Married Women and Property Law in Victorian Ontario.* Toronto: University of Toronto Press, 1997.

Christie, Nancy. *Engendering the State: Family, Work, and Welfare in Canada.* Toronto: University of Toronto Press, 2000.

Christie, Nancy, and Michael Gauvreau. *A Full-Orbed Christianity: The Protestant Churches and Social Welfare in Canada, 1900–1940.* Montreal and Kingston: McGill-Queen's University Press, 1996.

Coates, Colin. *The Metamorphoses of Landscape in Early Quebec.* Montreal and Kingston: McGill-Queen's University Press, 2000.

Cochrane, J.D. "Agricultural Societies, Cattle Fairs, Agricultural Shows and Exhibition of Upper Canada Prior to 1867." Unpublished paper, University of Guelph, 1976.

Cohen, Colleen Ballerino, Richard Wilk, and Beverly Stoeltje, eds., *Beauty Queens on the Global Stage: Gender, Contests, and Power.* New York: Routledge, 1996.

Cohen, Marjorie Griffin. *Women's Work, Markets, and Economic Development in Nineteenth-Century Ontario.* Toronto: University of Toronto Press, 1988.

Comacchio, Cynthia. *The Dominion of Youth: Adolescence and the Making of a Modern Canada, 1920–55.* Waterloo, ON: Wilfrid Laurier University Press, 2006.

– *The Infinite Bonds of Family: Domesticity in Canada, 1850–1940.* Toronto: University of Toronto Press, 1999.

– "Nations Are Built of Babies": Saving Ontario's Mother and Children 1900–1940. Montreal: McGill-Queen's University Press, 1993.
Cork, Campbell, ed. *A History of the Mount Forest Agricultural Society, 125th Anniversary, 1859–1984*. Durham, ON: I.B. Printing Co., 1985.
Cormack, Barbara Villy. *"Learn to Do by Doing": The History of 4-H in Canada*. Ottawa: Canadian Council on 4-H Clubs, 1971.
Corn, Wanda. *Women Building History: Public Art at the 1893 Columbian Exposition*. Berkeley: University of California Press, 2011.
Cott, Nancy F. *The Bonds of Womanhood: "Women's Sphere" in New England, 1780–1835*. New Haven, CT: Yale University Press, 1977.
Couckuyt, Jack. *The Aylmer Fair, 1846–1982*. Aylmer and East Elgin Agricultural Society, 1982.
Cowan, Ruth Schwartz. *More Work for Mother: The Ironies of Household Technology from the Open Hearth to the Microwave*. New York: Basic Books, 1983.
Craig, Béatrice. *Backwoods Consumers and Homespun Capitalists: The Rise of a Market Culture in Eastern Canada*. Toronto: University of Toronto Press, 2009.
Crowley, Terry. *Agnes MacPhail and the Politics of Equality*. Toronto: James Lorimer and Company, 1990.
– "Experience and Representation: Southern Ontario Farm Women and Agricultural Change, 1870–1914." *Agricultural History* 73, no. 2 (Spring 1999): 238–51.
– "J.J. Morrison and the Transition in Canadian Farm Movements during the Early Twentieth Century." *Agricultural History* 71, no. 3 (Summer 1997): 330–56.
– "Rural Labour." In *Labouring Lives: Work and Workers in Nineteenth-Century Ontario*, ed Paul Craven, 13–104. Toronto: University of Toronto Press, 1995.
– "The Origins of Continuing Education for Women: The Ontario Women's Institutes." *Canadian Woman Studies* 7, no. (Fall 1986): 78–81.
Cummins, Light Townsend. "From the Midway to the Hall of State at Fair Park: Two Competing Views of Women at the Dallas Celebration 1936." *Southwestern Historical Quarterly* 114, no. 3 (January 2011): 225–51.
Danbom, David. *The Resisted Revolution: Urban America and the Industrialization of Agriculture, 1900–1930*. Ames, IA: The Iowa State University Press, 1979.
Derry, Margaret. *Art and Science in Breeding: Creating Better Chickens*. Toronto: University of Toronto Press, 2012.
– *Bred for Perfection: Shorthorn Cattle, Collies, and Arabian Horses since 1800*. Baltimore, MD: Johns Hopkins University Press, 2003.

Bibliography

– *Horses in Society: A Story of Animal Breeding and Marketing, 1800–1920.* Toronto: University of Toronto Press, 2006.
– *Ontario's Cattle Kingdom: Purebred Breeders and Their World, 1870–1920.* Toronto: University of Toronto Press, 2001.
Devine, Jenny Barker. *On Behalf of the Family Farm: Iowa Farm Women's Activism since 1945.* Iowa City: University of Iowa Press, 2013.
Drummond, Ian. *Progress without Planning: The Economic History of Ontario from Confederation to the Second World War.* Toronto: University of Toronto Press, 1987.
Errington, Jane. *Wives and Mothers, Schoolmistresses and Scullery Maids: Working Women in Upper Canada, 1790–1840.* Montreal and Kingston: McGill-Queen's University Press, 1995.
Fair, Ross D. "Gentlemen, Farmers, and Gentlemen Half-farmers: The Development of Agricultural Societies in Upper Canada, 1792–1846." PhD diss., Queen's University, 1998.
Federated Women's Institutes of Ontario. *Ontario Women's Institute Story: In Commemoration of the 175th Anniversary of the Founding of the Women's Institutes of Ontario.* Toronto: Federated Women's Institutes of Ontario, 1972.
Ferry, Darren. *Uniting in Measures of Common Good: The Construction of Liberal Identities in Central Canada, 1830–1900.* Montreal and Kingston: McGill-Queen's University Press, 2008.
Fidler, Merrie A. *The Origins and History of the All-American Girls Professional Baseball League.* Jefferson, NC: McFarland and Company, 2006.
Fink, Deborah. *Agrarian Women: Wives and Mothers in Rural Nebraska, 1880–1940.* Chapel Hill: University of North Carolina, 1992.
Floyd, Janet. "Back to Memory Land? Quilts and the Problem of History." *Women's Studies* 37, no. 1 (2007): 38–56.
Fullerton, Gail. *Paisley Agricultural Society: 150 Years of History, 1856–2006.* N.P., 2006.
Gaffield, Chad. *Language, Schooling, and Cultural Conflict: The Origins of French-Language Controversy in Ontario.* Montreal and Kingston: McGill-Queen's University Press, 1987.
Gal, Andrea M. "Grassroots Consumption: Ontario Farm Families' Consumption Practices, 1900–45," PhD diss., Wilfred Laurier University, 2016.
Gates, Paul Wallace. *The Farmer's Age: Agriculture, 1815–1860.* New York: Holt, Rinehart and Winston, 1960.

Gidney, R.D., and W.P.J. Millar. *Professional Gentlemen: The Professions in Nineteenth-Century Ontario*. Toronto: University of Toronto Press, 1994.

Giesberg, Judith Ann. *Civil War Sisterhood: The U.S. Sanitary Commission and Women's Politics in Transition*. Boston, MA: Northeastern University Press, 2000.

Glassford, Sarah, and Amy Shaw, eds. *A Sisterhood of Suffering and Service: Women and Girls of Canada and Newfoundland during the First World War*. Vancouver: UBC Press, 2012.

Greenhalgh, Paul. *Ephemeral Vistas: The Expositions Universelles, Great Exhibitions and World's Fairs, 1851–1939*. Manchester, UK: Manchester University Press, 1988.

Grier, Katherine C. "The Decline of the Memory Palace: The Parlor after 1890," in *American Home Life, 1880–1930: A Social History of Spaces and Services*, ed. Jessica H. Foy and Thomas J. Schlereth, 49–74. Knoxville: The University of Tennessee Press, 1992.

Guard, Julie. *Radical Housewives: Price Wars and Food Politics in Mid-Twentieth-Century Canada*. Toronto: University of Toronto Press, 2019.

Halpern, Monda. *And on That Farm He Had a Wife: Ontario Farm Women and Feminism, 1900–1970*. Montreal and Kingston: McGill-Queen's University Press, 2001.

Handlin, David P. *The American Home: Architecture and Society, 1815–1915*. Boston, MA: Little, Brown, 1979.

Hann, Russell. *Farmers Confront Industrialism: Some Historical Perspectives on Ontario Agrarian Movements*. Toronto: New Hogtown Press, 1975.

Harbeson, Georgiana Brown. *American Needlework: The History of Decorative Stitchery and Embroidery from the Late 16th to the 20th Century*. New York: Coward-McCann, 1938.

Harding, Colin. "How to Spot a Collodion Positive, Also Known as an Ambrotype (early 1850s–1880s)." National Media Museum, entry posted 24 April 2013, http://blog.nationalmediamuseum.org.uk/find-out-when-a-photo-was-taken-identify-collodion-positive-ambrotype/ (accessed 21 June 2016).

Harris, Julie, and Jennifer Mueller. "Making Science Beautiful: The Central Experimental Farm, 1886–1939." *Ontario History* 89, 2 (June 1997): 103–23.

Harvey, Karen. "Introduction: Practical Matters." In *History and Material Culture: A Student's Guide to Approaching Alternative Sources*, ed. Karen Harvey, 1–23. New York: Routledge, 2009.

Hawkins, Daisy Waterhouse. *Old Point Lace and How to Copy and Imitate It*. London: Chatto and Windus, Piccadilly, 1878.

Bibliography

Heald, Henry F. *The Ottawa Fair, 1903–1996*. Nepean, ON: Ottawa Winter Fair, 1998.

Heaman, Elsbeth. *The Inglorious Arts of Peace: Exhibitions in Canadian Society during the Nineteenth Century*. Toronto: University of Toronto Press, 1999.

Hewitt, Nancy A. *Women's Activism and Social Change: Rochester, New York, 1822–1872*. Ithaca, NY: Cornell University Press, 1984.

High, Steven. *Oral History at the Crossroads: Sharing Life Stories of Survival and Displacement*. Vancouver: UBC Press, 2015.

Historical Atlas of Peel County, Ontario. Toronto: Walker and Miles, 1877.

Historical Atlas of the County of Wellington, Ontario, Compiled, Drawn and Published from Personal Examinations and Surveys. Toronto: Historical Atlas Publishing Co., 1906.

Hoffert, Sylvia D. *When Hens Crow: The Woman's Rights Movement in Antebellum America*. Bloomington: Indiana University Press, 1995.

Hoffman, Frances, and Ryan Taylor. *Much To Be Done: Private Life in Ontario from Victorian Diaries*. Toronto: Natural Heritage, 1996.

Holson, Maya. "'And in the good times, they buy flowers': Private Gardening in Suburban Southern Ontario in the 1950s." MA thesis, University of Guelph, 2008.

Holt, Marilyn Irvin. *Linoleum, Better Babies, and the Modern Farm Woman, 1890–1930*. Albuquerque: University of New Mexico Press, 1995.

Howard, Irene. "The Mother's Council of Vancouver: Holding the Fort for the Unemployed, 1935–1938." *BC Studies*, nos. 69-70 (Spring-Summer): 249–87.

Hunter, James Davison, and Joshua J. Yates. "Introduction: The Question of Thrift." In *Thrift in America: Capitalism and Moral Order from the Puritans to the Present*, ed. James Davison Hunter and Joshua J. Yates, 3–32. New York: Oxford University Press, 2011.

Inwood, Kris, and Phyllis Wagg. "The Survival of Handloom Weaving in Rural Canada circa 1870." *Journal of Economic History* 53, no. 2 (June 1993): 346–58.

Irvine, Leslie, and Jenny R. Vermilya. "Gendered Work in a Feminized Profession: The Case of Veterinary Medicine." *Gender and Society* 24, no. 1 (February 2010), 56–82.

Irwin, Thomas. "Government Funding of Agricultural Associations in Late Nineteenth-Century Ontario." PhD diss., University of Western Ontario, 1997.

Jahn, Cheryle. "Class, Gender, and Agrarian Socialism: The United Farm Women of Saskatchewan, 1926–1931." *Prairie Forum* 19, no. 2 (Fall 1994): 189–206.

Jeffrey, Julie Roy. *The Great Silent Army of Abolitionism: Ordinary Women in the*

Antislavery Movement. Chapel Hill: The University of North Carolina Press, 1998.

Jellison, Katherine. *Entitled to Power: Farm Women and Technology, 1913–1963.* Chapel Hill: The University of North Carolina Press, 1993.

Jensen, Joan M. "Cloth, Butter and Boarders: Women's Household Production for the Market." *Review of Radical Politcal Economics* 12, no. 2 (Summer 1980): 14–24.

Johnson, Samuel. *Dictionary of the English Language in which the Words are Deduced from their Originals, Explained in the Different Meanings, and Authorized by the Name of the Writers in Whose Works They Are Found.* Vol. 1. 7th ed. London: n.p., 1783.

Jones, Calvin, and Rachel A. Rosenfeld. *American Farm Women: Findings from a National Survey.* Chicago, IL: National Opinion Research Centre, University of Chicago, 1981.

Jones, David C. *Midways, Judges, and Smooth-Tongued Fakirs: The Illustrated Story of Country Fairs in the Prairie West.* Saskatoon, SK: Western Producers Prairie Books, 1983.

Jones, Robert Leslie. *History of Agriculture in Ontario, 1613–1880.* 1946. Reprint. Toronto: University of Toronto Press, 1977.

Kaethler, Marjorie, and Susan D. Shantz, *Quilts of Waterloo County: A Sampling.* Waterloo, ON: Marjorie Kaethler, 1990.

Kechnie, Margaret. *Organizing Rural Women: The Federated Women's Institutes of Ontario, 1897–1919.* Montreal and Kingston: McGill-Queen's University Press, 2003.

Kelly, Catherine. "The Consummation of Rural Prosperity and Happiness: New England Agricultural Fairs and the Construction of Class and Gender, 1810–1860." *American Quarterly* 49, no. 3 (September 1997), 574–602.

Kingery, W. David, ed. *Learning from Things: Method and Theory of Material Culture Studies.* Washington: Smithsonian Institution Press, 1996.

King-O'Riain, Rebecca Chiyoko. *Pure Beauty: Judging Race in Japanese American Beauty Pageants.* Minneapolis: University of Minnesota Press, 2006.

Kinnear, Mary. "Religion and the Shaping of 'Public Woman': A Post-Suffrage Case Study. In *Religion and Public Life in Canada: Historical and Comparative Perspectives,* ed. Marguerite Van Die, 196–215. Toronto: University of Toronto Press, 2001.

Kline, Ronald R. *Consumers in the Countryside: Technology and Social Change in Rural America.* Baltimore, MD: Johns Hopkins University Press, 2000.

Knapp, Neal. "Transforming the Political Bodies of Livestock: Anglo-American Livestock Expositions." Paper presented at the International Conference of the European Rural History Organisation (EURHO), University of Girona, Spain, 7–10 September 2015.

Knowles, Jane B., and Wava G. Haney, eds. *Women and Farming: Changing Roles, Changing Structures*. Boulder, CO: Westview Press, 1988.

Kubic, Wendee. "Farm Women: The Hidden Subsidy in Our Food." In *Canadian Woman Studies: An Introductory Reader*, eds. Brenda Cranney and Andrea Medovarski, 234–43. Toronto: Inanna, 2006.

Larsen, Elizabeth Anne. *Gender, Work, and Harness Racing: Fast Horses and Strong Women in Southwestern Pennsylvania*. Lanham, MD: Lexington Books, 2015.

Lasser, Carol, and Carol Robertson, *Antebellum Women: Private, Public, Partisan*. New York: Rowman and Littlefield, 2010.

Lee, John B. *Head, Heart, Hands, Health: A History of 4-H in Ontario*. Ontario 4-H Council, 1994.

Levenstein, Harvey A. *Revolution at the Table: The Transformation of the American Diet*. Oxford: Oxford University Press, 1988.

Levine, Lawrence W. *The Unpredictable Past: Explorations in American Cultural History*. Oxford: Oxford University Press, 1993.

Lewis, Frank. *Myles Birkett Foster, 1825–1899*. Leigh-on-Sea, UK: F. Lewis, 1973.

Lewis, R.M. *From Traveling Show to Vaudeville: Theatrical Spectacle in America, 1830–1910*. Baltimore, MD: Johns Hopkins University Press, 2003.

Little, Jo. "Constructions of Rural Women's Voluntary Work." *Gender, Place & Culture* 4, no. 2 (1997): 197–210.

Liverant, Bettina. *Buying Happiness: The Emergence of Consumer Consciousness in English Canada*. Vancouver: UBC Press, 2018.

Loudon, Mrs. *Gardening for Ladies; and Companion to The Flower-Garden*. New York: Wiley and Putnam, 1843.

MacDonald, Cheryl. *Splendor in the Fall: Norfolk County Fair: An Unbroken Heritage*. Simcoe, ON: Norfolk County Agricultural Society, 1990.

MacEwan, J.W.G., and A.H. Ewen. *The Science and Practice of Canadian Animal Husbandry*. Toronto: Thomas Nelson and Sons, 1945.

Madill, A.J. *History of Agricultural Education in Ontario*. Toronto: University of Toronto Press, 1930.

Marks, Lynne. *Revivals and Roller Rinks: Religion, Leisure, and Identity in Late-Nineteenth-Century Small-Town Ontario*. Toronto: University Press, 1996.

Marling, Karal Ann. "'She Brought Forth Butter in a Lordly Dish': The Origins of Minnesota Butter Sculpture." *Minnesota History* 50, no. 6 (Summer 1987): 218–28.

Marti, Donald B. *To Improve the Soil and the Mind: Agricultural Societies, Journals, and Schools in the Northeastern States, 1791–1865*. Ann Arbour, MI: University Microfilms International, 1979.

Marwick, Arthur. *Beauty in History: Society, Politics and Personal Appearance, c. 1500 to the Present*. London: Thames and Hudson, 1988.

May, Zita Barbara. *Canada's International Equestrians*. Toronto: Burns and MacEachern, 1975.

McAleer, John, and John M. Mackenzie. *Exhibiting the Empire: Cultures of Display and the British Empire*. Manchester, UK: Manchester University Press, 2015.

McCalla, Douglas. *Consumers in the Bush: Shopping in Rural Upper Canada*. Montreal and Kingston: McGill-Queen's University Press, 2015.

– *Planting the Province: The Economic History of Upper Canada 1784–1870*. Toronto: University of Toronto Press, 1993.

– "Textile Purchases by Some Ordinary Upper Canadians, 1808–1861." *Material History Review* (Spring-Summer 2001): 4–27.

McCallum, Margaret E. "Prairie Women and the Struggle for a Dower Law, 1905–1920." *Prairie Forum* 18, no. 1 (Spring 1998): 19–33.

McKendry, Ruth. *Quilts and Other Bed Coverings in the Canadian Tradition*. Toronto: Van Nostrand Reinhold, 1979.

McLeod, Ellen Easton. *In Good Hands: The Women of the Canadian Handicraft Guild*. Montreal and Kingston: McGill-Queen's University Press, 1999.

McManus, Sheila. "Gender(ed) Tensions in the Work and Politics of Alberta Farm Women, 1905–29." In *Telling Tales: Essays in Western Women's History*, ed. Catherina A. Cavanaugh and Randi R. Warne, 123–46. Vancouver: UBC Press, 2000.

McMurray, Sally. "Women's Work in Agriculture: Divergent Trends in England and America, 1800 to 1930." *Comparative Studies in Society and History* 34, no. 2 (April 1992): 248–70.

McPherson, Kathryn, Cecilia Morgan, and Nancy M. Forestell. "Introduction: Conceptualizing Canada's Gendered Pasts." In *Gendered Pasts: Historical Essays in Femininity and Masculinity*, ed. Kathryn McPherson, Cecilia Morgan, and Nancy M. Forestell, 1–11. Don Mills, ON: Oxford University Press, 1999.

Melder, Keither. *The Beginnings of Sisterhood: The Women's Rights Movement in the United States, 1800–1840*. New York: Schocken, 1977.

Bibliography

Milkman, Ruth. "Women's Work and Economic Crisis: Some Lessons of the Great Depression." *Review of Radical Political Economics* 8 (1976): 179–97.

Miller, Audrey Saunders, ed. *The Journals of Mary O'Brien, 1828–1838*. Toronto: Macmillan, 1968.

Miller, Bruce A. *The Fairest of Them All: A History of the Arran-Tara Fall Fair, Published on the Occasion of the 150th Consecutive Fair, 1858–2007*. Tara, ON: Great Books of Tara, 2006.

Mink, Nicolaas. "Cooking in the Countryside: The Rural Reform of Taste and the Wisconsin Farmers' Institute's Cooking Schools." *The Wisconsin Magazine of History* 92, no.2 (2008): 2–13.

Mizener, David. "Furrows and Fairgrounds: Agriculture, Identity, and Authority in Twentieth-Century Rural Ontario." PhD diss., York University, 2009.

Monture, Ethel Brant. *Canadian Portraits: Brant, Crowfoot, Oronhyatekha: Famous Indians*. Toronto: Clarke, Irwin, 1960.

Morgan, Cecilia. *Public Men and Virtuous Women: The Gendered Language of Religion and Politics in Upper Canada, 1791–1850*. Toronto: University of Toronto Press, 1996.

Mosby, Ian. *Food Will Win the War: The Politics, Culture, and Science of Food on Canada's Home Front*. Vancouver: UBC Press, 2014.

"Mrs. Hamlin's Family History Wreath." The Lost Art of Sentimental Hairwork, Victorian Gothic. Entry posted on 4 February 2012. http://www.victoriangothic.org/the-lost-art-of-sentimental-hairwork/ (accessed 22 April 2015).

Nasby, Judith. "Historical Views of Guelph." MacDonald Stewart Art Centre. http://www.msac.ca/downloads/publications/Historical%20Views%20of%20Guelph%20(Judith%20Nasby).pdf (originally published at msac.ca 2002) (accessed 21 June 2015).

Neth, Mary. *Preserving the Family Farm: Women, Community, and the Foundations of Agribusiness in the Midwest, 1900–1940*. Baltimore, MD: Johns Hopkins University Press, 1995.

Nurse, Jodey. "Reaching Rural Ontario: The County of Peel Agricultural Society and the Peel County Fall Fair, 1853–1883." MA thesis, University of Guelph, 2010.

Ontario Association of Agricultural Societies. *The Story of Ontario Agricultural Fairs and Exhibitions, 1792–1967, and Their Contribution to the Advancement of Agriculture and Betterment of Community Life*. Picton, ON: Picton Gazette, 1967.

Osterud, Nancy Grey. *Bonds of Community: The Lives of Farm Women in Nineteenth-Century New York*. Ithaca, NY: Cornell University Press, 1991.

Palm, Regina Megan. "Women Muralists, Modern Women and Feminine Spaces: Constructing Gender at the 1893 Chicago World's Columbian Exposition." *Journal of Design History* 22, no. 2 (2010): 123–43.

Parker, Rozsika. *The Subversive Stitch: Embroidery and the Making of the Feminine.* London: Women's Press, 1984.

Parkins, John R., and Maureen G. Reed. "Introduction: Toward a Transformative Understanding of Rural Social Change." In *Social Transformation in Rural Canada: Community, Cultures, and Collective Action*, ed. John R. Parkins and Maureen G. Reed, 3–20. Vancouver: UBC Press, 2013.

Parr, Joy. "Gender History and Historical Practice." *Canadian Historical Review* 76, no. 3 (September 1995): 354–76.

Petrzelka, Peggy, and Susan E. Mannon. "'Keepin' This Little Town Going': Gender and Volunteerism in Rural America." *Gender and Society* 20, no. 2 (April 2006): 236–58.

Pierson, Ruth Roach. *"They're Still Women After All": The Second World War and Canadian Womanhood.* Toronto: McClelland and Stewart, 1986.

Pini, Barbara, Berit Brandth, and Jo Little. "Introduction." In *Feminisms and Ruralities*, ed., Barbara Pini, Berit Brandth, and Jo Little, 1–12. Lanham, MD: Lexington Books, 2015.

Pini, Barbara, and Belinda Leach. "Transformations of Class and Gender in the Globalized Countryside: An Introduction." In *Reshaping Gender and Class in Rural Spaces*, ed. Barbara Pini and Belinda Leach, 113–128. Surrey, UK: Ashgate, 2011.

Prown, Jules David. *Art as Evidence: Writings on Art and Material Culture.* New Haven, CT: Yale University Press, 2001.

Quaile, Meredith. "Dairy Pin-Up Girls: Milkmaids and Dairyqueens." *Material Cultural Review* 66 (Fall 2007): 19–31.

Raizman, David, and Ethan Robey. *Expanding Nationalisms at World's Fairs: Identity, Diversity, and Exchange, 1815–1915.* London: Routledge, 2017.

Ready, J.C. *A Manual in Canadian Agriculture.* Toronto: The Ryerson Press, 1930.

Reaman, G. Elmore. *A History of Agriculture in Ontario.* Vol. 1. Toronto: Saunders, 1970.

– *A History of Agriculture in Ontario.* Vol. 2. Toronto: Saunders, 1970.

Rennie, Bradford James. *The Rise of Agrarian Democracy: The United Farmers and Farm Women of Alberta, 1909–1921.* Toronto: University of Toronto Press, 2000.

Rockwell, F.F., and Ester C. Grayson. *The Complete Book of Lilies: How to Select, Plant, Care for, Exhibit, and Propagate Lilies of all Types*. Garden City, NY: Doubleday, 1961.
– *Flower Arrangement*. 1935; repr. New York: The MacMillan Company, 1938.
– *Gardening Indoors: The Enjoyment of Living Flowers and Plants the Year Round, and New Opportunities for Home Decoration*. New York: The Macmillan Company, 1938.
– *Flower Arrangement in Color*. New York: Wise, 1940.
– *The Complete Book of Flower Arrangement, for Home Decoration, for Show Competition*. New York: American Garden Guild, 1947.
– *The Complete Book of Bulbs: A Practical Guide Manual on the Uses, Cultivation, and Propagation of More than 100 Species, Hardy and Tender, which the Amateur Gardener Can Enjoy Outdoors and in the Home*. Garden City, NY: American Garden Guild and Doubleday, 1953.
– *The Complete Book of Annuals: How to Use Annuals and Plants Grown as Annuals to Best Effect, Out-of-Doors and In, with Cultural Information and Other Pointers on More than 500 Species and Varieties*. New York: American Garden Guild, 1955.
– *The Rockwells' Complete Book of Roses*. Garden City, NY: Doubleday, 1958.
– *10,000 Garden Questions Answered by 20 Experts*. Garden City, NY: Doubleday, 1959.
– *New Complete Book of Flower Arrangement*. Garden City, NY: Doubleday, 1960.
– *The Rockwells' Complete Guide to Successful Gardening*. Garden City, NY: Doubleday, 1965.
Rollings-Magnusson, Sandra. "Canada's Most Wanted: Pioneer Women on the Western Prairies." *Canadian Review of Sociology and Anthropology* 27, no. 2 (May 2000): 223–38.
– "Hidden Homesteaders: Women, the State and Patriarchy in the Saskatchewan Wheat Economy." *Prairie Forum* 24, no. 2 (Fall 1999): 171–83.
Root, Waverly, and Richard de Rochemont. *Eating in America: A History*. New York: William Morrow, 1976.
Rosenberg, Gabriel N. *The 4-H Harvest: Sexuality and the State in Rural America*. Philadelphia: University of Pennsylvania Press, 2016.
Rowe, Rochelle. *Imagining Caribbean Womanhood: Race, Nation and Beauty Contests, 1929–70*. Manchester, UK: Manchester University Press, 2013.

Ryan, Mary P. *Civic Wars: Democracy and Public Life in the American City during the Nineteenth Century*. Berkeley: University of California, 1998.
– *Cradle of the Middle Class: The Family in Oneida County, New York, 1790–1865*. Cambridge: Cambridge University Press, 1981.
– *Women in Public: Between Banners and Ballots, 1825–1880*. Baltimore, MD: Johns Hopkins University Press, 1990.
– *World of Fairs: The Century-of-Progress Expositions*. Chicago, IL: University of Chicago Press, 1993.
Rydell, Robert. *All the World's a Fair: Visions of Empire at American International Expositions, 1876–1926*. 1984. Reprint. Chicago, IL: University of Chicago Press, 1987.
Sachs, Carolyn. *Gendered Fields: Rural Women, Agriculture, and Environment*. Boulder, CO: Westview Press, 1996.
– *The Invisible Farmers*. Totowa, NJ: Rowman and Allanhead, 1983.
Salant, Priscilla. *Farm Women: Contributions to Farm and Family*. Washington, DC: USDA, Economic Research Service, Economic Development Division, Agricultural Economics Research Report No. 140, 1983.
Samson, Daniel. *Spirit of Industry and Improvement: Liberal Government and Rural-Industrial Society, Nova Scotia, 1790–1862*. Montreal and Kingston: McGill-Queen's University Press, 2008.
Sandwell, Ruth. *Canada's Rural Majority: Households, Environments, and Economies, 1870–1940*. Toronto: University of Toronto Press, 2016.
Sangster, Joan. "Mobilizing Women for War." In *The First World War in Canada: Essays in Honour of Robert Craig Brown*, ed. David Mackenzie, 157–93. Toronto: University of Toronto Press, 2005.
– *One Hundred Years of Struggle: The History of Women and the Vote in Canada*. Vancouver: UBC Press, 2018.
Sanmiya, Vibeke. "A Spirit of Entreprise: The Western Fair Association, London, Ontario, 1867–1947." PhD diss., Wilfrid Laurier University, 2002.
Schlereth, Thomas J. *Victorian America: Transformations in Everyday Life, 1876–1915*. New York: HarperCollins, 1991.
Scott, Guy. *A Fair Share: A History of Agricultural Societies and Fairs in Ontario, 1792–1992*. Toronto: Ontario Association of Agricultural Societies, 1992.
– *Country Fairs in Canada*. Markham, ON: Fitzhenry and Whiteside, 2006.
Scott, Joan Wallach. *Gender and the Politics of History*. Rev. ed. New York: Columbia University Press, 1999.

Bibliography

Selden, Steven. "Transforming Better Babies into Fitter Families: Archival Resources and the History of the American Eugenics Movement, 1908–1930." *Proceedings of the American Philosophy Society* 149, no. 2 (June 2005): 199–225.

Sheumaker, Helen. *Love Entwined: The Curious History of Hairwork in America.* Philadelphia, PA: University of Pennsylvania Press, 2007.

Smith, Andrew, and Shelley Boyd. "Talking Turkey: Thanksgiving in Canada and the United States." In *What's to Eat? Entrées in Canadian Food History*, ed. Nathalie Cooke, 116–144. Montreal and Kingston: McGill-Queen's University Press, 2009.

Smith, William W. *The Elements of Livestock Judging.* Philadelphia, PA: J.B. Lippincott, 1927.

Smith-Rosenberg, Carrol. *Disorderly Conduct: Visions of Gender in Victorian America.* New York: A.A. Knopf, 1985.

Spike, Sarah. "Sights Worth Looking at: Agricultural Exhibitions in Nova Scotia," NiCHE: Network in Canadian History & Environment, entry posted 20 August 2014. http://niche-canada.org/2014/08/20/sights-worth-looking-at-agricultural-exhibitions-in-nova-scotia/ (accessed 3 April 2015).

Strasser, Susan. *Never Done: A History of American Housework.* New York: Pantheon Books, 1982.

Street, Kori. "Bankers and Bomb Makers: Gender Ideology and Women's Paid Work in Banking and Munitions during the First World War in Canada." PhD diss., University of Victoria, 2001.

Summerfield, Penny. "Patriotism and Empire: Music-Hall Entertainment, 1870–1914." In *Imperialism and Popular Culture*, ed. John M. MacKenzie, 17–48. Manchester, UK: Manchester University Press, 1986.

Switzer, June. *Erin Fall Fair since 1850, Written by June Switzer in Recognition of the 150th Anniversary of the Erin Agricultural Society.* Erin, ON: Erin Agricultural Society, 2000.

Taylor, Jeffrey. *Fashioning Farmers: Ideology, Agricultural Knowledge and the Manitoba Farm Movement, 1890–1925.* Regina, SK: University of Regina, Canadian Plains Research Center, 1994.

Tenneriello, Susan. *Spectacle Culture and American Identity, 1815–1940.* New York: Palgrave Macmillan, 2013.

Tessier, Marc. *Spencerville Fair, 1855–2005: Our Past, Our Present, Our Future.* Brockville, ON: Henderson Printing for Spencerville Agricultural Society, 2005.

Thompson, Paul. *The Voice of the Past: Oral History*. 3rd ed. Oxford: Oxford University Press, 2000.

Tippett, Maria. *By a Lady: Celebrating Three Centuries of Art by Canadian Women*. Toronto: Viking, 1992.

Tobin, Beth Fowkes, and Maureen Daly Goggin. "Introduction: Materializing Women." In *Women and Things, 1750–1950: Gendered Material Strategies*, ed. Maureen Daly Goggin and Beth Fowkes Tobin, 1–14. Farnham, U.K.: Ashgate, 2009.

Tosaj, Nicholas. "Weaving the Imperial Breadbasket: Nationalism, Empire, and the Triumph of Canadian Wheat, 1890–1940." *Journal of Canadian Historical Association* 28, no. 1 (2017): 249–75.

Traill, Catherine Parr Strickland. *The Backwoods of Canada: Being Letters from the Wife of an Emigrant Officer, Illustrative of the Domestic Economy of British America*. London: C. Knight, 1846.

– *The Canadian Emigrant Housekeeper's Guide*. Montreal: James Lovell, 1861.

Tumblety, Joan. "Introduction: Working with Memory as Source and Subject." In *Memory and History: Understanding Memory as Source and Subject*, ed. Joan Tumblety, 1–16. New York: Routledge, 2013.

Turkle, Sherry. *Evocative Objects: Things We Think With*. Cambridge, MA: The MIT Press, 2007.

Tye, Diane. *Baking as Biography: A Life Story in Recipes*. Montreal and Kingston: McGill-Queen's University Press, 2010.

Ulrich, Laurel Thatcher. *The Age of Homespun: Objects and Stories in the Creation of an American Myth*. New York: Alfred A. Knopf, 2001.

Valverde, Mariana. *The Age of Light, Soap, and Water: Moral Reform in English Canada, 1885–1925*. Toronto: University of Toronto Press, 2008.

Vamplew, Wray, and Joyce Kay, *Encyclopedia of British Horseracing*. New York: Routledge, 2005.

Van de Vorst, Charlotte. *Making Ends Meet: Farm Women's Work in Manitoba*. Winnipeg: University of Manitoba Press, 2002.

Van Die, Marguerite. "Revisiting "Separate Spheres": Women, Religion, and the Family in Mid-Victorian Brantford, Ontario." In *Households of Faith: Family, Gender, and Community in Canada, 1760–1969*, ed. Nancy Christie, 234–63. Montreal and Kingston: McGill-Queen's University Press, 2002.

Varon, Elizabeth. *We Mean to Be Counted: White Women and Politics in Antebellum Virginia*. Chapel Hill: University of North Carolina Press, 1998.

Bibliography

Victoria and Alberta Museum. "Mrs Loudon & the Victorian Garden." Victoria and Alberta Museum: The World's Greatest Museum of Art and Design. http://www.vam.ac.uk/content/articles/m/mrs-loudon-victorian-garden/ (accessed 20 April 2015).

Volz, Candace M. "The Modern Look of the Early-Twentieth-Century House: A Mirror of Changing Lifestyles." In *American Home Life, 1880–1930: A Social History of Spaces and Services*, ed. Jessica H. Foy and Thomas J. Schlereth, 25–48. Knoxville: The University of Tennessee Press, 1992.

von Baeyer, Edwinna. *Rhetoric and Roses: A History of Canadian Gardening, 1900–1930*. Markham: Fitzhenry and Whiteside, 1984.

von Baeyer, Edwinna, and Pleasance Crawford, eds. *Garden Voices: Two Centuries of Canadian Garden Writing*. New York: Random House, 1995.

Walden, Keith. *Becoming Modern in Toronto: The Industrial Exhibition and the Shaping of Late Victorian Culture*. Toronto: University of Toronto Press, 1997.

Walker, Marilyn I. *Ontario's Heritage Quilts*. Toronto: Stoddart, 1992.

Wall, Sharon. *The Nurture of Nature: Childhood, Antimodernism, and Ontario Summer Camps, 1920–55*. Vancouver: UBC Press, 2009.

Wardle, Patricia. *Victorian Lace*. 1968; repr. Bedford, UK: Ruth Bean, 1982.

Watson, Elwood, and Darcy Martin, eds. *There She Is, Miss America: The Politics of Sex, Beauty, and Race in America's Most Famous Pageant*. New York: Palgrave MacMillan, 2004.

Watson, E.H.A. *Ontario Red Cross, 1914–1946*. Toronto: Canadian Red Cross Society, 1946.

Weimann, Jeanne Madeline. *The Fair Women: The Story of the Woman's Building, World's Columbian Exhibition, Chicago, 1893*. Chicago, IL: Academy, 1981.

Wessel, Thomas, and Marilyn Wessel. *4-H: An American Idea 1900–1980: A History of 4-H*. Chevy Chase, MD: National 4-H Council, 1982.

Westover, Ruth Ann, ed. *Fair Days and Fair People: 130 Years of Fall Fairs*. Durham, ON: Harriston-Minto Agricultural Society, 1989.

Whatmore, Sarah, Terry Marsden, and Philip Lowe, "Introduction: Feminist Perspectives in Rural Studies." In *Gender and Rurality*, ed. Sarah Whatmore, Terry Marsden, and Philip Lowe, 1–10. London: David Fulton, 1994.

Wilkie, Rhoda M. *Livestock/Deadstock: Working with Farm Animals from Birth to Slaughter*. Philadelphia, PA: Temple University Press, 2010.

Wilson, Catharine Anne. "A Manly Art: Plowing, Plowing Matches, and Rural Masculinity in Ontario, 1800–1930." *The Canadian Historical Review* 95, 2 (June 2014): 157–86.

- "Reciprocal Work Bees and the Meaning of Neighbourhood." *Canadian Historical Review* 82, no. 3 (September 2001): 431–64.
- "Shaking Out the Quilt and Quilting Bee: Ontario Farm Diaries and the Dynamics of Household and Neighbourhood Production." Paper presented at The Local is Global: Gender and Rural Connections across Time and Place, Rural Women's Studies Association Triennial Conference, San Marcos, Texas, United States of America, 19–21 February 2015.

Woodward, Sophie. "Looking Good: Feeling Right – Aesthetics of the Self." In *Clothing as Material Culture*, ed. Susanne Küchler and Daniel Miller, 21–40. New York: Berg, 2005.

Woolum, Janet. *Outstanding Women Athletes: Who They Are and How They Influenced Sports in America*. Phoenix, AZ: Orynx Press, 1992.

Wynn, Graeme. "Exciting a Spirit of Emulation among the 'Plodholes': Agricultural Reform in Pre-Confederation Nova Scotia." *Acadiensis* 20, no. 1 (Autumn 1990): 3–51.

Yano, Christine R. *Crowning the Nice Girl: Gender, Ethnicity, and Culture in Hawai'i's Cherry Blossom Festival*. Honolulu: University of Hawai'i Press, 2006.

Yee, Shirley J. *Black Women Abolitionists: A Study in Activism, 1828–1860*. Knoxville: The University of Tennessee Press, 1992.

Yellin, Jean Fagan, and John C. Van Horne, eds. *The Abolitionist Sisterhood: Women's Political Culture in Antebellum America*. Ithaca, NY: Cornell University Press, 1994.

Young, Nancy. "The Reins in Their Hands: Ranchwomen and the Horse in Southern Alberta, 1880–1914." *Alberta History* 52, no. 1 (January 2004): 2–8.

Young, W.R. "Conscription, Rural Depopulation, and the Farmers of Ontario, 1917–19." *The Canadian Historical Review* 53, no. 3 (September 1972): 289–318.

Zeller, Suzanne. *Inventing Canada: Early Victorian Science and the Idea of a Transcontinental Nation*. Montreal and Kingston: McGill-Queen's University Press, 2009.

Index

4-H Clubs: 49, 196–204, 207–8

agriculture: appreciation for, 45, 208, 235; capitalist modes of, 21, 139; definition of, 11, 20; improvement of, 40; as a profession, 20–2, 32, 37, 208; scientific, 186

Agricultural Association of Upper Canada: establishment of, 22; Provincial Exhibition, 22, 62, 77, 106, 109

agricultural fairs in Ontario: Acton, 169, 229–31; Albion, 144; Ancaster, 205; Arran-Tara, 119, 122–5, 129, 132–3, 158–62, 195; Arthur, 153; Aylmer, 120, 130, 197, 202, 214, 216–17, 220, 222, 236, 238–9; Bolton, 215; Brampton, 63, 74, 91. *See also* Peel County; Brooke and Alvinston 85–6, 122; Brussels, 96; Burford, 45, 158; Burwick, 193, 219; Caledonia, 170; Central (Walter's Falls), 93–4; Centre Wellington, 193, 220; Collingwood Township, 85, 88–9, 95, 156–8; Cooksville, 193, 227, 235; Drumbo, 122; East Grey County, 217; East Huron, 122; Egremont, 119, 154; Elgin County, 36, 197, 216, 227; Eramosa, 72, 74, 91, 144, 188–9, 219; Erin 59, 61, 71, 74, 77–80, 96–100, 109–10, 119–21, 129, 131, 134–5, 140, 142–3, 148–9, 152, 156, 164–5, 167–9, 195–6, 198–9, 205–7, 217, 221, 224, 241–2; Etobicoke Township, 144; Dufferin County, 74; Georgetown, 85, 225, 227; Grand Union Exhibition, 79; Guelph Central, 67, 152; Guelph Township, 91; Harriston, 148; Harrow, 99; Hespeler, 110; Ingersoll, 122; Lakehead Exhibition, 65, 68–9; Leamington, 99; Luther, 176; Markham, 73, 92, 112, 219, 224, 231. *See also* Union; Minto Township, 62, 219; Mount Forest, 217, 236; Murillo, 237; Newmarket, 226; Nichol, 73; Norfolk County, 212–13, 220–3, 228–9, 235, 237, 243; Orono, 222–3, 229; Peel County, 24, 62–4, 74, 79–81, 91, 112, 117–18, 144, 193, 240–1; Perth County, 73; Pinkerton, 94–5; Puslinch, 75, 140, 188–9, 227; Renfrew, 65; Ridgetown, 90; Rockton, 238–9; Rodney, 30, 93, 166, 203–4; Schomberg, 95, 163; Seaforth, 122, 154–5; Shedden, 30–1, 99, 101, 119, 202, 213–14, 232; Smithville, 47, 22–5; Springfield, 65, 194, 237; Thorold, 227; Toronto Township, 112, 139, 213; Union, 67, 79; Wallacetown, 66, 102, 156, 194, 214, 224, 236, 238. *See also* West Elgin; Waterdown, 163, 165; West Durham, 92; West Elgin, 73, 144, 156, 194, 240; Western, 115, 181; Weston, 63, 226; Woodbridge, 62, 213, 215. *See also* agricultural societies in Ontario

agricultural halls, 7, 57, 213

agricultural societies in Ontario: administration of, 18–56; Ancaster, 18, 51; Bayham,

67; Beeton, 34, 193; Binbrook, 159; Brant, 26; Brooke and Alvinston, 122; Dundalk, 47; Eramosa, 144; Erin, 36, 52–3, 55, 70, 177; Forest, 122; Georgetown, 51, 53, 82; Harriston, 34; Niagara, 20; Normanby, 34; Paisley, 47; Perth and District, 46; Picton, 68; Puslinch, 189, 227; Rockton, 49–50, 238; Smithville, 47; South Brant, 45; Southwold and Dunwich, 30; Teeswater, 47; Toronto Township, 67; West Elgin, 36, 41; West Riding and Vaughan, 74, 212; Wyoming, 47. *See also* agricultural fairs in Ontario
agrarian feminism. *See* feminism
Ambrose, Linda, 13
animal husbandry: gendered roles, 20, 42, 184–8, 191–2, 199, 209–10; improved, 4, 21, 144. *See also* livestock
Aoewentaiyouh, 219–20
arts and crafts, 153–68
athletics. *See* sports

baby shows, 233–40
Bake Queen, 111
baked goods. *See* food
Bartlett, Gail, 159, 181
beauty: association with femininity, 6, 107, 109, 136, 145, 147–8, 150, 153, 248; concept of, 105, 126; physical characteristics of, 183, 194, 207, 222–3, 227–33, 241, 248
beauty contests, 227–33
Benton, Glenda, 51, 53, 85
Britain, 20, 184, 217. *See also* England
Bushman, Richard L., 150

Cairns, Kathleen, 168
Calvert, Fanny Colwill, 169–70, 173
Canada Packers Ltd., 96
Canadian Jersey Cattle Breeders' Association, 186
Canadian Legion Ladies' Auxiliary, 214–15
Canadian National Exhibition, 53, 223, 230
Canadian Red Cross, 66, 154, 157–8, 212, 214, 216–17
canned goods. *See* food
Carbert, Louise, 13
carnival: acts, 193; rides, 216; shows, 217

Carroll, J.A., 36–8, 43–4
cattle. *See* livestock
cheaters, 69–70
children: caring for, 51–3, 141, 153; children's exhibits, 29, 47, 65, 196; clothing for, 153–4, 156, 158–62, 168; as fairgoers, 90, 191, 216; keeping them on the farm, 113; labour of, 76–7, 82, 116, 147; proper rearing of, 23, 109, 234–40. *See also* girls, 4-H Clubs, Girls' and Boys' Clubs, and baby shows
church groups: decorations for, 125–6; fundraising efforts of, 30, 212, 214–16; public presence, 21
Cohen, Marjorie Griffin, 184
concerts, 201, 217, 219, 223, 244
consumer culture, 138, 145
cooking, 60, 87, 99, 212. *See also* food
courting, 241–3
craftwork. *See* fancywork
Craig, Beatrice, 140
Creelman, Mr G.C., 27
crops: committees about, 24; exhibits of, 21, 106; improved, 8, 26, 76, 113, 238; production and harvest of, 20, 42, 57, 72, 76, 198
Crystal Palaces, 68, 216, 236
culinary arts, 87, 103–4, knowledge of, 60; men's, 249; skill in, 212. *See also* domestic science

dairy cattle shows, 196–7, 202–3, 208. *See also* 4-H Clubs.
dairy farming, 26, 186–8, 207–8
Dairy Princess. *See* Queen Contests
dairy products. *See* food
dairywomen, 22, 43, 184, 186–8, 191, 207–8
dances, 217, 221
DeVonda, Dorothy, 220, 222
Dickenson, Velda, 18–20, 55
domestic science, 27, 153, 156, 199; committees and department of, 64, 93, 103; exhibits, 64, 66, 75, 84, 99; exhibitors of, 83. *See also* food
domesticity, 6, 13, 92, 164, 195
domestic manufactures, 137–53, 156, 159, 165, 181
domestic products. *See* food

Index

domestic work, 8, 10, 26, 45, 60
Dominion Grange, 24
Donaldson, Elizabeth Ann, 175–8

empire, 113. *See also* Imperial Order Daughters of the Empire
England, 169, 181, 189, 191
entertainment: acceptable forms of, 218, 224; professional, 220–3; racialized, 219–20
equestrian sport, 184, 192, 205, 222. *See also* horses

fair banquets, 19, 211–13
fair boards. *See* agricultural societies in Ontario
Fair Queens. *See* Queen Contests
fair ribbons, 171–3, 175–8, 180–1, 193, 210, 239
fancywork, 6, 15, 17, 43, 62–3, 137–8, 141, 144–53, 159, 162–70, 181–2, 222, 241, 245
femininity, 6, 10–13, 112, 127, 148, 150, 218, 222–4, 230, 233, 248, 250
feminism: rural feminism(s), 11–12, 250; social feminism, 19; second-wave feminism, 53
Ferry, Darren, 11
fine art, 17, 62, 113, 137, 151, 164, 180
Fink, Deborah, 12
First Nations, 14, 219–20
First World War, 26, 28, 30, 43, 66, 75, 120, 153, 158, 214, 235
flour companies, 92, 95–100
flowers, 17, 54, 63, 77, 105–36, 145, 147, 149, 151, 159, 191, 222
food: baked goods, 70, 87, 90–103; booths, 31, 52, 214–16; canned goods, 11, 66, 83–90, 100, 103, 191; dairy products, 22, 35, 58, 63, 66–7, 71–5, 103; fruit, 22, 24, 58–9, 61, 63, 67, 75–84, 103, 106, 119, 127, 131, 191, 214, 242; garden vegetables, 10, 22, 58–9, 61, 63, 75–87, 103, 106, 116, 119, 127, 131, 181; manufactured, 99; midway 7, 57, 215–16
friendship, 8, 51, 177, 243, 249–50
fundraising, 17, 30, 52, 211, 214, 216–17, 243, 248. *See also* church groups

garden vegetables. *See* food

gender: ideas about, in relation to farming, 13, 53, 60, 108, 209
Girls' and Boys' Clubs, 196–200, 209. *See also* 4-H Clubs
Girls' Garden Clubs, 82
Goodfellow, Miss K., 36, 64
Grayson, Ester C., 122, 126–7
Great Depression, the, 35, 66, 121–7, 156–7

Halpern, Monda, 12
Harkin, Dianne, 53
Heaman, Elsbeth, 7
Historical County Atlas, the, 113, 115
Hodgins, Miss Ina, 42, 46
Holson, Maya, 128
homecrafts, 5, 47, 51, 144, 159, 249. *See also* domestic manufactures
horses: draft, 46, 198–9; professional riders of, 222; shows of, 43, 74, 183–4, 188–96; 205–7, 209, 248–9
horticultural societies, 63, 127
horticulturalists, 180, 117, 120. *See also* flowers

Indigenous peoples. *See* First Nations
immigrants, 14, 138, 181
Imperial Order Daughters of the Empire, 212, 214
improvers. *See* rural reformers.

Jackson, Sarah Wheeler, 168–71
Jameson, Jeanette, 49–50
Jellison, Katherine, 59
Jell-O, 99
Jones, Eliza Maria, 186–7
judges: criticism of, 70; female, 28, 36, 41–2, 87, 93, 102, 120, 164–5, 185–6, 208, 237, 240
Junior Farmers, 18, 31, 200

Kechnie, Margaret, 27, 199
Kelly, Catherine, 141
Kodak, 164

ladies' work, 76, 139, 144, 153–62, 180, 188, 219
Levine, Lawrence, 121
livestock: associations, 54; beef cattle, 197–8, 205; breeding of, 3–4, 8, 21, 42, 75, 184,

186–8; Calf Club, 49, 197, 201–4, 207–8; caring for, 184–8; chickens, 3, 184, 188, 196; *See also* poultry; dairy cattle, 75, 184, 186–8, 191, 197, 202, 207–8; exhibitors of, 183–4, 188–210; as a masculine domain, 21, 26, 46, 106; pigs, 183, 197, poultry, 183– 4, 188, 191, 197, 201; sheep, 49, 183, 202
Loudon, Jane, 106
Lovering, Margaret, 51–2, 55

MacMaster, Phyllis, 208–9
Mannon, Susan E., 51
material culture, 15–16
May, Minnie, 84, 91, 108
McCulloch, Mary, 116
McDermand, Miss Bess, 39–40
Merry, Katherine, 202
midways, 214–16, 244, 249
milking contests, 224
Milton, Jessie, 82, 85
Miss Canada, 221, 228
Monture, Ethel Brant, 37, 40–1
morality, 105, 109, 150, 222, 248
Morgan, Cecilia, 186
musical performances, 194, 211, 220–4
mutuality, concept of, 5, 12–13, 246

Neal, Christena Nelena McMillan, 178–80
Needham, Miss May, 64–6, 69

oldest person competitions, 224–5
Ontario Agricultural College, 119, 201
Ontario Association of Agricultural Societies (OAAS), 14, 18–19, 28, 32, 42, 229; Convention, 19, 32, 34, 36–45; 64–6, 69, 75, 82, 128; Women's Section, 18–19, 28, 37–9, 41, 44, 46–7
Ontario Association of Fairs and Exhibitions. *See* Ontario Association of Agricultural Societies.
Ontario Department of Agriculture, 54, 80, 84, 114, 196, 200, 209; Minister of Agriculture, 237; annual report of, 54; Agricultural and Horticultural Societies Branch, 129
Ontario Horticultural Association, 129

Ontario Ministry of Agriculture and Food. *See* Ontario Department of Agriculture
oral history, 15–16
orchards, 59, 82, 85, 107
ornamental plants, 106–9, 114–19. *See also* flowers
Osterud, Nancy Grey, 11

paintings. *See* fine arts.
Paris Exposition, the, 180
Parker, Rozsika, 150
Patrons of Industry, 24
Petrzalka, Peggy, 51
Philadelphia Centennial Exhibition, the, 180
photography. *See* fine arts
pigs. *See* livestock
Potts, Ada L., 120
poultry. *See* livestock
premiums, 3, 21, 27, 71, 74, 91, 106, 172. *See also* prizes
Preston, Isabella, 119–20, 127
prizes: company sponsors of, 95–6, 100, 159, 229; donors of, 65, 94, 158, 227; money, 30, 39, 53, 74, 96–7, 158, 227–8, 239; rules and regulations about, 69–1, 194, 230; special, 31, 65–6, 91, 94–5, 119, 122, 231, 233
prize lists: baked goods, 91, 96–8; canned goods, 85–6; domestic manufactures, 154–5, 159–62; domestic sciences, 99–100; fancywork, 145–8; fine art, 164; flowers, 109, 117–18; garden exhibits, 77–81; horse show, 189, 195; recognition of women in, 36, 41, 53, 59, 188; as sources, 14, 26; updating of, 34–5; wartime, 43, 162–3
Prospect House, 115
Putnam, George, 40

Queen Contests: Dairy Princess, 231–3; Fair Queen, 223, 227–33; Grape Queen, 231; National Dairy Queen Contest, 232–3; Queen of the Furrow, 231; Tobacco Queen, 231
quilts. *See* domestic manufactures and fancywork

refinement, 5–6, 24, 76, 107, 109, 111, 138, 144, 148, 150, 182, 187, 222, 248

Index

respectability, 4–5, 94, 141, 145, 147–8, 150, 182, 225
Reid, Myrtle, 52
religion. *See* church groups
Robertson, Clara Jackson, 169, 172
Rockwell, F.F., 122, 126–7
Rotary-Kiwanis, 214
Royal Horse Show, the, 205
Royal Winter Fair, the, 202–5
Rudd sisters (Bernice, Irene, and Edna), 173–5
rural: beautification, 105, 114, 121; definition of, 9–10
rural reformers, 10, 23–4, 27–9, 54, 61, 84, 105–9, 113–14, 136, 150, 156, 177, 218
rural school fairs, 49, 51, 196

Sandwell, Ruth, 9
Second World War, 9, 42–3, 62, 65, 82, 85, 121, 127, 158, 162, 200, 214
seed catalogues, 83, 129
sewing machines, 74, 141, 153, 156, 168–9, 175
sheep. *See* livestock
Silverman, Eliane Leslau, 168

Simcoe, John Graves, 20
sociability, 12, 19, 30, 54, 213, 217
Spike, Sara, 61
sports, 200, 226–7
Stephen, Laura Rose, 32
Strickland, Nancy, 152, 180–1
Switzer, June, 52–5, 70, 207–9

taste, concept of, 150
Toronto Industrial Exhibition, 58, 213, 228
Tye, Diane, 215

Ulrich Thatcher, Laurel, 141, 144
United Farm Women, 24
United States, 82, 85, 192, 226–7, 229, 233
University of Guelph, 209
urban populations, 8, 29, 45, 54, 115

visual knowledge, concept of, 58, 61–2, 71, 111
volunteerism, 11, 19, 30, 41, 55, 215, 247

Walden, Keith, 7, 58, 247
weather, poor, 63–4
Wilson, Loetitia, 127
Women for the Survival of Agriculture, 53
Women's Institutes, 12–13, 24, 27, 29–31, 34, 40, 84, 153–4, 156, 199–200, 212–14, 217
Wood, Eleanor, 49
Woodward, Sophie, 178